Managing
New Technology
Development

McGraw-Hill Engineering and Technology Management Series

Michael K. Badawy, Ph.D., Editor in Chief

Managing New Technology Development

Wm. E. Souder

J. Daniel Sherman
University of Alabama in Huntsville

McGraw-Hill, Inc.

New York   San Francisco   Washington, D.C.   Auckland   Bogotá
Caracas   Lisbon   London   Madrid   Mexico City   Milan
Montreal   New Delhi   San Juan   Singapore
Sydney   Tokyo   Toronto

Library of Congress Cataloging-in-Publication Data

Managing new technology development / edited by Wm. E. Souder
and J. Daniel Sherman.
     p.   cm.
    Includes index.
    ISBN 0-07-059748-0
    1. Technology—Management. 2. Technological innovations—
Management. 3. New products—Management. I. Souder, Wm. E.
(William E.) II. Sherman, J. Daniel.
T49.5.M347   1994
658.4'062—dc20

          93-20840
             CIP

1 2 3 4 5 6 7 8 9 0 DOC/DOC 9 9 8 7 6 5 4 3

ISBN 0-07-059748-0

*The sponsoring editor for this book was Robert Hauserman, the editing
supervisor was Nancy Young, and the production supervisor was
Pamela A. Pelton. This book was set in Century Schoolbook. It was com-
posed by McGraw-Hill's Professional Book Group composition unit.*

Printed and bound by R. R. Donnelley & Sons Company.

For all managers of new technologies

Contents

Contributors

Calantone, Roger.   Michigan State University (CHAP. 7)

Clarysse, B. J.   University of Ghent, Belgium (CHAP. 9)

de Meyer, Arnoud.   INSEAD, France (CHAP. 9)

di Benedetto, Anthony.   Temple University (CHAP. 4)

Gupta, Ashok K.   Ohio University (CHAP. 3)

Jelinek, Mariann.   College of William and Mary (CHAP. 5)

Lawless, Michael.   University of Colorado at Boulder (CHAP. 10)

Leonard-Barton, Dorothy.   Harvard University (CHAP. 6)

Litterer, Joseph.   University of Massachusetts (CHAP. 5)

Millson, Murray.   Syracuse University (CHAP. 8)

Moenaert, R. K.   Free University of Brussels, Belgium (CHAP. 9)

Montoya, Mitzi.   Michigan State University (CHAP. 7)

Rubenstein, Albert.   Northwestern University (CHAP. 2)

Sherman, Daniel.   University of Alabama in Huntsville (CHAPS. 1 AND 11)

Smith, Warren.   Harvard University (CHAP. 6)

Souder, Wm. E.   University of Alabama in Huntsville (CHAPS. 1 AND 11)

Wilemon, David.   Syracuse University (CHAP. 8)

Dr. Wm. E. Souder holds the Alabama Eminent Scholar Endowed Chair in Management of Technology at the University of Alabama in Huntsville (UAH). At UAH, Dr. Souder is also the founder and Director of the Center for the Management of Science and Technology (CMOST), and he holds positions as Professor of Engineering and Professor of Management Science. He has Adjunct Research Professorships at the University of Pittsburgh and at the Georgia Institute of Technology. Dr. Souder received the B.S. with Distinction in Chemistry from Purdue University, M.B.A with a concentration in marketing from St. Louis University, and Ph.D. in Management Science from St. Louis University. He has 12 years of varied industrial managment experience in the chemical industry, 7 years of government laboratory experience and over 20 years of experience in academe where he has initiated several courses and programs in the management of technology. He has founded three small start-up firms, and was for many years Professor of Industrial Engineering and Director of the Technology Management Studies Institute at the University of Pittsburgh. He has many years of consulting experience with numerous *Fortune* 100 firms and has a long track record of research in the management of R&D, engineering, and innovation. He has served as a referee, board member, and editor of several professional journals and is the author of over 150 publications and 6 books. Dr. Souder is the recipient of numerous awards, including one from the White House for service on the President's commission on industrial innovation policies.

Dr. J. Daniel Sherman received a B.S. degree from the University of Iowa, a M.A. degree from Yale University, and a Ph.D. in organizational theory/organizational behavior from the University of Alabama. In 1989–90 he was a visiting scholar at the Stanford Center for Organizations Research at Stanford University. He currently serves as chairman of the Management and Marketing department at the University of Alabama in Huntsville. He is the author of over 40 research publications and his research has been published in *Academy of Management Journal, Psychological Bulletin, Journal of Management, Personnel Psychology, IEEE Transactions on Engineering Management*, and other journals. His current interests focus on the management of innovation.

Series Introduction

Technology is a key resource of profound importance for corporate profitability and growth. It also has enormous significance for the well-being of national economies as well as international competitiveness. Effective management of technology links engineering, science, and management disciplines to address the issues involved in the planning, development, and implementation of technological capabilities to shape and accomplish the strategic and operational objectives of an organization.

Management of technology involves the handling of technical activities in a broad spectrum of functional areas including basic research; applied research; development; design; construction, manufacturing, or operations; testing; maintenance; and technology transfer. In this sense, the concept of technology management is quite broad, since it covers not only R&D but also the management of product, process, and information technologies. Viewed from that perspective, the management of technology is actually the practice of integrating technology strategy with business strategy in the company. This integration requires the deliberate coordination of the research, production, and service functions with the marketing, finance, and human resource functions of the firm.

This task calls for new managerial skills, techniques, styles, and ways of thinking. Providing executives, managers, and technical professionals with a systematic source of information to enable them to develop their knowledge and skills in managing technology is the challenge undertaken by this book series. The series will embody concise and practical treatments of specific topics within the broad area of engineering and technology management. The primary aim of the series is to provide a set of principles, concepts, tools, and techniques for those who wish to enhance their managerial skills and realize their potentials.

The series will provide readers with the information they must have and the skills they must acquire in order to sharpen their managerial performance and advance their careers. Authors contributing to the series are carefully selected for their expertise and experience. Although the series books will vary in subject matter as well as approach, one major feature will be common to all of them: a blend of practical applications and hands-on techniques supported by sound research and relevant theory.

The target audience for the series is quite broad. It includes engineers, scientists, and other technical professionals making the transition to management; entrepreneurs; technical managers, and supervisors; upper-level executives; directors of engineering; R&D and other technology-related activities; corporate technical development managers and executives; continuing management education specialists; and students in engineering and technology management programs and related fields.

We hope that this series will become a primary source of information on the management of technology for professional managers, researchers, consultants, and students, and that it will help them become better managers and pursue the most rewarding professional careers.

Dr. Michael K. Badawy
Professor of Management of Technology and Strategy
Virginia Polytechnic Institute and State University
Falls Church, Virginia

Preface

This book is part of the modern management of technology (MOT) movement. The MOT movement is the result of the recognition by scholars and practitioners alike that new models, concepts, and paradigms of management are required to guide the management of the development of new technologies and new product innovations.

This book is designed to respond to the needs of practicing managers, researchers, and students in the area of the management of new technology development and new product development. It is of particular interest to practicing R&D managers and marketing managers seeking information on how to manage the process of new product innovation. For researchers in areas relevant to the management of technology, this book contains valuable information reviewing current research in a number of MOT areas and identifies directions for future research. This book is also designed to be a textbook for either a graduate-level or upper-level undergraduate course in the management of technology, new product development, product design, or engineering management.

This book takes the reader on an exciting journey. It is a journey through the fascinating life-cycle development of a typical new technology. This cycle begins with the origin of a new technology and ends with its successful application as a product in the customer's environment. During the journey, this book presents a combination of practice and research, experience-based advice, and state of art MOT methods for effectively managing each step in the new technology development cycle.

While many of the topics are addressed in other texts, few books address the process systematically as a life cycle. The management of the new product development process is a major theme in U.S. industry today. Thus, the need for books focusing in this area is rapidly emerging. In addition, this book presents information on new areas such as cross-cultural comparisons with Europe, issues associated with managing telecommunications, and directions for future research.

A final important aspect of this book is the quality and reputation of the contributors. This book brings together in one place the latest knowledge and practices of some of the foremost experts in the management of technology. The MOT area is just now beginning to emerge as a systematic discipline of research and management practice. We hope you, the reader and user of this book, will find it a stimulating state-of-the-art guide to the best new product MOT practices and concepts.

Wm. E. Souder
J. Daniel Sherman
Huntsville, Alabama

Critical Front-End Decisions: Idea Generation, Concept Development, and Market Definition

Part 1 of this book introduces the reader to the new product/technology development process and focuses on the "critical front-end decisions" in that process. The front-end decisions are "critical" because they largely determine the conduct of the remaining steps in the process and their effectiveness.

In Chap. 1 the new technology and new product development process is discussed in terms of its life cycle. In this introductory chapter, the logic underlying the organization of the book and the major stages in the product development life cycle are addressed. Chapter 1 also presents the sequencing rationale and an overview of the chapters in this book.

In Chap. 2 Professor Albert Rubenstein addresses the critical initial stage of the new product development cycle: idea generation. The quality of the new product and its market success are dependent on the quality and innovativeness of the ideas generated in this initial, critical stage. In this chapter, critical organizational roles in the idea-

generation process are described. Intrapreneurship, centralized versus decentralized R&D, the role of cooperation between functionally distinct groups, and the appropriate use of outsourcing versus internal venturing for new product ideas are discussed. Most importantly, this chapter addresses questions of how new ideas can be stimulated and how their development should be managed.

Professor Ashok Gupta addresses the critical issue of developing new product concepts in Chap. 3. The product concept is formed by linking the initial technological ideas with the customer's needs and requirements. Thus, at the concept development stage of the cycle, critical decisions must be made to minimize the likelihood of new product failure. Gupta presents several market research techniques which can be utilized to assess customer needs and to tailor the product concept to the customer's requirements. These include the use of conjoint analysis for selecting among alternative product concepts, perceptual mapping to assess customer perceptions of the product concept relative to competing products, and the use of focus groups for concept development and testing.

Building on the ideas presented in Chap. 3 by Professor Gupta, Professor Anthony di Benedetto focuses on defining target markets in Chap. 4. Di Benedetto discusses the importance of user needs and outlines methodologies for screening candidate new product concepts. Following screening, he discusses the development of prototypes, prototype use tests, and pretest marketing. Pretest market models like ASSESSOR, TRACKER, COMP, and LITMUS are presented and discussed in terms of their potential utility. Based on customer feedbacks from such testing, subsequent product modifications can be made before the full-scale product launch.

1

Introduction

Wm. E. Souder

J. Daniel Sherman
University of Alabama in Huntsville

The Challenge

New technologies, products, and services are the heartbeat of the world. They create employment, economic growth, high standards of living, high qualities of life, and world progress. Our daily existence is facilitated by hundreds of products and services that were unknown a decade ago. They have become so commonplace in our standard daily routines that we seldom think about how difficult it may have been to create them.

The successful development of new technologies, products, and services remains a major challenge for today's managers—all over the world. It will undoubtedly continue to be a major challenge for tomorrow's managers. The challenge is simple to state: how to transform technologies into useful products and services in a timely and economical way. The challenge has never been easy to meet.

Some organizations in some countries, with some technologies, have met this challenge better than others at some times. No one can say they have been universally successful. In part, this is because the challenge is dynamic. It changes with technologies, people, cultures, processes, organizations, groups, customers, products, knowledge, past experience—the list of contingencies is nearly endless. The challenge seems to outrace even the best experience and knowledge curves. Yet within all these ever-changing challenges, the notion persists that there must be some baseline best practices, principles, and theories that can be used to guide the effective management of new technology developments. This book focuses on those practices, principles, and theories.

Managing the New Technology/New Product Development Process

It seems oversimplified to speak in terms of *the* new technology/product development process, as if there were one universally successful process. There is not. There *is* a generally accepted *template* for successfully moving a technology from its birth to its final implementation in customer environments. This stepwise process is shown in Fig. 1.1. The template in this figure is a summary of a much more detailed process that contains many other steps and many dynamic interactions between these steps. Important choice decisions lie behind each of these steps. These decisions often involve irreversible commitments of scarce organizational resources and significant risks to both the organization and the careers of the managers who make them. But, for those who succeed, considerable financial and personal rewards exist.

Research and experience have repeatedly demonstrated that following the general process outlined in Fig. 1.1 is important to the successful development of most new technologies. However, there are many successful variations of this model. They are necessarily contingent upon the nature of the technology, the customer, and many other factors. The challenge for managers is to understand when variations from the template in Fig. 1.1 are allowable. For example, many of the steps in the figure consume time and resources. Resources are invariably scarce and time is money. When should any of these steps be bypassed? Concept testing is the quintessential example. Because testing is expensive, time consuming, difficult to design, and therefore prone to yield incomplete results, a strong impetus exists to go directly from concept to production. The answers to questions about optimal test strategies depend on many variables. All such questions should involve a careful analysis of the risk-cost trade-offs and the reward-regret outcomes of various actions.

Another common variation is to compress the duration of the entire new technology development process by accelerating some steps and/or combining adjacent steps. In today's fast-paced global economy, cycle time reduction (reducing the total time to get through the entire process in Fig. 1.1) has become a major pursuit. Several methods are available for compressing various steps within the process (e.g., applying additional resources, subcontracting some of the work to suppliers, using quality management methods,

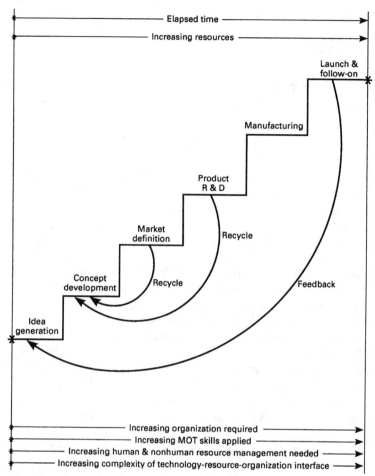

Figure 1.1 A stepwise view of the new technology development cycle.

and conducting some of the steps in parallel). For example, 3-D graphics, virtual reality, stereolithography, and other computer technologies can be applied to combine concept development with market testing. Stereolithography can be used to develop a working prototype of the product from a sketch in the user's presence. The user's comments and thoughts can thus be solicited and incorporated in real time during the design process. Virtual reality techniques allow the user to gain hands-on experience with realistic computer models of the product in ultrarealistic but simulated environments. When should some steps be accelerated? When should the above methods be used? The answers depend on the cost-effectiveness of these techniques in each particular situation and the risk-reward factors in using them.

Focus and Origins of This Book: The Management of Technology (MOT) Movement

This book is focused on the process outlined in Fig. 1.1. The various steps that are summarized in the figure are detailed. The many factors behind those steps are discussed, and the best management practices are presented for each step. This book prescribes rules of thumb, policies, and procedures for deciding which steps are most important, which steps should receive the most attention, and when some steps may be skipped. It discusses numerous variations in the process outlined in Fig. 1.1, with an eye to designing the best process under various circumstances. Making risk-resource trade-offs between various choices within the process is discussed (when to test, when some steps can be skipped, etc). Throughout, real-world examples and organizational experiences are cited to illustrate the principles and concepts. The objective is to impart a better understanding of the emerging paradigm for managing new technology/product development. Better understanding means better management of the new technology/product development process. Better management means greater efficiencies within the process and therefore better quality outputs (products and services) from the process.

This book is part of the MOT movement. This movement is an outgrowth of the demands of modern technologies. Modern technologies will continue to evolve along their established trends. They will exhibit increased complexity, involve multiple sciences and disciplines, create many uncertainties, have multicultural origins, exhibit many externalities, and involve complex interfaces with other technologies. Traditional, classical management theories are inadequate to handle such complexities. New management theories and concepts will be needed. This need has recently been recognized by scholars and practitioners throughout government, academe, and industry. Their joint efforts have culminated in a management theory revisionist movement. This emerging effort has been dubbed the MOT movement. Through research, studies of best practices, and real-world experimentation, the MOT movement is currently developing empirically based management theories and concepts for effectively managing tomorrow's technologies. The new technology/new product development paradigm is an integral part of that MOT movement.

Because the MOT and new technology/new product development paradigms are only now emerging, this book has necessarily taken a very eclectic approach. The reader will find the chapters to be a mixture of theory, concepts, proven practices, rules of thumb, and prescriptions for effective management. The main focus is on industrial new products. However, the distinction between industrial and consumer products is often blurred (e.g., the personal computer). Thus, the reader will find that many of the concepts and examples are crossovers. That is, they present lessons that are equally applicable in industrial and consumer arenas. The chapters represent the latest thoughts of world experts on each step within the new technology/new product development process. Some differences in writing style and approach will be noted among the chapters, as one might expect from different experts. These differences have been intentionally preserved in order to convey the realities of an emerging discipline. The reader will also find that this book uses the terms "technology," "product," and "service" almost interchangeably. Indeed, there are some important differences in the way the new technology development process is carried out for products and for services. These differences will be pointed out in the various chapters. The reader should also note that the static picture of the new technology development process used here (e.g., the linear stepwise process depicted in Fig. 1.1) is created only to facilitate our analysis of the new technology development process. In reality, many loopbacks, feed-forward loops, and other dynamics constantly occur between all the steps in the process. It must be noted that successful new technology/product development processes for some organizations may not culminate in the very last step shown in Fig. 1.1. For example, technologies may be successfully sold or licensed at the end of their concept development step, or their production processes may be purchased by other developers. Some organizations exist for the sole purpose of developing and selling ideas to other organizations, who then take these ideas on to further stages of their development.

This book is necessarily limited to discussing the general new technology/new product development process with some of its more common variations and to presenting prescriptions for improving those processes. It is impossible to cover all the many variations of this process or to try to particularize the prescriptions for every process. It is hoped that the reader will find relevancies to his or her own world and will be able to think by analogy as necessary to use the material in this book.

TABLE 1.1 The Chapter Plan

Parts	Chap. no.	Intro. to the book	Idea generation	Concept development	Market development	R&D & product development	Mfg. & operations	Product launch & follow-on	Summary of the cycle	Supporting functions & mgt. oversight	International considerations	Emerging issues	Directions for future research
				The new technology/product development cycle						Challenges and needs over the cycle			
			Critical front-end decisions			Decision implementation							
Part 1: Idea, generation, concept development, & market definition	1	X											
	2		X										
	3			X									
	4				X								
Part 2: Managing development, manufacturing, and launch	5					X				X			
	6						X						
	7							X					
	8								X				
Part 3: Emerging issues, cross-cultural issues, and directions for future research	9									X	X		
	10											X	
	11												X

X = this chapter focuses on this element.

Rationale for the Flow of the Chapters in This Book

Table 1.1 presents the rationale for the order of the chapters in this book. The chapters are grouped into three parts, based on their alignment with the theme presented in Fig. 1.1. Part 1 contains Chaps. 1 through 4. These chapters give an overview of the new technology/ product development process that is depicted in Fig. 1.1 and discuss the three steps (idea generation, concept development, and market definition) within the critical front-end decisions portion of this process. These functions are critical in the sense that their outcomes will largely determine the remainder of the technology cycle. New product development managers often refer to this front end as the *fuzzy front end* because they encounter considerable uncertainties in making the basic decisions about products and markets that are critical to determining the direction of the remaining steps in the process.

Part 2 of this book contains Chaps. 5 through 8. These chapters discuss the decision implementation portion of the process, in which the decisions made during the front-end decisions portion are implemented and followed through. This grouping into critical front-end and implementation activities is not meant to imply that decision making is not a mainstay of the implementation portion of the process. Management decision making is the primary focus of this book, and it is the major challenge of all the steps in the process depicted in Fig. 1.1. Rather, the distinction between the critical front-end and the implementation portions is a matter of focus. The front-end activities lay the groundwork for the subsequent implementation activities. These implementation activities include carrying out the actual product development and manufacturing, launching the new product, and providing for its service and follow-on, as discussed in Chaps. 5 through 8 (see Table 1.1).

Part 3 of this book contains Chaps. 9 through 11. These chapters discuss several key issues, emerging trends, and directions for future research in new technology/product development. These chapters are included because the process depicted in Fig. 1.1 and discussed in Parts 1 and 2 of this book is undergoing some evolution. Although the cycle model in Fig. 1.1 is accurate and relevant, the advent of a global economy, the rapid emergence of complex new technologies, and revisions in management theories are leading to changes in the way the cycle is managed. Thus, the contents of Part 3 are designed to alert the reader to new horizons and vistas that are now occurring with respect to the process depicted in Fig. 1.1.

Overview of the Chapters in This Book

As Table 1.1 notes, Chap. 2 begins the discussion of the front end (Part 1 of this book) of the new technology/product development process. This chapter covers the first step in the process: idea generation. Like any process, the final outcome of the new technology/product development process is heavily dependent on its origins. The final transfer and acceptance of the developed technology will depend on the source and quality of its origins. The care and feeding of the original idea for the technology is a vital part of the overall process. In this chapter Professor Rubenstein shares many valuable lessons that are based on many years of research, consulting, and personal experience. He shows how individual and organizational roles, tensions between corporate and divisional operations, cooperation between functionally specialized groups, organization cultures, the proper use of outside agencies as idea sources, and internal venturing are all matters that demand high-quality management attention.

Chapter 3 takes the reader to the next major step in the overall new product development process: concept development. This is another vitally important step in the front end of the new product development process. New product failures can often be traced to an inadequate or improper product concept. The old adage that new product success is a matter of staying close to the customer is epitomized in concept development. A concept that does not fill the customer's expectations will never become a completely successful product. In this chapter, Professor Gupta provides an in-depth look at the need to carefully transform technological ideas into powerful product concepts. He discusses several techniques for assessing customer needs and desires and for tailoring the product to the user. His treatise is carefully punctuated with examples and anecdotal stories that effectively illustrate the ideas he presents. The lesson is simple: Take the time to develop high-quality product concepts. As Professor Gupta points out, this rule is easily violated.

Chapter 4 concludes the discussion of the front end of the new technology/product development process (Part 1 of this book). This chapter continues the theme that was begun in Chap. 3, while making a bridge for the reader to the next step: understanding the user and defining markets for the new technology. In this detailed and illustrative chapter, Professor di Benedetto stresses the importance of having a clear conceptualization of the new technology as a basis for targeting the market and the user. He forcefully reiterates the importance of conducting a thorough user needs analysis, conducting product use tests with prototypes, and carrying out careful market tests. Several important technology screening, technology fitting, and user needs analysis techniques are presented. Iteratively adjusting the technology to fit the user, often in an extended trial-and-error fashion, is an essential part of the new technology development process. The reader will find that this chapter contains the latest state-of-the-art thoughts about how to determine the optimum testing strategy.

Chapter 5 moves the reader on to Part 2. This part of the book focuses on implementing the decisions and foundations for new technology/product development that were fixed in Part 1 (see Table 1.1). Chapter 5 deals with aspects that pervade the physical development/R&D step in the process: organizing and managing (see Fig. 1.1). Organizational, motivational, and general management issues begin to become of paramount concern at this step, as efforts ramp up to meet the challenges of new technology development. Professors Jelinek and Litterer demonstrate the inadequacies of traditional management and organizational methods as vehicles for modern new product development. The authors show how structuring for effective management of technological innovation requires entirely new directions in the management of strategic change.

Moving on into Part 2, Chap. 6 takes the reader to the next step in the process: product engineering and manufacturing. Collectively, these activities are often referred to as the "operations," or "process," phase of new technology development. This chapter is an exciting new look at a traditional topic. The theme that Professors Leonard-Barton and Smith advance is subtle yet profound: Manufacturing and processing know-how represent a vital repository of organizational knowledge. New technology developments must thus be viewed in light of the organization's repository. The repository is the foundation for effective new products, and it must be maintained in a state of readiness, consistent with the nature of the technology being developed and the nature of the markets that the technology is expected to serve. Many real-world anecdotes and examples are cited by the authors to buttress their cogent points and recommendations.

Chapter 7 completes the discussion of all the steps in the new technology development process. This chapter focuses on the last step in the process: new product launch and follow-on. Professors Calantone and Montoya detail the traditional principles of taking a technology to market. These principles involve managing customer interfaces, setting distribution strategies, transferring the completed technology, and providing responsive customer service. In their discussions, the authors provide many examples and illustrations to crystallize their conclusions and recommendations.

Chapter 8 completes Part 2 by summarizing the materials and themes in Chaps. 2 through 7, while also detailing the emerging dynamic MOT paradigm of new technology development. Professors Wilemon and Millson contrast the traditional approach with the new approach to new technology development. The traditional perspective emphasizes a strict recipe for new product development. The new perspective emphasizes fast, flexible, and close-to-the-customer MOT development operations. Professors Wilemon and Millson vividly demonstrate a statement made earlier in this book: It is not so much how the process is prescribed on paper that makes it effective. Rather, it is how the dynamics of the process are actually managed. In this chapter, the reader will be able to vividly see how earlier statements made in this book about the emergence of an eclectic MOT paradigm are in fact occurring. Professors Wilemon and Millson show how the emerging paradigm is built on some of the traditional principles that were detailed in Chaps. 2 through 7. The reader will also see how the emerging paradigm combines these traditional principles with several new thoughts. Finally, the reader will see how the emerging paradigm contains some radically new approaches that are required to cope with tomorrow's expected MOT challenges. Thus, Chap. 8 serves as a close-out summary of the materials in Part 2 on the implementation portion of the new technology/product development process. Chapter 8 also serves as a bridge to Part 3, where the focus turns to emerging issues, challenges to the management of new technologies and products, and directions for future research in the management of new technology development.

In Part 3, Chaps. 9 through 11 summarize several important aspects of concern for the future. Chapter 9 introduces the reader to cultural differences in how the state of art in managing the new technology/product process is practiced. The authors discuss important cultural differences between American and European new product processes and their management. This is an important topic for tomorrow's managers, as the world rapidly becomes one large globally connected economy. Understanding how technologies are managed differently in different cultures and, especially, the reasons for these differences are becoming vitally important for survival in the new global economy.

Chapter 10 presents the emerging theory and concepts of the strategic management of technology (SMOT). Coverage of the new technology development process and the MOT movement is not complete without a discussion of how this process dovetails with an organization's strategic management processes. But SMOT is not yet a fully developed discipline. Although many important notions and general lessons for managers have been defined, it is not yet possible to prescribe the detailed steps that must be taken to integrate new technology development processes with strategic technology management philosophies. Thus, this chapter is a progress report on accomplishments to date. It presents materials that managers and researchers should know to prepare them for future developments and thoughts about SMOT.

Chapter 11, the last chapter in this book, is the clean-up batter in this journey through the life cycle of a typical new technology development. It focuses on the unresolved challenges for tomorrow's mangers. Chapter 11 is a summary of the remaining issues to be resolved in creating an effective MOT paradigm that managers can routinely use for managing modern new technology developments. It is thus an agenda of directions for future research.

A Comment on the Chapters

The reader may have noted that one step in Fig. 1.1, the physical product development, or R&D, step, has not been accorded a separate chapter in the discussions in this book. This all-important step has not been slighted. The R&D function is critically important to MOT and new technology development. Entire books have been written on the R&D function.

But the new product R&D portion of the new technology/product development process is actually very diffuse. It plays a critical role in idea development, concept development, new process development, new manufacturing initiatives, and many other finite steps within the overall new technology/product development process. Thus, a conscious decision was made to treat the product R&D step in its proper role: as an integrated part of other steps. The reader should note that Chap. 2 includes a discussion of the relevant R&D functions, Chap. 6 includes a discussion of many product R&D aspects, and Chap. 8 includes a summary of the product R&D step. For a more thorough discussion of R&D functions, the reader should consult standard textbooks on R&D management.

The reader will also note that several themes run through all the chapters in this book, and some topics are repeatedly treated in more than one chapter. For example, management concepts, human resource management, project management, and organizational learning issues appear in several chapters. This is a result of artificially segmenting a process that is not easily segmented. The integrated new technology development process has necessarily been segmented into its several steps in this book in order to better examine them in detail.

The overlap in topics is also a natural artifact of the universal importance of these issues and of the diversity of perspectives that experts often have on complex issues that remain incompletely resolved. The reader is challenged to reflect on the occasional differences that the various experts who wrote the chapters in this book have taken on these issues. One of the experts is not necessarily wrong. Rather, it is not unusual to have more than one "good" solution to a common management problem.

Summary

This book is a unique new product. It is an eclectic, omnibus collection of theories, expert opinions, state-of-the-art practices, and immediately useful advice for improving new technology development processes. Thus, it may have many applications and serve many needs.

This product (book) sprung from the notion that a gap existed in the emerging market for MOT literature. A recognition of a gap is a typical origin of many new products. The development of this book then followed the new technology development process, almost exactly the way that process is detailed in this book. As the development proceeded, it became apparent that eclecticism and breadth was necessarily required of a book like this.

The new technology development field consists of many highly effective traditional methods and principles. Yet, newer approaches, philosophies, practices, and expert opinions are rapidly emerging as a result of changes in the world. In some cases, the newer methods are quickly rendering the traditional approaches obsolete. In other cases, they are combining with them to create even more powerful approaches. Therefore an eclectic book is necessary to convey the state of the art in this changing field, but it is also a necessary artifact of the emerging nature of the field. It is hoped that this product will thus serve many different readers in many different ways.

2

Ideation and Entrepreneurship

Albert H. Rubenstein
Northwestern University

Ideas are the foundation for the new technology development process. The new product or service that is derived from the process is necessarily limited by the quality and origin of the idea. How can new ideas be stimulated? How can their quality be maximized? Are there any rules for their proper care and feeding? Are ideas generated outside the firm better than ideas generated inside the firm? How do effective managers keep good ideas from being killed by the natural forces that exist within most organizations? Are special organizational arrangements necessary to generate good ideas? These are some of the questions answered in this chapter.

WM. E. SOUDER AND J. DANIEL SHERMAN

Introduction

This chapter deals with the front end of the R&D/innovation (R&D/I) process—the generation and communication of "ideas." This early stage is critical to the whole R&D/I process since it is the starting point for technical activities or projects which are intended to provide new and improved products, processes, services, and know-how to support the technology base of the firm's current and potential future activities. It is also, interestingly, critical for the "downstream" end that encompasses the application, adoption, implementation, transfer, and effective use of the results of the firm's technology activities. That is, if "bad" ideas form the basis for the firm's R&D/I "project portfolio," the results are not likely to be winners in either technical or commercial terms. Good technical work on bad ideas has led many industrial R&D labs into a situation of missed opportunities, wasted resources, lost credibility, and, eventually, decline or dissolution. The old question that many Ph.D. graduates ridicule when it is posed by recruiters—"if you had a lot of money and time, what would you work on?"—is no joke. Firms which hire scientists and engineers for R&D/I

functions are not just looking for project-*doing* capabilities; they are looking for project-*originating* capabilities—that is, idea generation, promotion, and development. Relatively well-trained R&D technical people working on irrelevant, unfeasible, insignificant, me-too projects which hold little promise of providing the firm with a comparative advantage in the market can be and have been a great waste of resources and a rationale for cutting back and reducing the role of, for example, the corporate research lab. It seems trite, but it is true that ideas are the technical life blood of the firm, whatever their sources. Without a steady stream of high-potential ideas, the firm's technical efforts in R&D, engineering, product development, manufacturing methods, market development, and customer and factory service can fall far short of needs and expectations.

Theoretically, one can imagine a set of all possible ideas of potential relevance to a particular field of technology, market, product line, or business. This initial set is based on the state of the arts involved and the underlying theory and empirical experience in that field. Seldom can a given firm generate such a "total" list, let alone pursue or exploit a significant percentage of all potentially relevant and useful ideas. This is because of the many constraints under which the firm operates—time, resources, skills, openness to ideas, patterns of cooperation within the firm, availability of outside sources of expertise, and many more. Despite this limitation, a firm that is in a position to take advantage of technology to differentiate its products and support them in the factory and the market, to seize new technology-based opportunities early in the game, and to fend off competitors with strong technology programs is likely to be one which has paid a lot of attention to this "front end" of R&D/I—idea generation and development. The importance of idea generation to the acceptance, use, and transfer of the results of R&D/I was mentioned above. More will be said about that in the following section on the R&D and marketing interaction. At this point, however, it is important to note that the actual source or sources of ideas that turn out to be successful technically—that is, lead to successful R&D projects—can have significant impacts on the acceptance of the R&D results by potential external clients or users inside the firm. Although the term not-invented-here (NIH) has been overworked in attempting to explain the failure of R&D/I projects and programs, it is clear that joint authorship by R&D and marketing (or at least early warning of or involvement with an idea coming down the pike) can have a significant impact on the efforts devoted by both groups and its ultimate commercial or financial success (e.g., as a cost-saving manufacturing innovation or a profit-producing market innovation).

Many individuals and groups, both inside and outside the firm, have roles to play at the front end, or idea stage, of R&D/I. The following sections examine several of them: R&D, marketing, corporate research labs (CRLs), operating unit or divisional technology groups, special entrepreneurial teams, and a variety of sources outside the firm itself.

Individual and Organizational Roles at the Front End of the Idea Flow Process

From the viewpoint of research on the R&D/I process, it would be nice to think that all ideas for new and improved products and processes have a clear time- and space-specific origination point. Many of them do, and it has been possible to identify both the originator and his or her organizational location (function, department) as well as the approximate time of origination. This identification has been helped by the existence and careful use of research notebooks, internal correspondence, and idea origination or proposal forms in many of the field sites where idea flow studies have been collected (e.g., see Rubenstein, 1963a and 1963b). The original notation or version of many of the ideas that have been identified and traced forward in time bears little resemblance to the final form of the idea as manifested in actual products, manufacturing processes, or spin-offs from the original idea. However, parentage can be established and the flow and transformation can be tracked for several hundreds of ideas. In these several studies (Rubenstein, 1963a and 1963b), these ideas originated, primarily, in the formal R&D functions of the firms studied. However, transformations, modifications, add-ons, destruction, and (in some cases) reconstruction of the original idea came from many additional sources along the way toward implementation, storage, or scrapping. A lesser number of ideas that originated in manufacturing and related activities (manufacturing engineering, industrial engineering, etc.) were also traced, and under the heading of "suggestions," a number originated among manufacturing personnel who were not officially charged with or expected to "innovate" (Bonge, 1968; Martin, 1967).

For almost a decade in the 1960s and 1970s, faculty and graduate students at Northwestern University pursued studies of idea flow in industry. This research, supported by the National Science Foundation (NSF), National Aeronautics and Space Administration (NASA), the Office of Naval Research (ONR), and several industrial firms, produced over a dozen dissertations and masters theses and over a dozen publications in the open literature. The research drew on a broad range of theories in the management and social sciences, as well as on the accumulated industrial experience of the research team members. It both generated and tested propositions about idea flow which derived from and were, in turn, incorporated into a series of conceptual flow models. Tracing of idea generation, communication, and disposition (acceptance, rejection, storage, recycling) was accomplished by a number of methodological tools at the behavioral level—general interview protocols, various forms of questionnaires, self-administered record keeping, observation techniques, and records analysis (notebooks, files). Results of this research are reported in detail in the references and bibliography at the end of this chapter. In this chapter, the opportunity is presented to look back on the results of this research program and speculate on their implications for management of the R&D/I process in the firm.

Since those specific studies of idea flow in the 1960s and 1970s, there has been a trail of "ideas" in connection with research in such contexts as: technical entrepreneurship in the firm, R&D/marketing interface, R&D/production interface, the software development process, technology networking in the decentralized firm, make or buy of technology, mergers and acquisitions, and others. This variety of experiences with the R&D/I process (in several hundred firms) suggests that ideas for technical work can originate in many different places. They can come from both inside and outside the firm, and they can range from formal, systematically formulated proposals for R&D work containing both a clearly stated market or technical need and a first cut at an identified or suspected means of approach to hunches or notions of directions in which the firm might look for technical and market opportunities. It is not clear which end of the continuum contains the larger proportion of "winners" in commercial and financial terms, so in this chapter the whole spectrum of sources is considered.

In particular, this chapter deals with issues such as the various individual and organizational roles involved at the front end of the idea flow process, the decision imperatives and constraints under which they operate, the potential for increased cooperation in idea generation, and the increasing trend in many firms to look outside for ideas as compared to encouraging and exploiting internally generated ideas. This latter aspect is a matter of mounting concern, since the internal technology base of many firms seems to be shrinking and even disappearing. In addition to "outsourcing," the shrinking is occurring due to factors such as drastic restructuring, giving ground to external competition, "leaning and meaning," and even further decentralization into smaller and smaller and even more narrowly focused business units. This chapter also draws upon a number of insights gained through consulting engagements in a wide variety of firms, both domestic and foreign, in the 2 decades since the formal research on idea flow was concluded. Those insights, captured in much looser form than the findings in the above-mentioned research program, add to our knowledge of the barriers which interfere directly with the front end of the idea flow process, as well as decreasing opportunities for facilitating that part of the process through new and improved management practices.

People and groups at the intersection of R&D and marketing

Very frequently, relations between R&D and marketing in connection with idea generation and development are at arm's length and linear. That is, one group offers or suggests ideas to the other for new or improved products or services or the technology related to them. Acceptance or approval decisions are made unilaterally ("we'll accept that idea from marketing" or "we'll go along with that idea from R&D") or only nominally made in a cooperative or truly collaborative mode. Hence, many ideas are attributable to (or "blamed on") one source or the other and often do not benefit from the joint experience and conceptual capabilities of both groups. This R&D/marketing interface is one of the most critical stages in the idea generation and development process and can easily make or break the product or process based on the idea (e.g., see Souder, 1987; Gupta, et al., 1985).

Much attention in the literature on R&D/I has been paid to this interface at the downstream end of the R&D/I process—when a new or improved product or service is coming up for market testing or commercialization. Good working relationships between these critical company functions are certainly important at that stage. However, they are also critical at the front end of the process when ideas for new improved products or services are being generated and communicated prior to detailed technical work being done to develop them and convert them into concrete results.

A critical aspect of this early stage involves obtaining some accurate insight into or educated guesses about "what the customer *really* needs now and is likely to need in the future." On the face of it, this seems like a natural task for R&D and marketing to undertake cooperatively so that customers' *needs* and technology *means* can be properly combined to formulate ideas that have a good chance of satisfying those needs. Due to classical differences in makeup and approach and styles of operation, such close cooperation is rare in many firms.

Classical recriminations between the two functional areas are generally phrased in terms of the deficiencies of the "other" group, such as:

- Too short-sighted or too far into the future (blue sky)

- Indifferent to customers' expressed needs or overreactive to customers' whims

- Too cavalier about costs and difficulties of development and manufacturing or too cautious and engrossed in technical detail

- Too willing to promise far-out features or too resistant to providing new values to customers

These and other attitudes suggest that R&D and marketing are often coming from "different places" with respect to ideas for new and improved products and their commercialization. Very often the attitudes struck by the two parties vis-à-vis a particular idea put forth for R&D work on a particular product line or for a particular market are the result of one or both parties being frustrated at their inability to get at "the *real* needs of the customer" or of possible future customers.

It is important to clear the air in order for a more fruitful working relationship to develop at the front end of the R&D/I process so that more and better ideas can be generated and developed. This requires that several things occur on both sides of the interface. Marketing and their associated sales people in the operating units must look beyond immediate and obvious consumer *demands,* which can often be unstable and short term. They must try to discover future and more robust needs and market opportunities. Understanding the differences between wishes, wants, needs, and market demands must be improved and common definitions shared at all stages of the R&D/I process, including market introduction and full-scale commercialization. Customers' and potential customers' expressed needs must be probed and analyzed in more depth than is usually done and in terms of the direction those needs are likely to take in the intermediate and longer term. Radical approaches and significantly different product and technology projects should be staffed by "entrepreneurial" types of people who are risk accepting and credible enough in the organization to command resources and cooperation, given the proper incentives.

A general approach to the accomplishment of the above state of affairs might include the following steps and activities. R&D and marketing people at all levels should form a partnership of a general and continuing type, signaling to all concerned that it is okay to work closely together. Their time horizons should be reconciled with a "mixed portfolio"—short-term focus on expressed customer demands; midterm, focused on likely customer needs in the next, say, 2 to 5 years, depending on the industry, technology, and market; and long-term potential or "conceivable" needs. A methodology should be worked out jointly which can probe a current or potential customer's likely midterm and potential long-term needs, based on a deep understanding of the customer's business, products or services, problems, and underlying technology base. Rather than focus on the obvious and immediate expressed needs of the customer, the methodology should probe into "the current and anticipated problems and opportunities you, the customer, have in developing, making, selling, and maintaining the products and services you provide to *your* customers or in using the products and services *we* provide to you." From a deeper understanding of the customers' problems and opportunities, a better picture of their needs—short-, intermediate-, and long-term—can evolve. It will take longer and cost more than a quick "what do you need that we have or can develop quickly?" approach but has the potential for longer-term and more substantial benefits to the seller. Part of this methodology involves attempting to identify the "real" customers or users in the customers' organizations. They may be production, engineering, marketing, sales, or service people, and their "real" needs may not be evident to the part of the organization that does the actual buying from you. Although this is not easy, it is important to try to involve the real users in this "needs methodology." The two parties must recognize that their differences in objectives, style, time horizons, view of the customer, view of the technology, and other aspects are, indeed, different but that both are real and relevant and deserving of respect. Reconciliation of the differences does not require either party to give up any deeply held beliefs or decision premises but only to disclose them openly and completely so that joint efforts can accommodate the differences and converge upon a joint strategy for addressing customer needs. Differences could include perceived urgency, priority, or value of the target (customer, market, individual sale, entry into a new market, maintaining market position). Differences could also involve the importance of establishing and maintaining company reputation for quality, reliability, responsiveness, integrity, competitiveness, stability, innovativeness, etc.

Such a partnership requires incentives—both positive and negative—ranging from clear signals by top management that such cooperation is expected and will be rewarded to actual material rewards for cooperation and censure for noncooperation. In many firms, marketing is less likely to receive recognition or material rewards for the ideas leading to new or significantly improved products, product lines, or services. Sales often receives tangible rewards (e.g., commissions and bonuses) and R&D often receives recognition through patent filings, publications, or "ownership" of the new product. Marketing needs equivalent recognition, assuming that they did, indeed, contribute to the original idea and/or its transformations on the way to the final product or service. Mixed teams are needed to do the "needs" probing, allowing direct access to customers for R&D people—a situation which is often resisted by marketing and sales people. Mixed teams are also needed to do product planning at the concept level as well as the actual product or service performance specification level. Teams of marketing and R&D people must follow up and follow through on concepts and opportunities rather than let them sink to the bottom of the in basket or pending file. Note that the customer probe should include the following kinds of questions: What technology-related issues, problems, or opportunities do you see over the next N years in your industry? your customers' organizations? your own organization? What new and emerging technologies are likely to affect all three? What are the hot buttons in each case—that is, what keeps your executives and managers awake at night and tied to their desks or the plants for long days and weeks or tied up in endless rounds of meetings? What are the hot buttons of your technical people, your operating people, your marketing people, and other important groups in the firm?

This R&D/marketing partnership, aimed at providing both *real needs* as targets for ideas and realistic *technical means* of approaching the needs, can greatly improve the typical R&D/I front end, or the idea generation and development stages, in most firms, which are often subjected to divisive and self-defeating R&D/marketing relationships. This plea for closer and more systematic cooperation between R&D and marketing needs to be viewed in the light of a number of studies of sources of ideas that seem to show that most successful ideas come "from the customer or user"—that is, they are market or *pull* ideas, rather than technology-*pushed* ideas (e.g., see Von Hippel, 1988).

The evidence on that issue is persuasive for shorter-term and "present"-type needs. However, it is also clear that whole new technologies and industries, some of which took decades and huge amounts of money to develop and bring to market, were, indeed, the product of insight and foresight and the pressure of technical feasibility and the momentum of advancing technology rather than the direct response to expressed market needs by customers. Whatever balance between push and pull is most effective in generating ideas for R&D/I, it is clear that a combination strategy is needed by firms who are attempting to lead or anticipate market trends. The convergence of insight into "real" needs and the technological means to satisfy them profitably makes a formidable combination.

The roles of CRLs and operating unit labs/groups in idea generation

A number of things have happened to CRLs since the decades-long research on R&D in decentralized firms (Rubenstein, 1964 and 1989; Rubenstein and Radnor, 1963) and the idea flow studies in the 1960s which grew out of the decentralization study. Although some new CRLs have been established and some have increased in size and scope, a large number have been downsized, abolished, or transformed into entities which are expected to become more closely tied to operations. In addition, many more continue in an atmosphere of uncertainty about their current and future roles in the firm and how they can continue to provide general technical support for their firms' wide range of products and businesses (Rubenstein, 1985). An aspect of this trend or set of trends is that, by necessity, and in order to survive, some CRLs have become much more responsive than they have been to the current and near-term needs of their operating units and the markets and technology they represent. They are beginning to come up with ideas that are less esoteric, closer in time and scope to current products and processes, and more ready for implementation and transfer to operations. A potential downside to this increased responsiveness may be that many CRLs, once the repositories of much of their firms' technology base and source of ideas for future technology directions, are on the path to becoming adjuncts to the operating units: groups, divisions, or strategic business units (SBUs). They are reflecting, in their skills mix, recruiting patterns, idea generation, and project portfolio, the near-term and narrow technology "slices" that they perceive as more likely to satisfy their clients and founders—the operating units—rather than their earlier long-range and more esoteric portfolio (from the operating units' viewpoint).

This trend, not yet dominant but rapidly moving in that direction across many U.S. industries, leaves the firms in which it is occurring in a highly vulnerable position versus their current and future competitors and the rapidly changing state of the arts on which the firms' businesses are based. Traditionally, operating divisional labs and engineering functions have been closely tied to current operations—products, markets, manufacturing technologies, materials. The vast majority of such units in most industries, except for some very large ones in high-technology industries such as electronics and chemicals, have been quite small—on the order of tens or, at most, scores of engineers and scientists (where there are any). This becomes a huge number in industries such as machinery, primary materials, engineered products, and food, which typically have professional staffs of less than 25 and very few with advanced degrees. In addition, as the range of hiring has declined due to the restructuring and thinning out of "marginal" activities such as R&D, the average age of the professionals in these operating unit labs or departments has risen in an almost linear pattern, paralleling the situation in most federal R&D laboratories over the past couple of decades.

This means that the opportunity is equally diminished for an infusion of new ideas and new technology, which often transfers with new graduates who have advanced degrees. Taken together with the narrow, short-term focus of most such divisional labs, this throttling back of the input of new, advanced, and divergent ideas and technology constrains many divisional labs from becoming familiar with, let alone exploiting, the newer technologies. Some of their competitors may be pushing such new technologies as a means of entry into the firm's markets or "building a technology wall" around their own products and manufacturing processes. If the project portfolio and skill mix of the CRL converges on that of the firm's operating divisional labs or groups, there may be no place within the firm for radical ideas for new technologies to emerge, enter, take hold, and lead the way to new products, processes, and markets.

The "solution" may not be to attempt to reestablish the previous leadership and technologically proactive role of the CRL, where it has declined significantly or disappeared. Much of the damage is essentially irreversible. A good, broad, forward-thrusting CRL can take decades to establish. This is not just because of hiring lags and the time and resources needed to reestablish research programs. It can also take a long or "infinite" time to reestablish its credibility as a leading-edge function that can identify and help open up new markets and even industries and to reestablish the patterns of collaboration and technology transfer which may be quite different from the ones it is currently using to provide near-term support for operating units. Further, the general organizational climate or culture in most large decentralized firms—the vast majority of all U.S. firms—may not permit this return to a past role. The cutbacks, eliminations, and downgraded roles of many CRLs are the result of a sea of changes in attitudes and practices of corporate management vis-à-vis staff activities of all kinds which are focused on the future and vague role of "preserving and enhancing corporate assets" as compared to "pitching in and helping with the bottom line in the current accounting period." Thus, Human Resources, Long-Range Planning, Management Science, "Conceptual" Marketing, Organizational Development, and other such staff groups have taken big budget hits. Many have been eliminated or moved down to the operating units where their time horizons and scope are narrowed in the interests of helping the unit to enhance profits in the short term. In the climate of lean and mean, it is very difficult for such staff activities and, especially, for corporate research to justify their existence as corporate, future-oriented activities.

Given this set of circumstances, corporate managers or even some forward-looking or desperate division managers who want to or need to come up with ideas for radical improvements in their products, processes, and know-how are forced to go outside for it or do without (discussed later in this chapter). At this point, an essential issue is that where there is a poor climate and skill level for *originating* radical ideas within the firm, there is also likely to be a poor climate for accepting and implementing such ideas from outside. In fact, the probability of successful exploitation of outside ideas may be even lower than that for internal ideas which come from an internal "foreign body" like the CRL or some other division's technical people who "really don't know our business and technology."

Given the trends affecting CRLs and, subsequently, the firmwide capability for radical idea generation, is nothing to be done to repair the internal damage to the firm's technology base and its internal idea-generation capability? There are some things that can be done, but most of them are counter-current and will take major efforts by top managers to remedy the situation. Some of these remedies are listed below and discussed in more detail elsewhere (Rubenstein, 1989).

1. Restructure the funding basis for existing CRLs so that all or most of their funds will not be essentially fee-for-service to the operating units. A reasonable ratio may be about 20 to 30 percent corporate funded with no specific deliverables and the rest funded either entirely or jointly by the operating units, focused on their important intermediate-term problem areas.

2. Provide the chief technical officer (CTO) with some venture funds from which he or she can support longer-term, broader-scope, deeper, and between-the-chairs projects and programs aimed at *preventing* the firm from being technologically blind-sided and

helping the firm to gain comparative technical advantages over its current and possible future competitors.

3. Provide incentives for divisional and other operating unit managers to set aside and protect a minimum level of resources (people and funding) that can help assure the unit's longer-term survival and prosperity (e.g., escrow a *significant* percentage of each operating unit manager's personal bonus and pay it out on a contingency basis as a reward for longer-term results than the current year). Despite lack of widespread acceptance of this approach (a few firms have tried some version of it), there is hope that it might counter the trend toward near-term evaluation and reward to the exclusion of longer-term impacts on the operating unit and the firm.

4. Develop and nurture a true "open-shop" situation between CRL or staff corporate technology groups and the technology groups in the operating units. Work hard to reduce or eliminate the traditional barriers to communication and cooperation between these two kinds of entities that have grown up by virtue of differences in formal education and degrees, hands-on experience, time horizon and focus, styles of R&D, and other factors which create and perpetuate gulfs between them.

5. Encourage and, if necessary, force both parties to adjust their time horizons and to focus on technology and projects of intermediate- and longer-term interest to the operating units through funding, cross-assignments of personnel, shared incentives, and jaw-boning methods by top corporate and divisional managers.

6. Persuade operating unit managers that their best interests lie in a mixed technology portfolio which meets current market, customer, and factory needs and also provides for survival and profitability in the intermediate and longer terms.

7. Infuse a moribund or fading CRL with new blood from universities and other organizations which are active in new fields of technology of potential importance to the firm.

8. *Do* the usual "good things" to coordinate and network the operating units and the CRL into a technology capability that can address both the present and the future technology needs of the firm. This is in contrast to merely *saying* the right things via rhetoric and good intentions. For example, a technical advisory council intended to coordinate R&D and other technology programs across the company should have strong leadership, fund-

ing, and a charter that allows them to take action to fill gaps in the company's technology one way or another, including outsourcing, when necessary.

Special Arrangements for Innovation—The Internal Entrepreneurship Process (Intrapreneurship)

The internal entrepreneurship or intrapreneurship issue is vividly portrayed in the following personal anecdote. While lecturing recently on entrepreneurship, this author was approached by a middle manager in the group, who works for a relatively new high-tech company. It depends for almost 80 percent of its revenue on government defense contracts. His question related to a new area of commercial communication technology mentioned during the lecture. He wanted to "find out about this area," which he had heard about in general but was not very familiar with technically. He said it sounded like an area his firm might diversify into as defense spending wound down and as the firm looked for new opportunities. When asked "is this the kind of idea you are expected to come up with or propose for the firm as part of your job?" he said no, since he was in a technical function of a fairly routine nature, but he said that the founder and CEO was a dynamic entrepreneurial type who might well be receptive to such an idea for diversification, whatever its source. At first it appeared that this was a fanciful "notion" rather than an idea for technical work that might lead the company into new fields of technology and new markets. That is, his "idea" did not seem to fit the "standard" definition of an idea: a suggestion, recommendation, or proposal that includes both a *need* and a technical *means* of addressing that need.

Perhaps at the very front end of the idea process, it is not really necessary to have both a clear need and a clear technical means at hand in order to address a new business or technical opportunity. There might be, in this situation, more than a superficial "notion" for a venture. That is, the whole organization (less than a decade old) exists and has grown because of its ability to come up with ideas and approaches to complex communication and control problems. The company's leadership and rate of progress to date suggests that at least the idea would be given a hearing and, if it made any sense to the CEO, some support for early explorations. The field, although in its infancy, was already flooded with both large and small firms seeking a niche in it, and many of them had comparative advantages that this firm did not currently possess. It might take significant time and resources to develop or acquire such an advantage. This did not seem to discourage the idea generator. It will be interesting to see what happens in this situation.

This approach to stimulating and accelerating innovation (intrapreneurship) has not worked very well in most large companies. It also encounters difficulties in medium-size and smaller companies beyond the "one great idea" on which some of these companies were founded. There is a fundamental question about whether the *special* arrangements needed to nurture internal technical entrepreneurship are *too special* to be fit into or accommodated by the modern U.S. corporation. Based on more than a dozen cases examined over the past dozen years, both critical barriers and facilitators to the effective use of intrapreneurship have been studied (Rubenstein, 1990). These cases include two types of internal ventures: (1) those in which the original idea came from a would-be entrepreneur who proposed and fought for his or her idea against the usual tide of opposition, indifference, skepticism, or "it can't be done that way" resistance and (2) those which were proposed and accepted at upper-management levels and then assigned to an individual or small group to "get it organized" and "run with it." With such a limited sample of cases it is difficult to say which category is more likely to lead to success, especially since the overall failure rate was so high (perhaps two-thirds never achieved commercial success). Those ventures which were set up some time after the original venture-generating idea was proposed and staffed by people who did not include the idea originator or "idea champion" did not seem to fare as well as those where the original idea generator, often in the role of a "fanatic," kept pushing the idea and the venture arising from it. On the other hand, a dominant parameter in these situations was the degree and stability of support provided by top management—corporate or divisional—who provided the funding, encouragement, and protection of the fledgling venture. Even with an authentic "fanatic" at the helm (almost a necessary condition for success), such idea-driven ventures also needed management support and protection.

Where a venture did not include the originator or even any real "enthusiasts" for the idea ("it's just another project as far as I am concerned"), the road to success appeared to be much less certain and the tendency to give up in the face of adversity seemed higher (e.g., false technical starts, dead-ends, lack of interest by other functional areas of the firm, or loss of key people). Once such an internal entrepreneurial venture gains some momentum and begins to show promise of getting something out the door that may be of value to the firm, a whole set of "in-process" barriers and difficulties can (and often do) arise for this special project. Noses get out of joint at special resources, attention, and treatment (e.g., incentives) provided to the entrepreneurial team or its leader. Facilities and equipment become mysteriously backlogged and not available for use by the team in their time frame ("join the queue, buddy"). People in various functional areas are not easily assignable to the team. Haggling arises as to who pays for what (people, facilities, materials, equipment usage, market research, exploring legalities of intellectual property, and so on).

These issues and circumstances are addressed elsewhere (Rubenstein, 1989). This chapter, however, deals with such entrepreneurial teams as the *source* of ideas for not only their own venture project but for other areas of the company's business. When an entrepreneurial team is "hot," especially in the early stages before the routine and less exciting stages of a venture are encountered (getting space, people, equipment, experimental results, market testing, manufacturing procedures), ideas can come bursting forth in a gusher that can conceivably spread all over the organization. In the early "gee whiz" stage of a venture, nothing seems impossible or too far out technically to the people who originated the idea or are running with it. Elaborations of the original idea, spin-offs from it, and even ideas not clearly related to it can be given free rein in an orgy of creative expression.

Even in the rarified environment of an entrepreneurial project, sobriety soon sets in. Deliverables must be identified, practical limits must be identified, milestones must be set, and side trips must be either abandoned or put on hold in favor of the main event—getting technical and commercial results from the venture. In that atmosphere, "extraneous" ideas, especially those not clearly related to the venture itself, must be shelved or abandoned, most of them never to surface again, unless the venture becomes a full-blown operating unit and the need for "other shoe" ideas becomes pressing. In the ordinary course of events, such excess ideas are lost to the organization or at least are unavailable to it for some significant time during which market and technology conditions may have changed enough to make them less attractive than they originally seemed.

Focusing on intrapreneurial teams as sources of ideas beyond the specifics of the venture team's charter, we can see the necessity and possibility of systematically capturing additional ideas for other parts of the firm and other current and future projects. Some of these "extra" ideas are often generated through the kind of free-wheeling contacts which team members have with technical and marketing people, manufacturing people, suppliers, customers, competitors, and others, including university people. The following anecdote illustrates this point. During the writing of this chapter, a Northwestern University group was engaged in providing technical support and research for a venture group in a large firm which was seeking to enter a new field of technology with a potential worldwide market. During meetings and casual conversations between members of the team and the university researchers, ideas flew thick and fast. Few of the ideas that were not of direct, short-term interest to the venture team were well received by them. In fact, they were viewed as distractions from "the job at hand," which involved a lot of routine administrative activities and resource gathering that did not, in themselves, generate any "interesting" technical ideas. Fortunately for posterity, the motivation of the university researchers and the fact that they were also working with other organizations in the same field helped to preserve some of the ideas and give them another chance at exploitation.

Absent such special "idea-capturing" situations, other mechanisms must be used to assure that potential good and useful ideas are not lost in the heat of moving the venture along. This is especially true where the venture team is not embedded in the traditional organizational arrangement where other people and groups are available to catch and use "extra" ideas. Most typically, the venture team is relatively isolated from mainstream R&D and other technical activities—an almost necessary condition for the survival of some of them. "Idea-catching" colleagues are not at hand to investigate or preserve them. This leads to the following suggestion for the capture of these extra or spin-off ideas from venture teams. Each venture team should have a member who is assigned to do that—record, preserve, analyze, and pass on these extra ideas. If, as is usually the case, the prime team members are overcommitted and cannot or will not play this role, an additional person should be assigned to the team on a part-time or occasional basis to do the catching of ideas.

That person, with appropriate technical and/or marketing skills (or in some cases manufacturing knowledge, since many spin-off ideas deal with "how to make it" issues), should be charged with an initial examination of the ideas and their routing to places in the firm where they might be relevant to ongoing or future-oriented activities. This "idea transfer agent" should be charged with being, temporarily, a champion of this set of ideas until they are abandoned or find a receptive home somewhere in the R&D/I process of the organization.

"Irrelevant" ideas may be among the most exciting candidates for future projects in the firm because they have been generated in an atmosphere of excitement, entrepreneurship, and "suspension of disbelief" while the venture itself is being conceptualized. Even if the venture itself fails (a high probability in most firms), the spin-off of ideas may be well worth the investment in it.

Using External Sources for Ideas and Research Results

Many operating and top management members have become impatient and, in some cases, disillusioned with the number and quality of ideas for R&D/I that come from internal sources (e.g., the regular R&D, engineering, marketing, and other technology-related formal groups in the firm). They would like to skip or significantly compress the time from recognition of a need, threat, or opportunity to having, in hand, a piece of technology or a technical approach, sometimes even a solution to the problem. As a consequence, many firms are increasingly turning to outside sources for "technology packages" (e.g., product designs, manufacturing processes, and/or turn-key and other ready-to-use technology). They perceive a saving of "milling-about" time on the part of their own technology groups as well as a presorted set of technologies where the bad or infeasible ones have already been screened out. This relates to the old joke about the R&D manager who says, "The odds of getting a good product out of all the ideas and projects we have going are about 1 in 10 or 1 in 100." Fine, says the nontechnical manager—"I'll take the 1 and let's not waste time on the other 9 or 99." Of course, a premium must be paid for the presorting and all the up-front work which went on in the vendor's or other organization's own R&D functions so that the item offered for sale or license appears to be ready to go or at least highly promising.

The purchaser may also see an advantage in starting far down the spectrum of R&D/I by bypassing the early idea, exploratory, and feasibility stages so that the cycle time for new product or service development can be shortened considerably. These are compelling attractions for going outside for ideas, especially for those already embodied in product designs, equipment, or other forms. This approach often works and can satisfy certain needs for ideas and new technology. However, there are also some downsides. There may still be a significant and expensive learning period needed for the firm's own technology people to verify, further develop, and feel comfortable with the purchased or licensed technology. Ironically, this learning period can stretch out way beyond expectations since the internal people may find the technology or specific approach or package unfamiliar and may need to go "way back upstream" and learn about it from "scratch." This situation may contrast with that of internally generated ideas and technology, which arise from know-how and experience that already resides in the firm, especially among idea generators and developers. The NIH factor may present obstacles to acceptance of or even interest in the purchased items. Internal R&D and engineering people may resent, feel threatened by, or view with deep skepticism this "foreign body" that has been thrust upon them with the instructions to "make it work" (Sen and Rubenstein, 1989). The technology package from outside may not match the internal technology base and precise needs of the firm. Although such a package may appear to be focused on and intended to blend perfectly with the internal needs and technology of the firm, it is difficult for an outside group to know exactly what features and operating conditions have to be met to make the transfer a success. Some company secrets in product formulation, manufacturing, materials, quality control, etc., may either be withheld from the supplier for proprietary reasons, or their transfer at an early and appropriate stage may have been overlooked or deliberately blocked. "If that vendor is so smart, let them figure out the operating parameters of our manufacturing process." The purchased package—whether hardware or software—may not be fully documented and supported. Questions that arise some time after the purchase may not find a ready source of information in the vendor's organization; the people who worked on the package may be gone, otherwise occupied with new projects, or unwilling or unable to deal with hitherto unanticipated problems. In relation to this issue, a major study has been undertaken of user involvement in the information system development process (Hoffman, 1992). One measure of the effectiveness of this involvement (ranging from nonexistent to very heavy in the service industries) is the cost of software maintenance following transfer of or sign-off on the systems package. High maintenance costs can indicate an inadequate design, poor documentation, features overlooked in the development stage, or other mismatches of user needs and deliverables by system developers. The delivered package may constitute a "one-shot" effort and may not spin off ideas for enhancements, follow-ons, and diversifications stemming from the original idea and product or service. Again, one of the historical reasons for establishing and maintaining in-house technical capabilities (CRLs, operating unit development labs, and engineering groups) is to have a capability for *con-*

tinuing innovation in a given field or special area, based on accumulated experience with the relevant technologies, products, and manufacturing processes of the firm.

Given the above limitations, how can outside sources be effectively used for idea generation and early-stage development? This issue is discussed below in relation to particular kinds of outside sources: universities, vendors, consortia, and small high-tech companies.

Relations with universities

Smooth working relations and an effective flow of ideas and technology between two quite different entities—the university and the industrial firm—are rare. Many barriers exist that inhibit communication and effective transfer of information and ideas. Some of them, in addition, constitute "fatal flaws," which can seriously impede or destroy the relationship in terms of delivering cost-effective results to the firm. Some of the key words in these lists include "deliverables," "attitudes," "respect," "responsiveness," "relevance," "delays," "mismatch," and "incompatibility." However, some of these relationships *do* work and can be made to produce mutually satisfactory outcomes—both as part of the process itself and as part of the ultimate impacts on the goals of both organizations.

This section suggests an approach to the generation and transfer of early-stage ideas between the two kinds of entities in this "improbable" partnership. This approach concerns, again, the front end of the R&D/I process, where the firm is looking for good and relevant ideas for starting or redirecting projects and programs. The approach recognizes the different orientations of university faculty and graduate students versus those of industrial R&D and other technology groups in the firm. These differences are as they should be, in view of the differing missions and decision imperatives of the two kinds of organizations. An idea for a dissertation or scientific experiment may fully satisfy university people but leave their industrial counterparts cold. If the idea breaks new theoretical or experimental ground, it may be valued intrinsically by the university people and that is that. If it has no commercial or financial implications via cost savings, new market opportunities, or other competitive advantage, it may be looked on as useless by the industrial people *at this time.* Many ideas that were earlier rejected as being irrelevant by industrial firms have been eagerly pursued at a later date when the technology or the markets changed and competitive threats required new approaches and concepts.

The procedure suggested here concerns providing a setting or settings within which the university and industrial people can come together to generate and spin off ideas of mutual interest. The opening gambit should be a statement of perceived needs for technology and ideas by the industrial people, followed by timely responses by the university people. This is not the usual "show and tell" in which each party describes, in often numbing detail, their projects and programs. That kind of material can be presented in other ways (e.g., reports, publications, and working papers). One approach is to use technology exchange seminars and idea sessions—omit such formal presentations and get right to the hot buttons and longer-term concerns of the industrial people and the relevant knowledge and experience of the university people.

A large number of such sessions have started with the industrial people (usually a small group of under half a dozen in a particular market or technology area) briefly describing the technical and related marketing, financial, and management issues that are presently causing them concern and that are likely to cause them concern as they develop, grow, change, and otherwise pursue their R&D/I activities. The university people follow with specific ideas on how the issues might be approached—drawing upon theory, experimental results, direct experience, the literature, and their own intuition. Elaboration of the issues follows, and then the sessions are followed up with a series of "idea memos" which are not quite proposals for new projects but which contain the seeds of new projects which might be jointly undertaken. The design features for such sessions are important if the kinds of serious mismatches discussed earlier on CRLs and divisional technology relations are to be avoided. Since each group knows the role of the other, there is no need for a contest to see who is smarter, more erudite, more practical, or more creative. Mismatches do, in fact, occur. But these can be handled by subtle but firm chairing of the sessions, allowing different viewpoints to be expressed and diverting person-centered to task-centered differences or conflicts.

A major advantage of this type of interaction for the idea-generation process is that many of the ideas thrown onto the table and augmented in the discussion are essentially "free" for the taking and using. The objective of the university people, of course, is to propose ideas that need further R&D work in the form of contracts or grants to the university group. The ideas themselves are the starting point for problem solution and product or process generation rather than an end product. In some cases, an idea thrown out during such discussions causes a light to flash in the mind of one of the industrial participants and helps him or her to see a direct connection with a problem or gap in knowledge that can immediately, or in the near term, be helped by the idea with or without a great deal of further R&D work. Of course, good faith must be the basis of the sessions in that both parties want to gain something and the value of intellectual property must be respected and dealt with fairly.

Where equally competent people from both sides participate in a university-industry relationship, whether it is a seminar such as described above or a formal grant or contract arrangement, ideas can flow easily and be captured by both parties for individual and mutual advantage. Where too many of the "barriers" are salient and occupy too many of the resources of the collaborating group, the potential for idea generation may be devalued and the mechanics of the collaboration may come to dominate. In that case, the likelihood of a flow of good ideas being generated as part of the relationship may drop radically. As an example, our research group is currently involved in one such relationship which has degenerated to the point where technical idea exchange and development have given way to haggling over terms and conditions of the formal contract, deliverables, costs, accounting for time, and other "administrivia." No good ideas have emerged from the relationship for several months and are not likely to unless this undergrowth is cleared away and the relationship is refocused on idea generation and solving technical problems.

Of course, the distance between what many university researchers consider a "good, relevant" idea and what an industrial researcher or marketer is looking for may be sizable. Considerable translation and concessions may be needed on both sides to get an idea to converge upon a practical approach to a real problem. Likewise, if an idea generated by an industrial person is viewed as trivial or self-evident by the university people, it may be a long time before a happy marriage is consummated. Mutual respect for each other's vantage point (where they are coming from relative to the idea) is essential if the relationship is not to bog down in differences over language, time perspective, feasibility, resources required, etc. Again, a moderator (or moderators, perhaps one from each side) who respects and feels comfortable with both viewpoints can greatly facilitate idea generation and development. "Impractical" or "trivial" ideas may be the starting point, but good faith and good moderators can convert such early mismatches into exciting possibilities for project-initiating ideas. In one such relationship, the two parties bogged down for months on different perceptions of the meaning and value of "models" of a phenomenon of mutual interest. Some of the differences were nominal and lay in the different languages and time horizons of the two groups. However, the essential issue is that both wanted the same thing—a better, quantifiable understanding of the phenomenon that would help increase the effectiveness of the industrial process that was the focus of the project. Universities can be a great source of ideas for industrial R&D/I projects, but these ideas do not grow on trees that are easily harvested. Careful preparation, hard work, mutual respect, and good faith are needed to bring them to the point where they can, in fact, be used by industry.

Make or buy and the contracting out of technology

One of the current bandwagons in industrial management in both the manufacturing and service sectors is outsourcing. Although "make-or-buy" and "contracting out" alternatives to doing things internally have been available for many years, the subject has become hot again recently. This is due to the efforts of many firms to downsize, restructure, and otherwise reorganize many of their functional areas in the spirit of making them lean and mean. While these efforts often involve direct functional areas such as manufacturing and marketing, leading to heavier dependence on vendors and sales or marketing representatives and distributors, they are most often focused on "peripheral" or support activities. For example, one current new tendency in the service sectors (banking, insurance, trade, etc.) is to contract out the entire information systems activity. In some cases this includes the previously sacrosanct operation and management of the internal corporate computer and communication network. Of course, contracting out or buying telecommunication services has always been the dominant mode. In fact, after a period following deregulation and breakup of the AT&T system, when a large number of organizations seemed to be willing to "make their own" telecommunication systems, a slowdown or reversal may be imminent. Many of these self-sufficient organizations are again seeking to farm out their telecommunication and related services. Contracting out supporting services follows in the tradition of letting other firms provide services in law, health care, food service, advertising, public relations, security, janitorial, and even organizational development and human resources specialties. However, now that it is spreading to technological activities such as R&D, product development, manufacturing engineering, system design, software development, and other "knowledge-based" activities, it is time to take a closer look at the longer-term implications of pushing or allowing some of these functions to go to outside vendors and contractors.

There can be many advantages to such outsourcing for technology in the shorter run, when results are needed quickly, where a "fixed" cost is desired, where large outlays of capital for equipment and facilities are to be avoided, and where there is not much concern about losing know-how or proprietary information about the firm's operations. But in the context of the longer-run ability of the firm to "regenerate" itself and its products, services, or manufacturing technology in order to gain or maintain comparative technology advantage over its competitors, these practices may ill serve the firm. Discussions above about the size and diversity of the "idea pool" and its underlying skills base or "core technologies" in the firm come into play here. Many of the good technical ideas and those which are both technically sound and, ultimately, successful in a financial or commercial sense derive from an intimate familiarity with the business of the firm, its products, and the underlying technologies.

Although it *is* possible for an outside partner in a "strategic alliance" with the firm to get to know its technology over a period of time, the general barriers to idea generation and communication mentioned earlier and discussed below make it difficult for such an outside body to become steeped enough in the firm's technology and operations to come up with a steady stream of good and cost-effective technical ideas. This is especially true when we view the idea-generation process as a continuous, "improbable" process, depending on an accumulation of observation, experimentation, hands-on experience, market insights, familiarity with customers and clients both inside and outside the firm, as well as formal scientific and technical knowledge. "Mailing in" ideas to be further exploited by the remaining technical people in the firm may not be feasible if those remaining engineers and scientists constitute a less than critical mass in the quantity and/or quality that is needed to evaluate, further develop, adapt, implement, and maintain the resulting technological products, services, or manufacturing methods. If the firm's internal or embedded technology capabilities (Rubenstein, 1989) or set of "core competencies" are reduced below a certain quantitative and qualitative level, there is no longer the capability of feeding externally generated ideas into the innovation system in an effective manner. A brilliant outside inventor, consultant, vendor, strategic alliance partner, or other outside source requires counterparts in the firm who can understand, accept, elaborate on, and adapt the idea and its related technology to the realities of the firm's environment. By contrast, going outside to *augment, supplement,* or *gain synergy* for internal technology skills is a good strategy and should be encouraged. But when it is undertaken as a cost-reducing substitute for internal technology capabilities, it can kill the ability of the firm to either generate or implement significant ideas that have the potential for gaining and maintaining comparative technical advantage. Motivations for going outside and some of the pitfalls of doing so are more fully discussed elsewhere (Rubenstein, 1989, Chap. 6). Moves in the direction of going outside precipitously and for the wrong reasons, such as short-term cost reduction, should be weighed carefully in the light of the difficulties mentioned above when a firm is "technologically vulnerable" due to cutbacks and destruction of its internal technology base. Among the early and longer-term casualties of such cutbacks are internally generated and internally advocated ideas for innovation.

On the other hand, when the firm's own technology, people, and units show evidence of being "burned out," used up, or sterile in terms of idea generation, the only recourse may be to seek outside sources for an infusion or regeneration of the firm's idea-generating capability. In those cases, there are many outside sources of ideas. The need remains, however, to assure that once the outside ideas are delivered to or injected into the firm, there are people with the technical ability, enthusiasm, management support, and general capabilities to adopt, adapt, implement, and exploit the resulting products, services, or manufacturing methods. Although it may appear at first glance that inside ideas have a greater chance of being accepted and exploited than those from outside, that is not always the case. The next section of this chapter discusses barriers to such acceptance and exploitation within the firm by virtue of organizational structure, territorial conflicts, organizational politics, depth and direction of commitments, and other factors. These same barriers may block ideas from outside, even after they have "penetrated the skin of the firm" and have found a champion or at least some individual or group that is willing to promote them. A mitigating factor may be that ideas purchased from outside (either directly or through other contingency or deferred arrangements, such as royalties) generally have the attention of high-level management and may have some initial (or even continuing) push behind them. Under those circumstances, some outside ideas might even have a better chance of acceptance than some inside ones, despite the NIH factor mentioned earlier. However, it may still require some extraordinary efforts to make sure that the management-championed ideas do not bog down or die in the "bowels" of the organization, just as most internally generated ideas do, in terms of producing commercially and financially attractive results and impacts on the firm. One of these extraordinary efforts involves establishing mechanisms to get the R&D, engineering, marketing, and/or production people directly involved in the project that is formed from the idea infusion and providing an atmosphere of ownership, even though the idea itself was born outside.

Another approach is to give the idea sufficient visibility to make it attractive for people to be associated with a high-priority, management-supported idea and its related project. Most important, and sometimes hardest to achieve, is to arrange for early involvement of internal people in the idea before it is "dumped on them" or announced by upper management (Sen and Rubenstein, 1989). This could provide the internal people with the opportunity to make their inputs to the idea, thus providing some level of identification with it and also infusing it with the kind of reality discussed above in terms of the firm's particular characteristics, operating style, and constraints.

Organizational Constraints on Idea Generation, Development, and Exploitation

No matter what the original sources of ideas for R&D/I in the firm (inside/outside, formal/informal, technical/nontechnical, solicited/not solicited), there are many aspects of organizational life which can distort, derail, or destroy a potentially good or useful idea and keep it from leading to a significant new or improved product, service, or process. This is because ideas are improbable events in many organizations. They often disturb people, challenge entrenched positions, require intellectual and physical effort, distract people from their assigned and ordinary tasks, and carry great risks for both the originator(s) and those who become involved downstream in further developing them or exploiting their results. Even the potential rewards from ultimately "successful" ideas can generate barriers to their acceptance and smooth flow through the organization. Perceptions of the "zero-sum" nature of rewards for innovation may lead some people and groups in the organization to oppose or ignore ideas which are likely to provide rewards to others than themselves and even make them appear less innovative and worthy of support.

In reality, there are few people who are actively out to sabotage an idea as such. But there are many people who are "too busy," "not responsible," or "skeptical about" particular ideas so that the end effect is the same—blockage, delay, distortion, diversion of ideas. Indifference and lack of commitment—"it's not my job"—often kills or derails many ideas. Others are stopped or diverted because they challenge existing approaches and vested interests, despite lip service to the need for "more good ideas."

Of the many factors influencing the acceptance and disposition of ideas at early stages in the R&D/I process, two categories are described briefly below. More detail on all of them can be found in the references to this chapter.

1. *Organizational politics.*   Although this construct is relatively new in some literature (Prasad and Rubenstein, 1992), it is well known to practitioners of R&D/I. Most researchers and their managers, when they are being open about their perceptions and feelings, will attribute a lot of things that happen in the organization to "politics." It is not always clear to what extent they use this term in the pejorative or "naughty" sense as though it were an aberration in an otherwise formal and genteel environment.

Spinning off from a series of studies (Knapp, 1983; Fischer, 1987; Ginn, 1983), Prasad developed the notion of politics at the project level and identified many factors leading to and resulting from "political" behavior at the interface between functions along the R&D/I spectrum (Prasad and Rubenstein, 1992). Among the factors he identified are some which might act as barriers or facilitators to technology transfer (i.e., the exchange of ideas, information, and pieces of technology within the organization). They include uncertainty (technical and environmental), structural and behavioral norms in the organization, slack resources, interdependence of entities, salience of the issue (around which conflict arises), superordinate goals, power differentials between entities, and exit options for the participants in the interaction or exchange.

Some aspects of organizational politics can help the idea-generation, flow, and development process by providing alternative channels, "side payments" in terms of informal rewards, implicit encouragement without the need for full-dress proposals and approvals, and other "facilitators." Other aspects of politics can block, damage, or eliminate ideas, especially those in their early transitional or "fragile" stages. A nascent or "early-stage" idea, for example, might become "hostage" to a power play between individuals and/or departments in the firm. Divergent views on what the firm, the customer, the market, and the "world" needs in the way of new technology may be supported not on the merits of the idea but on the position, reputation, connections, and control of resources of the originator, proponents, and opponents of the idea.

2. *Decentralization philosophy, practice, and associated funding practices.*   One set of indicators of the impacts of organizational decentralization is the elements of the idea-generation and flow process (Rubenstein, 1964; Rubenstein and Radnor, 1963). That is, if we look beyond the stated philosophies and policies related to

decentralization (e.g., "moving the decisions down closer to the market"), we begin to identify perceptions and behavioral patterns that have a direct influence on what actually happens in the R&D/I process and some of the consequences of that behavior. Early findings on decentralization (Rubenstein and Avery, 1958) identified a shortening of time horizon and narrowing of scope of projects proposed and accepted by the operating units who either performed or funded the bulk of the total R&D in the firm. It became clear that the "decision imperatives" and the initial project selection and *continuation* decisions by divisional managements were significantly different from those of people in the CRLs. These decisions were also different from the leading technical edge decisions in their industries, where CRLs did not exist. In the idea flow literature (e.g., Baker, 1965; Pound, 1966; Utterback, 1965; Bolen, 1963; Martin, 1967), conceptual flow models and lists of factors are developed from both the literature and a large set of empirical studies. The latter involved obtaining the history of several hundred "ideas" for new and improved products and processes in over a dozen firms in several industry sectors.

Some of the factors in those models (Rubenstein, 1989, Chap. 6) were general enough so that they could have been observed in almost any kind of firm. Others were the direct result of or exacerbated by the extent of decentralization and divisionalization in the firm, with the associated intellectual and organizational walls between R&D/technology groups by virtue of the organizational structure and philosophy. Some of the key factors identified included the filtered perceptions of top management objectives and technology policy at the operating unit "lab" level; perceived relevance of the idea to the division's current product line; time pressures on members of the operating unit lab or technical departments; perceived rewards (positive and negative) for generating and pursuing ideas, especially unpopular ones or "distracting" ones; management receptivity to the ideas of particular people whose role was not viewed as "idea generating"; complexity of the review and approval process for "nonroutine" or potentially "radical" ideas; risk propensity of the various players (e.g., the originator and his or her boss and colleagues); market focus at the moment; perceived urgency of the problem, if any; perceived availability of human resources to assign to or "allow to work on" the idea; the cultural acceptance of "smuggled" ideas; and perceived time to fruition of the results of the idea. Although decentralization was not the "cause" of death, dismemberment, or burial of many of the ideas that were studied, the conditions it set up did have a direct impact on the idea-generation and flow process as barriers.

Among the most significant factors stemming from the decentralization mode of operation and philosophy has been the changing method of funding R&D and related activities, especially at the corporate level. Although there continues to be some "churning" on this issue of who pays for what and under what conditions of approval and control, the convergence has been toward more and more control of funding, project selection, and deployment decisions by the managers of the operating units. During the churning period in many firms, and carrying over into the convergence stage, a great deal of uncertainty was generated about continued funding of projects, programs, and radical new thrusts. Many of the large oil companies, for example, had programs in nonconventional energy sources, materials, biotech, and other "esoteric" areas. Most of them have been sold off, closed down, or attenuated. This has meant that ideas in those areas, which once were avidly sought, now find no takers in the organization. Projects which had been funded directly from corporate overhead were often "defunded" at the corporate level and the CRL people went to operating units to seek support. Few found such support, unless the projects were clearly of relevance to the operating units (e.g., a new way of finding oil), and many lines of investigation have disappeared from those firms' portfolios of projects and active ideas.

Extreme decentralization of R&D/technology (e.g., where there is no research capability at the corporate level) has led to virtual disappearance of projects, programs, and the motivation to pursue ideas which are not of "clear and present" relevance to at least one operating unit, preferably several. Even where there is latent interest or *should be* some interest, the venue for pursuing such ideas may have disappeared with the corporate labs and related corporate technology functions. Exceptions are in a few very large operating divisions with responsibility for supporting and promoting a broad line of products, markets, and technology. In those cases, such divisions are, in effect, integrated "companies," and very often their large labs and technology groups operate as though they were CRLs. Under the lean and mean philosophy of many large companies of the 1980s and 1990s, broad portfolios which allow for "far-out" ideas are becoming rare at both the corporate and divisional levels. The source of funding, the strings attached, and the criteria for project selection and continuation have become so restrictive that the opportunity to float and promote ideas not obviously of direct relevance to the existing businesses has become highly attenuated. Whether "better days" are ahead and whether such an attenuated R&D portfolio and its associated ideas can recover in the near future is an open issue.

Summary

The "engine" of idea generation and flow is critical to a healthy and productive R&D/I process. If the kinds of barriers described in this chapter are allowed to dominate the idea flow process, only poor ideas can result. This is especially the case for the "preemergence" phase—when an idea is still in the mind or the notebook of the originator. When many barriers to idea generation exist, the entire R&D/I process will suffer, and the fruits of R&D/I are likely to be much less than hoped for or expected by those supporting or participating in it.

References

Baker, N. R. *The Influence of Several Organizational Factors on the Idea Generation and Submission Behavior of Industrial Researchers and Technicians.* Ph.D. Dissertation, Northwestern University, 1965.

Bolen, F. *A Method for Real Time Measurement of the Flow of Ideas in Industrial Research Laboratories.* Unpublished Master's Thesis, Northwestern University, 1963.

Bonge, J. W. *Perception and Response to Major Change by Purchasing Agents.* Ph.D. Dissertation, Northwestern University, 1968.

Fischer, B. *Factors Influencing Success in the Introduction of New Production Processes.* Ph.D. Dissertation, Northwestern University, 1987.

Ginn, M. E. *Key Organizational and Performance Factors Reflecting on the R&D/Production Interface.* Ph.D. Dissertation, Northwestern University, 1983.

Gupta, Ashok, Raj, S. P., and Wilemon, David. "The R&D-Marketing Interface in High-Technology Firms," *Journal of Product Innovation Management,* 2, 1985, 12–24.

Hoffman, G. M. "User Participation in Application System Development." Center for Information and Telecommunication Technology, Northwestern University, 1992.

Knapp, C. M. *The Role of the Key Communicators in a Divisionalized Industrial Corporation.* Ph.D. Dissertation, Northwestern University, 1983.

Martin, R. B. *Some Factors Associated with the Evaluation of Ideas for Production Changes in Small Companies.* Ph.D. Dissertation, Northwestern University, 1967.

Pound, William H. *Communications, Evaluations, and the Flow of Ideas in an Industrial Research and Development Laboratory.* Ph.D. Dissertation, Northwestern University, 1966.

Prasad, L., and Rubenstein, A. H. "Factors Influencing Power and Organizational Politics During New Product Development." Working Paper, Northwestern University, 1992.

Rubenstein, A. H. "Studies of Idea Flow in Research and Development." Presented to the New York Chapter, TIMS, November 1963*a*.

———. "Field Studies of Project Selection Behavior in Industrial Laboratories." In B. V. Dean (ed.), *Operations Research in Research and Development.* New York: Wiley, 1963*b*, 189–206.

———. "Organizational Factors Affecting R&D Decision-Making in Large Decentralized Companies," *Management Science,* 10, July 1964.

———. "Trends in Technology Management," *IEEE Transactions on Engineering Management,* 32, November 1985, 141.

———. *Managing Technology in the Decentralized Firm,* New York: Wiley Interscience, 1989.

————. "Barriers to Successful Technical Entrepreneurship in the Large Firm: Some Case Studies." Presented at the Institute of Management Sciences, Philadelphia, October 1990.

————, and Avery, R. W. "Idea Flow in Research and Development." In *Proceedings of the National Electronics Conference, XIV, October 1958.*

————, and Radnor, M. "Top Management's Role in Research Planning in Large Decentralized Companies," *Proceedings of IFORS,* Oslo, Norway, July 1963.

Sen, F., and Rubenstein, A. H. "External Technology and In-House R&D's Facilitative Role," *The Journal of Product Innovation Management,* 6, June 1989, 123–138.

Souder, Wm. E. *Managing New Product Innovations.* Lexington, MA.: Lexington Books, 1987.

Utterback, J. M. *Accuracy of Perception and Enculturation of Researchers in an Industrial Laboratory.* M. S. Thesis, Northwestern University, 1965.

Von Hippel, Eric. *The Source of Innovation,* New York: Oxford University Press, 1988.

3

Developing Powerful Technology and Product Concepts

Ashok K. Gupta

Ohio University

After a technological idea has been generated (Chap. 2), concept generation is the next step in the technology development process. A product concept is an elaborated version of a technological idea that gives it shape, function, form, and defined benefits. The product concept thus links the technological idea to the customer's needs and wants. The concept development step is obviously a vitally important part of the new technology development process. How can powerful concepts be developed? What techniques are available to assist managers in understanding user needs and wants? How can these needs and wants be documented? How can they be used to guide the concept development process? How can managers know when a powerful concept has been developed? These are some of the many questions answered in this chapter.

WM. E. SOUDER AND J. DANIEL SHERMAN

Introduction

Every year thousands of new products are introduced in the marketplace. Most of them fail to reach their business objectives. Should these products not have been introduced? Could businesses have saved money if they had not brought these products into the marketplace? What could they have done differently? Did they learn any lesson? These are all good questions. However, there are no definite answers. In the world of intense competition and rapidly changing technologies, the cost of not acting could be as disastrous as the cost of proceeding further.

There are many ways to miss a market or lose in new product development. One sure way is to offer a product that customers do not want. While some companies consistently develop products that delight the customers, others do not come up to the mark. What differentiates them? Successful companies develop and introduce products with powerful concepts which provide high value to customers at a profit to the organization. How such concepts are developed is the topic of this chapter.

What Is a Product Concept?

A product concept is an elaborated version of the idea expressed in meaningful consumer terms (Kotler, 1991). Turning an idea into a concept means giving the idea form, substance, and shape (Kuczmarski, 1988). The concept must describe (Clark and Fujimoto, 1990):

What the product does. Defining the product in terms of its performance benefits, how well it meets customer needs, ease of use, safety, serviceability, disposability, and environmental concerns

What the product is. Describing its form in terms of its configuration and main component technologies

Whom the product serves. Target customers for the product

What the product means to customers. Highlighting a product's psychological benefits by describing it in terms of its character, personality, image, feel, and looks—the soft side of product design

Crawford (1991) summarizes a product concept as consisting of benefits plus either form or technology. Knowing benefits and form or technology permits us to talk to potential buyers to see if it fulfills their needs better than what is available in the market.

From a customer's point of view, powerful product concepts include all of the above dimensions. From a product developer's point of view, there are additional dimensions that should be included at the time of concept development. These include the considerations of manufacturability and assembly, product testing, future innovation, quick market adoption, disassembly, disposability and environmental concerns, and profitability and competitiveness. Traditionally, the subject of concept development has been looked at from the customer's point of view. In this chapter, we will break that tradition. We will also include producer's concerns that should be addressed at the concept development stage. Many times customer input is necessary to address producer's concerns.

Need for Developing Powerful Concepts

Provides a framework for decision making

A powerful product concept is like a vision statement for the product development team. It keeps the team's efforts focused and integrated.

It gives the team a goal to achieve by translating images into tangible product attributes. For example, as described in Clark and Fujimoto (1990), Honda successfully redesigned its Accord by developing a powerful product concept. In the early 1980s when Honda's engineers began to design the Accord, they did not start with a sketch of a car. They started with a concept—"man maximum, machine minimum." This is the way they wanted customers to feel about the car. Again in 1990 when the time for redesign came, Honda listened to the market, not to its own success. Market trends were indicating a shift away from sporty sedans toward family models. The new Accord model would have to send a new set of product messages. Therefore, a new product concept had to be researched.

Several brainstorming sessions later, involving close to 100 people, a new concept for the car as "an adult sense of reliability" was decided upon. The car would allow the driver to transport family and friends with confidence, regardless of weather or road conditions; passengers would always feel safe and secure. This message was still too abstract to guide the product and process engineers. So the next step was to find an image that would personify the car's message to consumers. The image the team agreed upon was "a rugby player in a business suit." It evoked rugged, physical contact, sportsmanship, and gentlemanly behavior—disparate qualities the new car would have to convey. The image then had to be translated into design details, such as new headlamps symbolizing the will of a rugby player looking calmly into the future with clear eyes.

The next step was to break down the rugby player image into specific attributes the new car would have to possess. Key words such as "tough spirit," "stress-free," and "love forever" captured the image. Tough spirit in a car, for example, meant maneuverability, power, and sure handling in extreme driving conditions, while love forever translated into long-term reliability and customer satisfaction.

These phrases helped people make coherent design and hardware choices in the face of competing demands. They were also a powerful spur to innovation and creativity. For example, in developing a stress-free Accord, Honda engineers had to work on reducing engine, wind, and road noises. In the process they also developed a four-cylinder engine that was as quiet as a V-6, designed a new electrically controlled engine mount to minimize vibration when the engine was idling, inserted paper honeycomb structures in the roof lining to reduce resonation from engine vibrations, and redesigned the body floor by creating a new sandwich structure of asphalt and sheet steel to reduce floor vibration and strengthen the body shell. Product concepts like the ones developed at Honda give the product development team a clear framework for finding solutions and making decisions.

Early planning pays off

In Booz, Allen and Hamilton's 1982 study of new product management, it was found that the companies that have excellent records of successful new product introductions are more likely to develop specific new product strategies. They conduct more analyses early in the process and focus on their idea and concept generation. Furthermore, they conduct more rigorous screening and evaluation of the ideas generated.

Among U.S. companies surveyed by Booz, Allen and Hamilton, the portion of expenditures in the first three steps of new product design (NPD) (i.e., expenditures on idea or concept generation, screening, and business analysis) has more than doubled from 10 percent in 1968 to 21 percent in 1981. There is still much room for improvement. In a recent study, Mahajan and Wind (1992) found that only 43 percent of the companies sampled conducted new product screening in all cases, while only 4 percent undertook detailed market studies for concept development and testing.

Also, in comparing product development in the United States with that in Japan, it is clear that the Japanese invest more time and attention to these early steps in the new product development process. Consequently, this approach has paid off well for Japanese companies (Table 3.1). They respond better to customer needs, waste less time on postlaunch debugging, and earn more from their NPD efforts.

Costs incurred versus costs committed

Another important reason to pay greater attention to the development of powerful product concepts and the early stages of NPD

TABLE 3.1   How Companies Spend Their NPD Efforts and With What Results

	Japan, %	U.S., %
Percent of total effort spent on screening, evaluating, and planning	40	25
Proportion of NPD efforts suffering development setback	28	49
Percent of total effort wasted on postlaunch debugging	5	15
Percent of profit contribution from products developed in last three years	42	22

ADAPTED FROM: *Fortune*, 1991; *Business Week*, 1991.

efforts is that the costs incurred during these initial stages are relatively small, but the decisions made could lock in the bulk of later spending. For example, data collected by Computer-Aided Manufacturing International, Inc., suggests that the cost incurred during the product conception stage ranges from 3 to 5 percent of the total development cost. However, the decisions made at this stage can commit 40 to 60 percent of the total costs in terms of product features to be provided, material to be used, and the manufacturing and assembly process to be employed (*Business Week,* 1990*a*).

The costs of making any changes in the new product are much lower at the conception stage than at the production stage. An example of the relative cost advantage of making changes early on can be obtained from the data compiled by Dataquest, Inc., on major electronic products (*Business Week,* 1990*a*). It is estimated that if the design changes are made during the design stage and it costs $1000 to make that change, the cost of making the same change would be $10,000,000 during the production stage. Paying attention to the quality, manufacturing, testing, and assembly issues at the concept stage could reduce the cost of product development and speed up the process.

Considerations in Developing Powerful Product Concepts

In this section we will discuss the factors that influence the creation of powerful product concepts and how some successful companies have profited from their use. Tools for understanding customers and for developing winning concepts are described in the next section.

Designing for customer delight

In a competitive environment where technology diffuses quickly, it is becoming increasingly difficult to compete on the basis of technical specification of products alone. Most products offer similar functional attributes. Customers are now looking for more than functionality in the products they buy. The product must feel right. It must look extraordinary. It must be ergonomically correct. It must be clean, crisp, and inviting. For example, the Japanese are working on creating sensuous cars. The car will provide delight and surprise by just opening the door, hearing the sound, and pressing the accelerator. Everything will be thought through mostly by emotion (Kotler, 1992). Esslinger of Frogdesign (one of the largest design firms in the world) notes that the secret of a good design is capturing a product's intangibles. In a similar way, Jay Wilson, design director at GVO in Palo

Alto, California, says, "We design a user experience, not just the envelope the product comes in" (*Business Week,* 1990*b*). A powerful product concept must therefore provide functional benefits to the customer and also the intangibles.

Successful product developers have been shifting away from their preoccupation with pure product performance to focus on how real people use things. The 1992 Industrial Design Excellence Award (IDEA) for winning products exemplifies this shift in emphasis (*Business Week,* 1992*a*). The winners are those that function well and solve specific work needs beautifully. These are the products that elicit an "I've got to have it" reaction. It is not just the looks that are important. Products that are primarily style-based will not make the grade with customers. The concept should have the right mix of engineering, ergonomics, and aesthetic appeal. At the same time, it should not be too difficult or expensive to produce.

An inviting design could provide a big boost to sales even in high-tech products. AT&T hired Frogdesign to give its new answering machine a special outward look. The company is convinced that no answering machine ever looked like its model 1337 and customers will be delighted to have it. Many start-up companies believe that one of the reasons Apple Computer has been so successful is because of its appealing designs. A good example of how a delightful design can help improve sales came from Sun Microsystems. Sun's products looked nerdy, space-hogging, and noisy. Frogdesign's Esslinger gave the workstation a cool grey casing and trim physique. By carefully positioning air vents for cooling, the machine's CPU could be packed into a unit the size of a pizza box. The machine received wide acceptance from consumers. Sun sold $2.5 billion worth in 1989 (*Business Week,* 1990*b*). Similarly, the development of Apple's PowerBook notebook computer is an excellent example of how a product can be made playful as well as practical (*Business Week,* 1992*a*).

Apple was already behind in portable computers when it began work on the PowerBook in 1990. Apple introduced its PowerBook in October 1991 in the background of its own flop, the Macintosh Portable, nearly 3 years after competitors shipped the first notebooks. PowerBook's special design has made it a standout in a dizzying field of 300-plus notebook machines.

PowerBook's success is partially dependent on its late entry. Apple used this time to study how people interacted with the existing notebook computers, using a *product-mapping* technique. Product mapping switches the focus of analysis from the product to the user. It analyzes how people interact with machines. For example, in the case of notebook computers, the design firm GVO was hired by Apple to compare the proposed PowerBook line with the competition. It used product mapping to measure 159 user-computer interactions from opening the box to computing on an airplane to a notebook-usability index. This index is a measure of peoples' experiences with notebook computers, including PowerBook, over a range of user-product interactions. The exercise helped develop the idea of (1) placing the trackball in a flat space just below the keyboard, which would make it easier to use for both lefties and righties, (2) a hot-dog-shaped screen hinge instead of the usual double hinges so that users can tilt the lid and adjust the screen's angle, (3) flip-down back feet which can set the computer's base at a slant for easier viewing and typing, (4) a floppy-disk drive positioned in a way that would not be blocked by airplane-seat arm rests, and (5) molded ribs on the outside of the case that not only give PowerBook its distinctive look but also keep it from getting scuffed or dinged while traveling (*Business Week*, 1992a).

Before going into production, the whole concept for PowerBook was subjected to six different studies involving 68 people to obtain their reactions on many things, including the color, the position of the on-off and brightness buttons, and the technical aspects. The result: a design that is practical but playful. In just 8 months following its introduction, PowerBook became the no. 2 notebook in U.S. computer stores.

Another emerging concept in delighting the customer is called "universal design." This involves adapting products for the elderly or physically challenged and, in the process, making them more useful to everybody (*Business Week*, 1992b). The universally designed products allow businesses to widen the consumer market. Kohler Co.'s new walk-in bathtub, Precedence Bath, takes the danger out of bathing by installing a door in the tub. This is truly a transgenerational idea, appealing to older adults and to younger children who do not feel secure stepping over a bathtub rim. OXO's good grip kitchen utensils are not only a delight to look at, they also are easy to use by people from 5 to 95 years old. The handles are big, made of rubbery material, and they have fins for fun. AT&T's new big button phone and Tide's Snap Top easy-to-open detergent box are other examples of universal designs which provide elegant solutions to all-age consumers.

Designing for the environment

Designing for the environment is no longer a choice; it is a necessity. Increasingly, new product concepts are emerging that require fewer resources to manufacture, consume less resources to use, incorporate less toxic but recyclable raw materials, and are easy to disassemble for improved recyclability. AT&T, among others, is adding design for environment (DFE) software to its computer-aided design systems (Table 3.2) to provide guidelines and options for making products easily recyclable, choosing the least harmful materials, and minimizing hazardous waste and energy use in manufacturing. In 1993, AT&T will market a new business phone based on the concept of environmental friendliness. The phone will be easier to disassemble

TABLE 3.2   Green Product Concepts

Corporations	Green product concepts
Dow Chemical	Replacing chlorinated solvents used for cleaning industrial equipment with less polluting, water-based systems
AT&T	Designing phones which will be easier to disassemble for recycling and using conductive plastic for interconnections to minimize use of lead solder
BMW	Using design for disassembly to build cars so they can eventually be taken apart and recycled more easily
GE Plastics	Using design for manufacture, service, and disassembly and materials compatibility approach to develop socially responsible products
Scott	Developing products that generate less waste and use postconsumer recycled fiber
3M	Pioneered Pollution Prevention Pays (3P) program by focusing on waste reduction strategies
XEROX	Implementing programs for equipment and parts recovery, remanufacturing, and recycling
P&G	Reduction in packaging by developing concentrates and refills
Pacific Gas & Electric	Implemented Green Lights Program by developing and encouraging the use of energy-efficient lighting products
NEC and Mercedes	Eliminating CFC use from all their products
Ford	Eliminating CFC from car air conditioners
Shell Oil	Improving conventional gasoline for cleaner burning
Audi	Manufacturing seat cushions without emitting CFC; also, designing Audi Duo, a full-size station wagon that is powered by gasoline for highway driving but can be switched to electricity in the city
Mazda & Toyota	Evaluating hydrogen-powered cars which will give only water in their exhaust
Michelin	Designing a new Green Tire which will reduce fuel consumption of vehicles by 5 % by using low rolling resistance technology

ADAPTED FROM: *Fortune,* 1992c; *Business Week,* 1992; *Newsweek,* 1992.

for recycling, and it uses conductive plastic interconnections to minimize the use of lead solder (*Business Week,* 1992c).

Another tool for developing environmentally friendly product concepts is to conduct a cradle-to-grave life-cycle analysis. The analysis involves taking an inventory of materials, energy consumed, and pollution emitted during a product's manufacture, use, and disposal. Conducting a life-cycle analysis can be a complex but useful exercise. Procter & Gamble's concepts of Tide Ultra and concentrated Downy fabric softener are some examples of products requiring fewer raw materials based on the life-cycle analysis. The life-cycle analysis has become an integral part of GE Plastics' approach to address the issues of environmental management in the conceptual phase of product design. To develop environmentally sensitive product concepts, GE Plastics is focusing on design for disassembly for effective service and recycling. The new concept refrigerator developed by GE Plastics uses modular design, compatible materials, and design for disassembly to create a product that is much easier to service in the field and is easy to recycle. To facilitate this process, the designers consider the following points:

- Use of recyclable and compatible materials
- Identification of materials on parts and time coding
- Use of two-way snap-fits and break points
- Reduced part size for ease of handling
- Avoidance of labels and nonrecyclable paints, coatings, or decorations

U.S. corporations currently spend $115 billion annually to comply with federal environmental regulations (*Fortune,* 1992a). The corporations have now realized that the best way to avoid air, water, and land pollution is to avoid creating pollution in the first place. How about the consumers? Will they be willing to pay more, if need be, to buy environmentally friendly products? In a survey conducted by Roper Organization for S.C. Johnson & Son, Inc., it was found that consumers are willing to pay between 6 to 7.4 percent more for products made of recycled paper and one-third more for less polluting cars, detergents, soaps, gas, aerosols and for biodegradable plastic packaging (*Fortune,* 1992a).

Designing for manufacturability and assembly

Even if the product concept meets customer satisfaction, what good is it if it cannot be cost-effectively manufactured to meet the corporate objective of profitability? By considering the manufacturing and assembly steps in the initial design equation, manufacturers can engineer both the part and the manufacturing process simultaneously. The secret of Japanese companies' ability to produce ever-higher quality at lower and lower prices is that they do not have separate design and manufacturing functions. They engineer quality into the manufacturing process instead of relying on assembly-line inspections to weed out defects. Most U.S. companies have now realized that it does not make sense to simply automate factories; no amount of factory automation can compensate for a poor design. Because 90 percent of production costs are preordained by design decisions made long before the blueprint reaches the shop floor, there is no point in efficiently producing a badly designed product (*Business Week*, 1990*a*).

Successful companies such as NCR, AT&T, Westinghouse, General Electric, TI, IBM, Ford, Eastman Kodak, Motorola, and Deere & Co, among others, are adopting the concept of "concurrent engineering" to bring R&D, manufacturing, marketing, customers, and suppliers together to reduce production costs, improve product quality, and trim cycle time. This helps bring in the critical manufacturing, assembly, and service issues at the product conception and design stage. Some examples of what is achievable are presented in Table 3.3.

How was it done? What did these companies do to produce products which delighted the customers and made them profitable to manufacture? They used teamwork, made organizational structural changes, talked to customers, adopted modular design, analyzed the designs, considered assembly requirements early on, upgraded manufacturing technology, and treated suppliers as partners. Some guidelines based on company experiences included the following aspects (*Business Week*, 1989; Dean and Susman, 1989; Whitney, 1988; *Fortune*, 1992*a*):

- Adopting simultaneous engineering/concurrent engineering or teamwork. Benefits include: 30 to 70 percent less development time, 65 to 90 percent fewer engineering changes, 20 to 90 percent reduction in time to market, and 200 to 600 percent higher overall quality.

- Making organizational structural changes. NCR brought designers on the shop floor to reduce the "me think, you do" mentality. Westinghouse put design and manufacturing under one manager.

- Thinking of customer requirements, skill level, service needs, safety, and intangibles early on.

- Considering modular design so that each module can be tested prior to assembly and different versions of a module can be made identical to facilitate assembly and handling.

TABLE 3.3   Profits by Design

Corporation	How did it benefit by incorporating manufacturability and assembly issues at concept development and design stage?
NCR	*For check-out counter terminal* Development time reduced to 22 months from 45. 85 percent fewer parts. Assembly time reduced to 2 minutes from 8 minutes.
AT&T	*For main phone-switching computers* Time from conception to production trimmed more than half to 3 years.
Deere	*For construction and forestry equipment* Cycle time dropped by 60%. Costs dropped by 30%.
Ford	Defects per vehicle dropped from 670 in 1981 to 150 in 1989.
TI	*For gun-sight component* Assembly time reduced by 85%. Number of parts reduced by 75%. Number of assembly steps reduced by 78%.
IBM	*For proprinter* Assembly time reduced to 3 minutes from 30 minutes.

ADAPTED FROM: *Business Week*, 1990, 1989.

- Analyzing the design. Are tolerances too tight? Can the number of parts be reduced? Can we use standard parts? Can we use proven parts? Can we eliminate the use of jigs and fixtures, or can we use common jigs and fixtures for different models? Can we eliminate certain operations? Use of design analysis tools such as value engineering, mechanical computer-aided engineering (MCAE), or design for manufacturability and assembly (DFMA) could be helpful.

- Considering assembly requirements earlier. Consider assembly sequence, ease of testing, ease of replacing faulty parts, access to fasteners and lubrication parts, and risk of damaging while assembling. For example, VW paid 18 percent extra for cone-shaped tip screws to speed up assembly.

- Upgrading manufacturing technology for flexibility and quality.

- Treating the suppliers as partners. Hold them to high standards of quality and timeliness. Work with them and help them perform better. Involve them in design. Show long-term commitment to them. Reduce the number of suppliers. For example, Ricoh helped

Axon in the United States to become a leading supplier of fax paper, Motorola told its suppliers to shape up or lose their business, and Ford asked its suppliers to reduce prices and improve quality by 5 percent every year.

Assessing Consumers' Reactions to New Product Concepts

Understanding consumer preferences and perceptions of the marketplace is the key to generating powerful product concepts. Consumers' receptivity to the concept determines their purchase intentions, which in turn influences their purchase behavior. The objective of this section is to discuss several tools that can be used to answer the following questions:

1. How do consumers react to the concept in terms of their purchase intention, importance of the product, need satisfaction, relative advantage over others, product quality, etc.?
2. How can the concept be improved?
3. How will the product do if it is introduced in the market as described by the concept?
4. How could the product be positioned in the marketplace against competing brands?

In conducting a concept test to gauge consumer reactions, there are two key issues that need to be considered (Dolan, 1992):

1. How should the concept be presented to the users? Should it be described in words only or visual only or a combination of the two?
2. What questions should be asked of the users to measure their reactions?

Presenting the concept

A product concept could be presented in a factual, nonemotional way. Or it could be presented in the context of persuasive communications, as would surround the product in a typical market situation. Crawford (1991) offers a good example to illustrate the difference in the tone of presentation of two concept statements.

Statement A. Light Peanut Butter, a low-calorie version of natural peanut butter that can provide a tasty addition to most diets (also called "core concept").

Statement B.   A marvelous new way to chase the blahs from your diet has been discovered by General Mills scientists—a low-calorie version of the ever-popular peanut butter. As tasty as ever and produced by a natural process, our new Light Peanut Butter will fit most weight-control diets used today (also called "positioning concept").

These statements have few substantial differences. However, consumers will react differently to them. Consumers are more likely to see concept statement B in the marketplace. Therefore, consumer reactions to concept statement B will be more realistic and would yield better behavioral predictions. However, their reactions are being biased by good or poor advertising copywriting.

The second communication issue is whether to use words only or to add an illustration, a model, or a film. In some situations, for example food or designer clothes, the use of a sketch is necessary. It has been found that purchase intention scores go up as one moves from core concept (statement A) to positioning concept (statement B). Similarly, visual cues added to words generally produce purchase intention scores higher than either alone. As the reliability of the concept test increases, the more concrete and physical the stimulus (Kotler, 1991).

Questions to ask during concept testing

Generally, four types of data are collected during the concept testing phase: purchase measures, overall reaction to the product concept, diagnosis of specific attributes, and respondent profiles.

Purchase measures.   Purchase measures consist of purchase intentions and expected frequency of purchase. These two data can provide a good idea of the sales potential of the product based on the concept being tested.

Purchase intention (PI) is measured by asking the question, "Based on this product description, how likely would you be to buy this product if it were available at a store in your area?" Responses are usually recorded on a five-point scale consisting of:

- Definitely would buy
- Probably would buy
- Might or might not buy
- Probably would not buy
- Definitely would not buy

Purchase intention is a good indicator of trial. For forecasting sales volume, we also need purchase frequency data. Purchase frequency is assessed by asking, "How often do you think you would buy this product if it were available to you?" Generally, the responses are checked on the following scale:

- Once a week or more often
- Once every 2 to 3 weeks
- Once a month
- Once every 2 to 3 months
- Once every 4 to 6 months
- 1 to 2 times a year
- Less often
- Never

The amount purchased each time a purchase is made is also asked.

Purchase intention is the most important information gotten from concept testing. Interpreting the data, however, can be tricky. When do we say that the purchase intention scores we obtained are good or poor? A general rule of thumb on "good" purchase intention scores suggests that a concept statement should receive 80 to 90 percent favorable answers (sum of *definitely buy* and *probably will buy*) for subsequent work on product development to continue (Taylor, Houlahan, and Gabriel, 1975). However, there is a variation in good PI scores by product categories (Schwartz, 1987). Therefore, to interpret the goodness of PI scores, one might need to consult published sources, the company's own experience with similar products, or rely on the experience of research companies who conduct concept tests.

Another important issue concerning purchase intention is how well it correlates with the actual purchase. Research shows that there is a strong correlation between PI and trial. For consumer packaged goods, the rule of thumb is that the percentage of responses falling in the category "definitely will buy" (also called "top box") is a good indicator of the trial rate. However, PI is not a predictor of trial rate because satisfaction with the product determines repeat purchase, which is not directly measured in concept testing (Tauber, 1981). Kalwani and Silk (1982) found that PI scores do correlate with actual purchase behavior. However, their research suggests that PI scores and purchase behavior vary by product categories. For consumer packaged goods, their research findings support the rule of thumb. For consumer durables, they suggest that a weighted average of all box scores would be a better estimate of trial.

Overall reaction to the product concept. The second set of data one collects during concept testing is about the overall reaction to the product idea itself. The type of questions generally asked include:

1. Are the benefits clear and believable? If consumers do not understand what benefits the product provides or do not believe in the core benefits of the product, the concept must be revised.

2. How important is the new product in solving a consumer's problem? This is a measure of need level. The stronger the need, the greater will be the consumer interest in the concept. The benefits may be clear and unique, but not important.

3. How does the new product compare with current products in meeting consumer needs? This measures the satisfaction gap between current and new products. The greater the gap, the higher the consumer interest in the new product. A need-gap score could be computed by multiplying need level with gap level. A higher need-gap score would suggest that the new product satisfied an important need that is not satisfied with current products (i.e., the new product offers a big relative advantage).

4. Is the new product a good value for the money? Is the price reasonable? The greater the perceived value, the higher will be the purchase intention.

Diagnostics of specific attributes. The third set of data collected during concept testing attempts to answer the question, "How do individual product attributes or benefits contribute to the overall purchase intention?" This is achieved by assessing consumer perceptions of each attribute and their importance (Table 3.4). For example, consumers could be asked to rate the attributes of an electric car concept on perception and importance scales.

TABLE 3.4   Assessing Attribute Perception and Importance

	Attribute perception					Attribute importance				
	Excellent				Poor	Very important				Not at all important
Good for the environment	5	4	3	2	1	5	4	3	2	1
Ease of maintenance	5	4	3	2	1	5	4	3	2	1
Safety	5	4	3	2	1	5	4	3	2	1

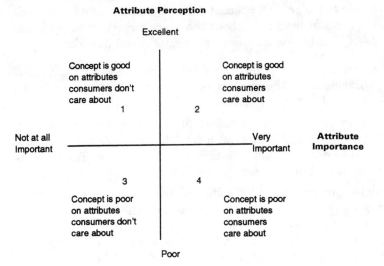

Figure 3.1 Quadrant analysis.

This data can be used to perform a quadrant analysis to see if the concept is strong on the attributes consumers care about. Each attribute will fall into one of the four quadrants in Fig. 3.1. Quadrant 1 contains those attributes where the concept is good, but consumers do not care. Quadrant 3 consists of those attributes where the concept is poor, and consumers do not care about them. Therefore, attributes in quadrants 1 and 3 cause no problems. Attributes falling in quadrant 2 are those where the concept is strong and at the same time consumers care about them. On the other hand, attributes falling in quadrant 4 are those which consumers care about, while the concept performs poorly. These are the attributes that need improvement for the concept to be a winner.

Respondent profile. Respondent profile data is helpful in market segmentation and in developing a targeted marketing strategy. The objective is to collect data on consumer demographics, psychographics, current brand purchase, benefits sought, etc., to see if there is some commonality in the clusters of those who responded differently about their purchase intentions.

Concept testing provides useful diagnostic information for refining the concept and gives an estimate of sales potential. The standard concept test falls short on providing scores if an attribute level of the product is changed (unless one tests all possible levels). Also, one does not get an idea of how consumers will make trade-offs among various attributes. Generally, all desirable attributes are rated "very important." To overcome some of these limitations, the conjoint analysis procedure is used to identify the most desirable product concepts.

Conjoint analysis

The conjoint analysis technique is used to measure consumer preferences for alternative product concepts. The technique yields the "utilities" of the various levels of the product features to the users. These utilities can be used to see how consumers make trade-offs between competing product features to choose the most satisfying product concept. Thousands of applications of conjoint analysis have been reported in a variety of industries, products, and services including condominium designs, snowmobiles, rug cleaners, credit cards, rural health care systems, performing arts series, computer software, aircraft, lift trucks, cellular telephones, and others (Cattin and Wittink, 1982; Wittink and Cattin, 1989).

In this section we will present a nontechnical introduction to the technique using the classical examples from Green and Wind (1975) and Johnson (1974). These examples highlight two typical approaches to data collection: the full-profile approach and the pairwise approach (Kotler, 1991).

Full-profile approach. Green and Wind have illustrated the use of the full-profile approach in connection with the design of a hand-held carpet cleaning brush. Suppose the product concept includes the following five design elements (product attributes) and their choice levels:

Design elements	Types/levels considered
Shape	A, B, C
Brand names	K2R, Glory, Bissell
Prices	$1.19, $1.39, $1.59
Good Housekeeping seal	Yes, No
Money-back guarantee	Yes, No

Using these five design elements and their choices, 108 possible product concepts could be generated for consumer evaluation ($3 \times 3 \times 3 \times 2 \times 2$). But it would be quite cumbersome for consumers to rate or rank 108 possible product concepts. Instead, these were reduced to 18 using fractional factorial designs (Urban and Hauser, 1980). Each of these 18 product concepts were presented on 3- by 5-inch cards for consumers to rank from the most preferred (or most likely to buy) brush for carpet cleaning to the least preferred (or least likely to buy). For example, card 5 would be:

```
                    Card 5
        Shape:                 B
        Brand name:            Glory
        Price:                 $1.59
        Good Housekeeping seal: No
        Money-back guarantee:  No

        Rank: _____
```

Table 3.5 shows how a consumer might rank all 18 cards (product concepts). This consumer ranked product concept 18 the highest, thus preferring shape C, brand name Bissell, a price of $1.19, having a Good Housekeeping seal, and money-back guarantee over all other concepts.

About 100 consumers were surveyed and provided rankings of 18 concepts. Conjoint analysis software was used to analyze these rank-

TABLE 3.5   A Consumer's Ranking of 18 Product Concepts

Card	Brush shape	Brand name	Price	Good Housekeeping seal	Money-back guarantee	Preference rank
1	A	K2R	1.19	No	No	13
2	A	Glory	1.39	No	Yes	11
3	A	Bissell	1.59	Yes	No	17
4	B	K2R	1.39	Yes	Yes	2
5	B	Glory	1.59	No	No	14
6	B	Bissell	1.19	No	No	3
7	C	K2R	1.59	No	Yes	12
8	C	Glory	1.19	Yes	No	7
9	C	Bissell	1.39	No	No	9
10	A	K2R	1.59	Yes	No	18
11	A	Glory	1.19	No	Yes	8
12	A	Bissell	1.39	No	No	15
13	B	K2R	1.19	No	No	4
14	B	Glory	1.39	Yes	No	6
15	B	Bissell	1.59	No	Yes	5
16	C	K2R	1.39	No	No	10
17	C	Glory	1.59	No	No	16
18	C	Bissell	1.19	Yes	Yes	1*

*Most preferred concept.

ADAPTED FROM: Green and Wind, 1975.

ings to derive consumer utility for each design element or product attribute. The derived utility functions may look like those presented in Table 3.6.

The utility functions indicate how sensitive consumer perceptions and preferences are to changes in product features. The greater the range in the utility, the more important that feature is and the less likely the consumers will be willing to trade it off for another feature. For example, by not having the Good Housekeeping seal, consumer's utility will decrease only by 0.1 units. By not having the money-back guarantee, the utility will drop by 0.5 units. Consumers will derive the highest utility from the concept with brush shape B, brand name Bissell, sale price of $1.19, a Good Housekeeping seal, and a money-back guarantee. Whether or not this concept can be profitably made and marketed is an issue which needs to be further analyzed.

Pairwise approach. Another method for collecting data for conjoint analysis is called the pairwise comparison, or trade-off, approach. In this method, consumers indicate their preferences for attribute levels, with two attributes taken at a time. Johnson (1974) provides the following example where consumers rank two attribute levels at a time for cars: price versus top speed, price versus seating capacity, price versus warranty, top speed versus seating capacity, top speed versus warranty, and seating capacity versus warranty. Rank order preferences for a respondent's trade-off data may look like those presented in Table 3.7.

The rankings show how the user will make the trade-offs. For example, the consumer will prefer to have a four-seat car with 130-mph speed. If this combination is not possible, the second preference will be for a car with 130 mph and two seats. That is, the consumer will trade off car capacity for speed. Similar matrices are complete by a large number of consumers. The conjoint analyzer software is then used to derive utility functions for each attribute. These utility func-

TABLE 3.6   Conjoint Analysis Utility Functions

Brush shape	Utility	Name	Brand utility	Price	Utility	Good House keeping seal	Utility	Money back guarantee	Utility
A	0.1	K2R	0.3	$1.19	1.0	No	0.2	No	0.2
B	1.0	Glory	0.1	$1.39	0.7	Yes	0.3	Yes	0.7
C	0.6	Bissell	0.5	$1.59	0.1				

ADAPTED FROM: Green and Wind, 1975.

TABLE 3.7   Ranking of Pairwise Comparisons

	Top speed, mph			Seating capacity			Months of warranty		
	130	100	70	2	4	6	60	12	3
Price, $									
8,000	1	2	5	2	1	3	1	3	4
12,000	3	4	6	5	4	6	2	5	6
16,000	7	8	9	8	7	9	7	8	9
Top speed, mph									
130				2	1	3	1	2	5
100				5	4	6	3	4	6
70				8	7	9	7	8	9
Seating capacity									
2							2	5	8
4							1	4	7
6							3	6	9

ADAPTED FROM: Johnson, 1974.

tions can then be used to design the car most preferred by consumers.

The conjoint analysis is a useful tool in selecting the most preferred product concept by understanding the consumer's utility functions. However, its application is limited to relatively few attributes and their levels because the potential product concepts for consumer ranking grow very fast as the number of features increases. Also, to conduct the conjoint analysis experiment, usually a personal interview survey is required due to the complex nature of the task.

Perceptual mapping for understanding consumer perceptions of your product concept in relation to the competitive offerings

Perceptual maps depict the positions of products on a set of key dimensions consumers use to evaluate them. For example, Urban and Hauser (1980) present a possible perceptual map for pain relievers (Fig. 3.2). The two dimensions considered most important by consumers in evaluating pain relievers are "effectiveness" and "gentleness." Perceptual maps could be useful in:

1. Identifying new product opportunities available due to "holes" in the map or the market niches that current competitors have overlooked. Also, this map can indicate the vulnerability of competitors by showing how consumers perceive their offerings.

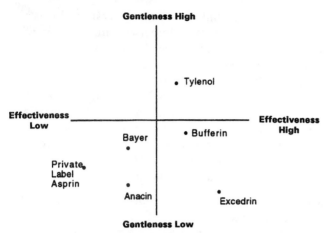

Figure 3.2 A perceptual map of pain relievers. The perceptual map presented here is for illustration only and is not based on any empirical data. (*Adapted from: Urban and Hauser, 1980.*)

2. Testing whether or not consumers perceive the product concept the way the firm intended.

3. Providing guidelines for R&D and the marketing effort. By including the consumer's "ideal concept," R&D and marketing could work on refining the concept or communication strategies.

Perceptual maps can be created by two methods: (1) attribute rating method (AR) and (2) overall similarity method (OS). For greater statistical details on these methods see Urban and Hauser (1980).

AR method.   In the AR method, consumers are presented with a complete list of possible product attributes and are asked to rate each product or brand on each attribute. In the pain reliever example, a list of attributes could include:

1. It is easy to swallow.

2. It is effective.

3. It is gentle on the stomach.

4. It doesn't leave an aftertaste.

5. It doesn't make me drowsy.

Each attribute is rated on a five-point strongly agree/strongly disagree scale by the consumers for the concept being tested as well as for competing brands. Thus, if we have n brands and m consumers, we will have $5 \times n \times m$ data points. The data are subjected to a statistical technique called "factor analysis." The analysis reduces the original list of attributes to the two most important dimensions with all the brands positioned on the map. Since the original data are in five dimensions and the perceptual map is reduced to two, the map does not capture all of the variation in the data.

The AR method is the most widely used method. However, it requires that the researcher be able to list all the important attributes and the consumers must think in terms of those attributes when rating. For product categories that are driven by taste, odors, or aesthetics, such as a soft drink or a perfume, the AR method may not be useful. In these situations, the OS method is useful.

OS method. The OS method produces a perceptual map like the AR method. In OS, the researcher does not provide a list of attributes on which all products are to be rated. Instead, consumers are asked to make judgments about the overall similarity of pairs of products. Each judgment is recorded on a five-point (very similar to very different) scale. If there are n products, there will be $n(n - 1)/2$ pairs to be evaluated. For example, four soft drinks can be compared as presented in Table 3.8.

The overall similarity data are then analyzed by a statistical technique called multidimensional scaling (MDS). The statistical analysis attempts to find a map such that the distances between the products on the map match up the overall similarity data of the input matrix. The output is similar to the perceptual map produced by the AR method without the labels for the axes. A panel of consumers having knowledge of the category can help label them.

The OS method allows the researcher to map products as well as to infer the attributes used by consumers in making distinctions among competing brands. Depending on the product category, the researcher can choose between the AR and OS methods.

TABLE 3.8   Overall Similarity Data for Perceptual Mapping

Possible pairs of 4 soft drinks	Very similar				Very different
Coke/Pepsi	1	2	3	4	5
Coke/7UP	1	2	3	4	5
Coke/BrandX	1	2	3	4	5
Pepsi/7UP	1	2	3	4	5
Pepsi/BrandX	1	2	3	4	5
7UP/BrandX	1	2	3	4	5

Use of focus groups for concept development and testing

The focus group interview is a popular qualitative market research technique due to its low cost, quick turnaround, and flexibility. Mostly, focus group interviews are conducted at the idea-generation stage of the new product development. However, research in the adoption and diffusion process suggests that the focus group interview technique might be well suited to the concept development and testing stage. The advantage of the group interview technique is the in-depth exploration it affords of the reasons behind the likes and dislikes expressed and discussed among group members. It is worth considering when the focus group interview may provide better insight than individual interviews during concept testing.

A focus group typically consists of six to ten people and is led by a moderator with the aid of a discussion guide prepared in advance (Wells, 1974). The moderator manages the group dynamics and makes sure that the discussion stays relevant to the sponsor's concerns. McQuarrie and McIntyre (1986) suggest a six-stage process for evaluation of product concepts within a focus group. Each stage is associated with a research question or task to be accomplished as follows:

1. *Orientation.*   Identify unmet needs and unsolved problems associated with the product category. Also assess current product offerings. What new products in this area have caught their eye or would they like to see?

2. *Exposure.*   Determine initial reactions to the product concept. Engage in a detailed discussion of product features and capabilities.

3. *Evaluation.*   Discover what is liked or disliked about the product concept. Determine why the concept is accepted or rejected. Get global reactions first. Get feature-by-feature reactions. Rank the most/least attractive attributes.

4. *Pricing.*   Assess the value of the needs satisfied, in part, by determining the prices that seem appropriate for the product. Get reactions to suggested prices. Ask the group to suggest a price.

5. *Extensions.*   Assess options and extensions that might be added to the product either before or after it is purchased.

6. *Modifications.*   Assess improvements that might be suggested for the product.

Focus group results can be misleading and should be treated as only another set of data (Wells, 1974). In all focus groups, the sample size is small and is never selected by probability methods.

Questions are not asked the same way each time. Responses are not independent. Statistical techniques of data analysis cannot be applied. The conclusions depend on the interpretative skills of the analyst. It is possible that moderators may influence the results. However, meaningful insights into purchase intentions along with diagnostic data on concept modification can be obtained. This insight could be helpful in broadening the thinking of the product development team.

Lead user analysis for concept development

Traditional research methods for assessing consumer reactions to new product concepts (preference for A versus B) or understanding user needs of new products generally require a relatively large group of users. The research may not provide meaningful results if these users have not felt the need for the product the marketer is testing. However, in many industries like semiconductors, instrumentation and control, and printed circuit boards, it has been observed that the richest understanding of needed new products is held by just a few users rather than a large number of users. These few users who face this need before the majority of other users do are called "lead users." They expect to benefit significantly by obtaining a solution to those needs. Herstatt and von Hippel (1992) provide a methodology for the joint development of "next generation" product concepts with lead users who possess not only future need data but also solutions to those needs. The lead user approach takes the popular slogan of becoming customer focused beyond passive and reactive response to clearly articulated customer needs. It suggests that at times it might be necessary to lead the customers by anticipating the needs of the majority (based on the experience of a minority of lead users) and educating them on new functionalities (Hamel and Prahalad, 1991).

The lead user approach to developing future new product concepts involves:

1. Identification of important trends in the evolution of user needs and innovation activities at the user end, suggesting that they will benefit as a result of investment in these innovations.

2. Identification of the lead user sample with the help of the manufacturer and other companies in customer contacts.

3. Bringing the sample of lead users together with company engineering and marketing personnel to engage in group problem-solving sessions.

4. Testing whether concepts jointly developed by lead users and the manufacturer will be found valuable by the more typical users in the target market.

It is reported by Herstatt and von Hippel (1992) that the lead user methodology applied for the development of new generation pipe hangers was faster and less expensive than the conventional marketing research method. The lead user method took 9 months and cost $51,000 as compared to the conventional method which took 16 months and cost $100,000. An additional benefit of the lead user approach is the increased cooperation between technical and marketing people in understanding user product and performance requirements.

Communicating the Concept to the Technical People

Although technical people are (should be) involved during the concept development process, there might be a need to send an actual request to R&D for the technical development of the concept. Crawford (1991) notes that in most firms this request usually comes in the form of features desired in the product, such as color, size, material, or appearance. Crawford suggests that instead of asking for features, the request should include the benefits sought. Some required features or constraints could be included in the request, while primarily listing the benefits. How those benefits will be provided should be left to the creativity of the R&D people. Crawford calls it a "protocol" between R&D and marketing. The R&D group promises that they will deliver the product with needed benefits, while marketing promises to market the product successfully.

Conclusion

In this chapter, we discussed why companies need to develop powerful new product concepts, what makes a concept powerful, and how it is developed and refined. From a customer's point of view, the powerful product concepts provide both tangible and intangible values in excess of competitive offerings. On the other hand, from a product developer's point of view, there are some additional dimensions that should be included at the time of concept development (e.g., the considerations of manufacturability and assembly, product testing, future innovation, quick market adoption, disassembly, disposability, environmental concerns, profitability, and competitiveness). Above all, the product concept must fit with the corporate strategy and the objectives of the new product development program or the product innovation charter (Crawford, 1991). This ensures that the product resulting from the concept will take advantage of the company's strengths, meet corporate objectives of risks and returns, provide customer value, and be defensible against competitive threats.

In this chapter we have deliberately avoided the discussion related to idea generation and screening techniques. Rather, the focus has been on developing product concepts in partnership with customers. Thus, concept development and testing seem to occur concurrently rather than sequentially. Readers interested in getting detailed information on idea generation and screening techniques may refer to Crawford (1991).

References

Booz, Allen and Hamilton, Inc. *New Products Management for the 1980s,* New York: Booz, Allen and Hamilton, 1982.

Business Week. "Pssst! Want a Secret for Making Superproducts?" October 2, 1989, 106–107.

———. "A Smarter Way to Manufacture," April 30, 1990a, 64–69.

———. "Rebel with a Cause," December 3, 1990b, 130–135.

———. "Questing for the Best," Quality Imperative Issue, October 25, 1991, 16.

———. "Winners: The Best Product Designs of the Year," June 8, 1992a, 52–68.

———. "What Works for One Works for All" April 20, 1992b, 112–113.

———. "Sustainable Development," May 11, 1992c, 75.

Cattin, P., and Wittink, R. "Commercial Use of Conjoint Analysis: A Survey," *Journal of Marketing,* 46, Summer 1982, 44–53.

Clark, K. B., and Fujimoto, T. "The Power of Product Integrity," *Harvard Business Review,* November–December 1990, 107–118.

Crawford, M. C. *New Products Management,* Illinois: Irwin, 1991.

Dean, J., and Susman, G. "Organizing for Manufacturable Design," *Harvard Business Review,* January–February 1989, 28–36.

Dolan, R. J. *Managing the New Product Development Process,* Massachusetts: Addison Wesley, 1992.

Fortune. "Closing the Innovation Gap," December 2, 1991, 56–62.

———. "Environment and Industry: Harnessing the Power of the Marketplace," June 15, 1992a, 17–36.

———. "Brace for Japan's Hot New Strategy," September 21, 1992b, 62–72.

Green, P., and Wind, Y. "New Way to Measure Consumers' Judgements," *Harvard Business Review,* 53, July–August 1975, 107–117.

Hamel, G., and Prahalad, C. K. "Corporate Imagination and Expeditionary Marketing," *Harvard Business Review,* July–August 1991.

Herstatt, C., and von Hippel, E. "Developing New Product Concepts Via the Lead User Method: A Case Study in a Low-Tech Field," *Journal of Product Innovation Management,* 9:3, September 1992, 213–221.

Johnson, R. M. "Tradeoff Analysis of Consumer Values," *Journal of Marketing Research,* May 1974, 121–127.

Kalwani, M., and Silk, A. "On the Reliability and Predictive Validity of Purchase Intention Measures," *Marketing Science,* Summer 1982, 243–287.

Kotler, P. *Marketing Management: Analysis, Planning, Implementation, & Control,* Englewood Cliffs, N.J.: Prentice Hall, 1991.

———. "Future Marketers Will Focus on Customer Data-base to Compete Globally." *Marketing News,* June 8, 1992, 21.

Kuczmarski, T. *Managing New Products,* Englewood Cliffs, N.J.: Prentice Hall, 1988.
Mahajan, V., and Wind, Y. "New Product Models: Practice, Shortcomings and Desired Improvements." *Journal of Product Innovation Management,* 9:2, June 1992, 128–139.
McQuarrie, E. F., and McIntyre, S. H. "Focus Groups and the Development of New Products by Technologically Driven Companies: Some Guidelines," *Journal of Product Innovation Management,* 3:1, March 1986, 40–47.
Newsweek. "The Environment: Responsible Development in the 21st Century," June 15, 1992 (special advertising).
Schwartz, D. *Concept Testing,* New York: AMACOM, 1987.
Tauber, E. "Utilization of Concept Testing for New Product Forecasting: Traditional versus Multiattribute Approaches," in Wind, Y., Mahajan, V., and Cardozo, J. (eds.), *New Product Forecasting,* Lexington, MA: Lexington Books, 1981.
Taylor, J. J., Houlahan, J., and Gabriel, A. "The Purchase Intention Question in New Product Development: A Field Test," *Journal of Marketing,* January 1975, 90–92.
Urban, G., and Hauser, J. *Design and Marketing of New Products,* Englewood Cliffs, N.J.: Prentice Hall, 1980.
Wells, W. D. "Group Interviews," In Ferber, R. (ed.), *Handbook of Marketing Research,* New York: McGraw Hill, 1974.
Whitney, D. E. "Manufacturing by Design," *Harvard Business Review,* July–August 1988, 83–91.
Wittink, R., and Cattin, P. "Commercial Use of Conjoint Analysis: An Update," *Journal of Marketing,* 53, July 1989, 91–96.

Defining Markets and Users for New Technologies

C. Anthony di Benedetto
Temple University

Defining the market for and the users of the new product being developed is an essential early step in the new technology development process. In real-world processes, this step is often conducted in concert with concept development (Chap. 3). The outcome of this step may result in a revision in the established concept, as well as the original idea (Chap. 2). What constitutes a market? How can potential users be identified early in the development? Does the perceived nature of technology, radical versus incremental, make a difference? How can user needs be created? How can the product be matched to the user? These are some of the important questions answered in this chapter. Numerous techniques are presented, with examples that demonstrate their proper use.

WM. E. SOUDER AND J. DANIEL SHERMAN

Introduction

Conceptualizing new products and technologies

This chapter focuses on understanding the roles of technology and market needs in developing successful products. It opens with a discussion of the product innovation process and the factors that act to drive this process. A later section focuses on market analysis and definition: how broadly the market boundaries ought to be defined, assessing the characteristics of the market, and targeting a product to the market. The tasks and activities that tend to lead to more successful products are discussed in detail: concept evaluation, screening, product use testing, market testing, and customer value analysis. This chapter concludes by examining how customer needs can be actively brought into the product development process and the ultimate impact upon quality commitment and customer satisfaction.

What is a new product?

Although used in different ways by different individuals, a "new product" is usually defined as a "product (good or service) new to the firm marketing it" (Crawford, 1991, p. 539). The term "innovation" is often associated with new products. It has been broadly defined as "an idea, practice, or material artifact perceived to be new by the relevant unit of adoption" (Zaltman, Duncan, and Holbek, 1973). It can refer to a new product or a new process (Utterback and Abernathy, 1975) and implies "at least some degree of newness to the market, not just to the firm" (Crawford, 1991, p. 536).

Innovations can be radical or incremental in nature (Ettlie, Bridges, and O'Keefe, 1984). Thus, a minor reformulation of a detergent or a discontinuous change in product or production technology could both be considered innovations. "Innovation" can also be used to refer to the act of creating a new product or process, including invention, as well as the work required to bring the idea or concept into final form (Crawford, 1991, p. 536). An important distinction can be made between invention and innovation, in that the innovative process includes the work done by management, marketing, engineering, production, and others subsequent to the actual invention that leads to the development of a marketable product (Crawford, 1991, p. 537). Of course, the innovating firm may not have been the firm that actually invented the product or process. Competitors, suppliers, licensees, or other entities may have been the inventor but for some reason did not or could not bring the product successfully to market (Utterback and Abernathy, 1975).

A dynamic model of product or technology innovation

In an important early study, Utterback (1974) found strategic patterns in the types of innovations and the sources of innovative ideas in several industries. The drivers of innovation appeared to depend on the key objectives within the industry. For example, in the automotive industry, sales maximization is an important objective, so he found a predominance of easily implemented product innovations that would be obvious to final customers. By contrast, in transportation and communications, an overriding objective is cost minimization, and production line innovations of the cost-saving type were more common. Other industries appeared to have different objectives, which would drive different kinds of innovative strategies: In the pharmaceutical industry, performance maximization would predominate, while in mining, resource control is an appropriate objective (Utterback, 1974).

Based on this and other studies of innovation in industry, Utterback and Abernathy developed a dynamic model of product and process innovation which proposed that a firm's growth strategies, its competition, and its environment would lead it to pursue certain types of innovations (Utterback and Abernathy, 1975; Abernathy and Utterback, 1978). In the first stage of the innovation process, product performance maximization is the major objective. Most innovations are product innovations in response to unmet customer needs and advancing technologies, and production processes remain highly flexible to accommodate relative uncertainty in the product form required by customers. The second stage is marked by a focus on sales maximization and the emergence of a dominant product form. During this stage, "islands of automation" appear in the production process, which becomes less flexible. Major process innovations, required by increasing volumes, predominate here since most of the key product innovations will have already occurred. In the third stage, cost minimization becomes the overriding objective and incremental, cost-saving innovations of both the product and process type occur. By this time, production processes are rigid and capital-intensive. While they were not able to provide empirical evidence of firms or industries passing through these stages using longitudinal data, Utterback and Abernathy (1975) found partial support for the model in a cross-sectional study.

This dynamic model has been investigated and questioned in further studies. De Bresson and Townsend (1981) confirmed some, but not all, of the relationships in the model by applying it to a different database. Although partial support was found for it, they concluded that it did not provide an integrative framework to industrial innovation. Another study investigated the model in terms of the technical autonomy of the innovating firm (Calantone, di Benedetto, and Meloche, 1988). Technically autonomous firms were found to be successful at cost-saving process innovations, and their internally developed innovations usually took place early in the production process. This dynamic model has some parallels to the product life cycle as well. Moore and Tushman (1982) noted that strategies and organizational structures must evolve over time as changes in innovation types necessitate changes in competitive emphasis.

Revolutionary versus evolutionary product innovation

The dynamic model of Utterback and Abernathy (1975) is concerned with describing evolutionary changes in an industry as product form and production processes stabilize through time. What about discon-

tinuous or revolutionary product changes, which can suddenly desta-
bilize an industry and possibly cause current technology to become
obsolete overnight? Utterback (1982) developed an extension of the
original model that sought to describe conditions where revolutionary
product innovations are likely to occur. Utterback provided examples
of revolutionary innovations in the automobile, transistor, and type-
writer industries as illustrations.

According to this model, as an industry matures, market shares
become relatively stable, no substantial product or process innovation is taking place, and the industry thus becomes a target for
attack by a firm from outside the industry with new technology.
Discontinuous change of this nature happened in, for example, the
vacuum tube industry as transistors were used to replace vacuum
tubes. Similar discontinuous change occurred as manual typewriters
gave way to electrics, electrics with memory, and finally word proces-
sors and desktop publishers. Incumbent firms are thus put into a
position of having to react either by being forced to make improvements to the technology to try to survive in the same arena as the
new entrants (as was the case with the vacuum tube makers) or by
actually moving into the new technology and competing directly, possibly via acquisition (Cooper and Schendel, 1976). Consistent with
this model, Calantone, di Benedetto, and Meloche (1988) found that
successful revolutionary product innovations often emerged from outside the industry. Also, large firms may have an advantage over
smaller firms in being able to enter a new market with a radical
innovation. But this may be true only to a point: Very large firms
may actually be at a disadvantage to some extent because they may
not be able to respond quickly enough to changes in technology
(Scherer, 1980, Chap. 15; Ettlie and Rubenstein, 1987). These studies
imply that a key to success is to discover a market that will be receptive to the new technology; this can allow a firm to compete effectively against very entrenched competition with the older technology.

Drivers of product innovation

Many characteristics of the firm will have an impact on its ability to
innovate. One of the most important is the firm's human resources. A
poorly educated workforce can be a major hindrance to innovation
(Fernelius and Waldo, 1982), and American firms are sometimes dis-
appointed with the level of written and math skills of high school
graduates. The "MBA syndrome" has resulted in American companies being run by managers without science training and little experience in the firm's products (Hayes and Abernathy, 1980; Crawford
and Tellis, 1981). Other innovation drivers will include the firm's

access to capital and the policies of government to make capital available (Dickinson, Ferguson, and Herbst, 1984), the size of the firm's assets and its reputation within the industry (Ettlie and Rubenstein, 1987), and the attitudes of top management toward risk taking and failure (Hayes and Abernathy, 1980; McIntyre, 1982; Quinn, 1985). The information resources available to a firm can also play a significant role. For example, the Japanese seem to be better than their American counterparts at making overseas technical visits and attending industry and research consortia to gain key information (Peck and Goto, 1981).

The nature of the firm's products and its production processes are two additional drivers influencing corporate innovation. As has been seen before, the stage of the product in the dynamic model of product and process innovation can have an effect on the nature and extent of innovations implemented by the firm. The similarity or synergy between the innovative product and the firm's current product line may also be a factor, as may the level of cooperation between the firm and the new product venture team (von Hippel, 1977). Problems with the production process can lead to a search for innovative solutions. These may include product overengineering, too expensive components, problems with quality or meeting lead times, or the need to convert from batch processing to mass production (Avlonitis, 1985).

Further drivers of innovation are the firm's customers and competitors and the government. Users can initiate innovation by developing prototypes and seeking out manufacturers or by licensing a technology (Foxall and Johnston, 1987). Customers can also have important input into manufacturer-initiated innovation (von Hippel, 1988). In industries marked by a high level of competition, there often is quicker diffusion of innovation due to behaviors like price cutting that stimulate early sales (Buskirk, 1986). The government can influence the rate of innovation by granting research funds, sponsoring industrywide research programs that provide for sharing of key information and human resources, and providing subsidies or tax breaks to innovating firms (Schnee, 1978).

Opportunity identification and technological development

The dynamic model of product and process innovation, together with the familiar product life cycle (PLC), can be used as guides to technology development, forecasting, and opportunity identification.

The PLC itself, of course, has limitations. It can perhaps be used as a good depiction of the sales patterns through time of a product but is of less help in strategic planning or opportunity identification. A good rundown of the problems associated with the PLC and its role in strategic planning is provided by Rink and Swan (1979):

1. Few studies have actually attempted to test factors that cause patterns in the PLC.

2. Most PLC studies have addressed consumer goods, and its applicability in industrial settings is comparatively untested.

3. Researchers have focused efforts on minor product changes instead of on technology changes.

4. Little is known about the length and sequence of PLC stages.

5. Little is known about the impact of the firm's characteristics on the PLC.

6. Little is known about how widespread the use of the PLC is among practicing managers.

Similarly, Wind and Claycamp (1976) described the familiar "generic" recommendations typically found in discussions of the PLC (such as how to manage a product through growth, maturity, and decline stages) to be "vague, nonoperational, not empirically supported, and conceptually questionable," lacking in concern for profitability, market-share position, and competitive activity. Small wonder that some authors (notably Dhalla and Yuspeh, 1976) called upon marketers to "forget the product life cycle."

A more balanced approach was presented by Day (1986), who suggested that "the PLC has *descriptive* value…for explaining the evolution of markets that are subject to the…interplay of customers, competitors, and technologies" (p. 85). It has less value as a *predictive* model of when products will grow, enter maturity, or decline or as a *prescriptive* model of what strategies to use during each PLC stage. A main message of Day is that the PLC is not a fait accompli but is indeed shaped by the strategies implemented by the firm itself (and its competitors).

American firms have too often managed their products as if they were in maturity or decline stages of the PLC rather than revitalizing them for renewed growth, resulting in weaker global competitiveness (Gordon, Calantone, and di Benedetto, 1991). A recent study using the PIMS database (Anderson and Zeithaml, 1984) supports this argument. Among firms in industries with slowing growth rates (presumably an indication of market maturity), many common traits were found: declining prices, marketing mix strategies centered around advertising differentiation, low relative investments in product or process research and development, and low emphasis on market share growth. These are traits found among firms that are managing mature products for the short term rather than trying to revitalize them for the long term. Gordon, Calantone, and di Benedetto (1991) provided four reasons why American managers frequently take this short-term view: stagnant technology resulting from an unwillingness to commit to long-term R&D efforts and investments, less customer-supplier interaction resulting in emphasis on low cost and product standardization, slower innovation rates due to narrowing of product scope, and a corporate culture shaped by managers with MBA degrees but lacking in technical training. Companies that have successfully revitalized products, on the other hand, share several tendencies: They focus on their target markets, adapt to environmental changes, keep their divisions small, tolerate failure and mistakes on the part of management, and ensure understanding and appreciation of technical aspects of the firm through job rotation (Maidique and Hayes, 1984). The Gordon et al. (1991) article proposes a five-step plan for revitalizing products:

1. *Customer analysis:* Searching out potential new technology and product enhancements being performed by customers

2. *Supplier analysis:* Including suppliers in hurry-up research

3. Developing contractual arrangements through which technology can be licensed

4. Pursuing strategic alliances

5. Exploring other means of cooperative technology or resource sharing (p. 44)

It is possible to take the benefits of the PLC, as identified by Day (1986), a step farther. Perhaps the cumulative information provided by examining *several* PLCs can be used to identify strategic windows of opportunity for gaining competitive advantage (Calantone, di Benedetto, and Gordon, 1991*b*). Additional information can also be gained by examining the:

> Patterns of product and process innovation over time. The strategic posture (in terms of technology, innovation, and devotion to R&D) of a successful firm is cultured by the PLCs and (patterns of product and process innovation) in a given industry over a long period of time. One might expect this length of time to vary by industry: one could expect it to be shorter for computer chips or environmental services...than for, say, automobiles...Taking a long-term perspective over several years (or decades) and several PLCs provides management with an understanding of the patterns of windows of opportunity in its industry (Calantone et al., 1991*b*, pp. 64–65).

Consider, for example, the computer chip industry. Innovations in integrated circuits were predictable over the 1970s and 1980s. Improvements in capabilities could in fact be plotted as an exponential function of time, and future improvements were expected to continue in this way (Penfield, 1982). Having information like this can allow the firm to decide whether it should invest today in R&D (that is, whether the payoffs in terms of product or process innovations will occur before the window of opportunity closes). If it is feared that the window of opportunity will have passed, other strategies for staying competitive may need to be chosen, such as joint venturing or technology licensing. This relates to the recommendations of Day (1986), who suggests that there are two ways for late-entrant firms to enter a market successfully: out-imitate (serve the market better than earlier entrants did) or leapfrog ("out-innovate" or refine the product or process technology).

In sum, examining the patterns of PLCs through time in a given industry can aid in the understanding of technological opportunity identification. It can also be useful to examine the patterns of product and process innovation [as in the Utterback and Abernathy dynamic model (1975)] and to assess whether substantial innovations of either type are expected in the near future. This can also be a guide for where to invest R&D dollars, especially when keeping in mind the possibility of imminent revolutionary product innovation (Utterback, 1982). In addition, firms lacking the appropriate human resources, capital, asset base, or (importantly) the right attitude toward risk taking should know that these are important drivers to product innovation and as such are critically related to the firm's ability to be at the technological forefront.

Market Analysis and Definition

Defining the market boundaries

One difficult aspect of market analysis is deciding just how broadly to define the relevant market. For example, what should McDonald's consider as its market? In part, the answer to this depends upon the competition in the fast food business; as chicken, taco, Italian, and other categories of fast food gain shares of this market, the scope of McDonald's business has expanded:

> Rather than defining its business as strictly fast food hamburgers, McDonald's has gradually broadened its market to fast food in general, adding chicken and salad offerings, and...contemplating entry into the pizza business. Changing consumer preferences have made the chicken and pizza markets faster growing than the burger market. Additionally, new cooking technologies allow virtually error-free pizzas to be made in five minutes (Bremmer, 1989; quoted in Guiltinan and Paul, 1991).

In general, corporate-level managers are most likely to define relevant market boundaries broadly. In doing so, they keep a lookout for long-run growth opportunities or environmental threats that may arise, which the firm must defend against (Guiltinan and Paul, 1991, p. 56). As an example, mainframe computer manufacturers had to assess their options when it became feasible for business people to use personal computers at work and buy them for home use. Record makers had to envision the need to get involved in tape- and CD-based technologies early on in order to continue to compete into the 1990s. Even so, record companies have responded differently to the changes in this industry. Motown Records has deleted many of its LPs from its catalog, while larger firms like CBS/Sony and Warner/Elektra/Atlantic continue to make vinyl records as well as tapes and CDs, noting that there are still a lot of vinyl record lovers, while keeping open the possibility of eliminating LPs sometime in the future (Hisrich and Peters, 1991, p. 415).

Broad relevant markets are, in general, appropriate when (1) technological changes are likely to create alternative product forms (as with the record industry), (2) social or cultural changes will affect the frequency of use or usage situations for the product, (3) sales gains or losses tend to come from alternative product forms, not simply competing brands, and (4) the product is so innovative, there is no competition at the product form level (Guiltinan and Paul, 1991, p. 57). Guiltinan and Paul (1991) provide Procter & Gamble's low cholesterol laxative product as a prime example where the market must be defined broadly: technological change is occurring, demand for cholesterol-reducing products will increase as buyers become more concerned about intake of cholesterol (a social change), and there are no direct brand competitors.

By contrast, middle-level managers (product managers, brand managers, advertising managers, and similar positions) are more likely to define the relevant market more narrowly, as they are more focused on shorter-term decisions and the day-to-day business of managing their product. In general, a narrower relevant market focus is seen in situations where (1) brand competition is more critical than product form competition, (2) no great technological changes or changes in product usage are envisioned, at least in the near future, and (3) there are no easily substitutable product forms (Guiltinan and Paul, 1991, p. 57). Thus, while top management of a record company may have been anticipating entry into tape-based technologies as early as the 1970s, its middle-level managers would be more concerned with selling this year's and next year's LPs.

Characteristics of the market

In determining the likely demand for a new product form or product class (that is, the *primary demand* for the product), management must consider the characteristics of the market it will be serving. There are two categories of market characteristics to consider: the identity of the buyers of the product form (demographics, life styles, customer turnover, and so on) and the buyers' levels of willingness and ability to buy the product form (Guiltinan and Paul, 1991, p. 59).

In the case of a new product, buyer resistance may occur. It is up to the innovating firm to diagnose the reasons for this buyer resistance and, if possible, to overcome it. A useful framework for this diagnostic process is suggested by Guiltinan and Paul (1991, p. 62), who consider buyer resistance in terms of willingness and ability to buy. A customer may be unwilling to buy a new product for several reasons: perceived economic, convenience, physical, or performance risks; lack of compatibility with the customer's values or experiences; perceived problems in product use; or inadequacies in related products and services. Consider the adoption of microwave ovens by consumers. At first, many customers may have been reluctant to purchase an oven that cooked in such a radically different way from conventional ovens. They may have feared that food would not be prepared as well as in conventional ovens (performance risk). They may have found the price too high for the benefit it delivered (economic risk). They might have feared radiation leaks or somehow felt the ovens unsafe (physical risk). They may have found the early models too difficult to operate (problem in product use). They may have been skeptical of something that they (or their parents) never used (lack of compatibility with customer values). Or there might not have been enough microwave cookbooks to help the new user in the kitchen (inadequacy in related products). Any of these factors would lead to an unwillingness to buy. Being able to uncover the reasons for this unwillingness is the first step in overcoming the problem. As time went on, microwaves were made simpler to operate, safety ceased to be a problem, and cookbooks written by well-known chefs were sold, leading to greater product acceptance. Note that these kinds of improvements do not necessarily sell any one brand of microwave oven. They all are geared toward increasing the primary demand for microwaves in general. Without stimulating this primary demand, none of the producers of microwave ovens will be able to make reasonable sales or profits.

It is also important to know if slow adoption of a new product by the market is a result of a lack of ability to buy, not a lack of willingness. Guiltinan and Paul (1991, p. 64) provide several factors limiting ability to buy, many of which are not under the control of the firm: cost factors, packaging and size factors, and spatial availability (that is, the product is not available in some areas or regions of the country). For example, getting solar heating panels or thicker insulation may be great cost savers in the long run, but when the economy is bad, home budgets are squeezed and alternatives that cost less in the short run will be preferred. Discretionary purchases (such as a new car or an expensive vacation) may be put off altogether. The demand for supermarket checkout scanners was expected to grow substantially in 1978 and 1979, but high interest rates forced retailers to make do with many mechanical cash registers until the early 1980s. Not anticipating this postponement in purchases, the scanner manufacturers overproduced in the late 1970s (Day, 1986). The purchase of some large products, such as some home computers or televisions, may be limited by a very fundamental consideration: lack of space in the customer's home (Guiltinan and Paul, 1991, p. 64).

As was the case with willingness-to-buy factors, a lack of ability to buy a new product form on the part of customers must be understood before the firm can appropriately respond. If price is an issue for a big-ticket household item, perhaps a no-obligation trial period or rent-to-own plan could overcome the problem. This would permit the customer to become aware of the benefits delivered by the product, and the customer would then value it more highly. On the other hand, as the above checkout scanner example illustrates, there are times when there is little the firm can do. It is debatable whether these firms could even have foreseen the phenomenal growth in interest rates that led to the difficulties they faced. One could still argue that they became so wrapped up in the potential for their new technology that they failed to develop contingency plans or, if they did, chose to ignore the "worst-case scenario."

Targeting a new product or technology to a market

In the case of a new product or new technology, anticipating user demand can be an especially challenging task. Not only must the number of years (or months) to maturity be predicted but also the level of sales at maturity. Without accurate forecasts, greatly misleading sales and marketing plans may result. Wildly inaccurate growth market forecasts have been reported. For example, in the late 1960s, such product forms as plastic housing, laser three-dimensional movies, family helicopters, home dry cleaning systems, and tooth decay vaccines had been predicted to be commonplace before 1980 (Berenson and Schnaars, 1986). To their credit, growth market forecasters are often closer to the mark. But misleadingly high forecasts continue to be made. Berenson and Schnaars (1986) noted that more than half the new product forecasts fail. They attribute this unfortunate observation to two main reasons: (1) unbridled optimism—the innovating firm overvalued the technology without considering the size of the potential buying market—and (2) product costs or lack of competitive advantage were not considered by the innovating firm. By contrast, better forecasts tended to be based on market factors rather than on how advanced or glamorous the technology was. Berenson and Schnaars provided several common sense recommendations to aid in growth market forecasting, including checking assumptions, staying flexible, and avoiding being "dazzled" by new technology. It is important to note that the two major reasons for misleading growth market forecasts each could have been avoided if marketing and R&D personnel had cooperated better. Marketing is in an ideal position to provide important information to R&D on user needs, competitive activities and likely competitive response strategies, and other environmental changes. The need for marketing to provide this key information, in a way that is usable to R&D, cannot be overstated. If it works independently of marketing, R&D may well come up with bold new technologies but will probably not have the customer knowledge required to bring the technology to the market in a way that customers will want to buy it. While the consumer market rejected Picturephones in the 1960s, for example, other products using the technology of sending pictures over phone lines (such as fax) have been successful in the industrial market. There are two important implications here: (1) new product development is a two-stream process—activities within the realms of marketing and production/R&D are essential and neither stream should be overlooked—and (2) the two streams need to be coordinated in their activities such that each really does help the other in developing successful new products. This is often easier said than done, as differences in training and background and, occasionally, a mutual suspicion (or mistrust) can exist between marketing and R&D. Both of these implications are discussed at length in the next section.

User-Need Analysis

What makes a new product successful?

New product development can be an important source of competitive advantage for firms making either consumer or industrial goods. Yet, as any manager responsible for developing or launching a new product into the market knows, it can be a very costly and risky activity, and some product failures (such as the Convair intermediate-range jet and the RCA videodisc) have cost their makers hundreds of millions of dollars. All things being equal, some risk is probably desirable; a firm without any failures may not be aggressive enough and may be missing out on emerging opportunities (di Benedetto and Calantone, 1990). The goal is to identify costly failures early and to manage the investment into new products for long-term profitability and competitive advantage. A firm that has adopted the credo of the marketing concept (identify and strive to meet customer needs and wants while meeting organizational goals) must, therefore, be able to analyze user needs and use these needs as input into the product development process.

To manage the risk undertaken by firms involved in new product development, one must first understand the extent of this risk. What proportion of new products fail? One hears failure rates as low as 30 percent and as high as 90 percent bandied about. Crawford (1991) conducted a meta-analysis on failure rates to set the record straight. Most high-end claims are unsupported by empirical evidence and appear to have come out of thin air. He found only seven carefully conducted, recent empirical studies and notes that the average failure rate was 38 percent—a little higher for consumer goods and food products, a little lower for industrial goods. (The studies used financial criteria to assess success or failure of products.) These studies were also very consistent: Except for two studies that focused on food products, all failure rates reported were between 24 and 39 percent. So the risk of product failure is there, although perhaps less than might have been anticipated by some. The first step in managing the process is to understand the factors that lead to success.

Early studies into successful or unsuccessful new products concentrated on identifying factors that correlated with product success or failure. A key finding emerging from most of these studies is the importance of both marketing and technical activities to product success. The National Industrial Conference Board (1964), in its study of what determines product failures, identified several factors, many of which seem to put the blame for product failure on the marketing department: inadequate market analysis, product deficiencies, high costs, poor launch timing, inadequate marketing effort, and competitive strength. Other studies of failed products (Hopkins and Bailey, 1971; Hlavacek, 1974) also pointed an accusing finger at the marketing department.

Project SAPPHO (Rothwell, 1972) compared successes and failures and attempted to identify discriminating factors between them. Five key factors were found. Firms with successful products (1) sought out user needs, (2) paid more attention to marketing publicity, (3) had more efficient product development, (4) made better use of outside technology and scientific communication, and (5) gave responsible managers more authority. The first two factors are marketing department related, the next two have to do with technical or production issues, and the last is related to managerial autonomy. Later comparison studies (Rothwell, 1974; Kulvik, 1977) provided consistent findings.

Building on these and other studies, Cooper (1979, 1980) conducted Project NEWPROD, an exploratory study of about 200 industrial product launches (half successes, half failures). He found that factors differentiating between success and failure included proficient launch execution, meeting customer needs better than competitors, good product quality, adequate test marketing and prototype testing, and understanding the customer's decision-making process. In brief, three key distinguishing factors were identified: (1) having a unique, superior product in the eyes of the customer, (2) having strong market knowledge and performing marketing research and launch tasks well, and (3) having technological synergy and performing the technical and production tasks well (Cooper, 1980). These marketing tasks include idea scanning, market investigation, market research, sales forecasting, and marketing planning. The technical activities include technical feasibility studies, product design, and prototype construction and testing (Cooper, 1986). A more complete list appears in Table 4.1.

Later versions of Project NEWPROD confirmed and extended these findings. In NEWPROD II (Cooper and Kleinschmidt, 1987a, b), a wider assortment of measures of success were studied than in previous studies (profit, payback, market share, sales relative to competitors, and so on). Critical success factors found were: having a superior product, good performance on preliminary activities prior to prod-

TABLE 4.1   New Product Development Activities

Marketing activities	Technical/production
Marketing-derived idea generation	Technically derived idea generation
Idea scanning	Preliminary technical feasibility
Preliminary market investigation	Product design
Market research	Prototype construction
Preliminary sales forecasting	In-house prototype testing
Development of marketing plan	Prototype trials with customer
Detailed sales forecasting	Trial production
Test marketing	Acquisition of production facilities
Final business analysis	Full production
Revision of launch plan	
Market launch	

SOURCE: Cooper, 1986.

uct development, having a project definition or protocol, and having key proficiencies and synergies. Marketing synergies refer to the degree of fit between the new product and the firm's distribution, advertising, market research and intelligence, and customer service systems and capabilities. Technical synergies have to do with the amount of fit between the product project and the firm's research, product development, engineering, and production resources (Cooper and Kleinschmidt, 1990). The most recent effort in this research stream, NEWPROD III (Cooper and Kleinschmidt, 1990), identified eight key factors underlying success, which overlap somewhat with previous findings: (1) a unique, superior product, (2) well-defined product and project prior to development, (3) technological synergy, (4) good execution of technological activities, (5) good execution of predevelopment activities, (6) marketing synergy, (7) good execution of marketing activities, and (8) market attractiveness. Some of these factors bear further explanation. Technological activities refer to product development, in-house prototype testing, pilot production, and production start-up. Predevelopment activities are defined as initial screening, preliminary market and technical assessments, market study or marketing research, and business analysis. Marketing activities include in-depth market studies, customer tests (field trials), test markets, and market launch. Attractive markets are also viewed as large, high-growth markets in which customers exhibit high need for the product and consider the purchase to be important (Cooper and Kleinschmidt, 1990). These and other studies (see, for example, Stern, 1966; Mansfield and Rapoport, 1975) emphasize the duality of the new product development process in which the marketing and technical development are depicted as parallel processes.

Several recent studies confirm the importance of these factors to product success (Hopkins, 1981; Cooper and de Brentani, 1984; Maidique and Zirger, 1984; Link, 1987; Davis, 1988; Calantone and di Benedetto, 1990). Calantone and di Benedetto (1988) explicitly modeled the duality of the product development process with a path-analytic model. In their study of industrial firms, marketing, technical, and launch activities specific to the new product were all found to be determinants of product success. The possession of critical marketing and technical resources and skills helped the firms conduct the specific activities and thus had an indirect effect on their success.

Once the framework had been built by these studies, later works sought to make recommendations to managers on how to improve their firm's rate of product success. One recommendation common to almost all of these studies is the need for a better linkage or interface between the marketing and technical/R&D departments. The importance of this linkage is implied, of course, in the dual process of product development. But getting adequate linkage can be difficult in practice since these departments are often adversarial. Souder (1980, 1981) and Souder and Chakrabarti (1980) found four key problems at the marketing-R&D interface: (1) lack of communication, (2) lack of appreciation, (3) mutual distrust, and (4) not enough objective criticism across departments, leading to tolerance of inadequate performance. Gupta, Raj, and Wilemon (1985a, b) and Gupta and Wilemon (1988) identified several barriers to effective linkage, including lack of support from top management, a credibility problem, and others that paralleled those found in the work of Souder. Crawford (1984) notes several failure scenarios that typify poor marketing-R&D linkages. In one of these, marketing does not provide enough information to R&D, leaving the latter to guess what the marketplace wants. In another, marketing dictates product specifications to R&D, leading to resentment.

The importance of an effective marketing-R&D linkage cannot be overstated. Gupta, Raj, and Wilemon (1985a) listed almost 20 key product development activities that require integration of the two departments (see Table 4.2). The rewards are apparent for firms that can encourage a healthy marketing-R&D interface. Some of the methods suggested by researchers to foster a good linkage include keeping projects small, involving both sides early in the process, keeping both sides' power and status about equal, having interdivision steering committees or teams (Souder, 1980, 1981; Gupta, Raj and Wilemon, 1985a, 1987; Golden, Huerta, and Spivak, 1985), and establishing a protocol (or agreement between the two parties) on benefits and performance specifications (Crawford, 1984).

Using this information as a guide to new product development

The literature on new product success stresses the importance of both the marketing and the technical/R&D departments and the need for cooperation. By virtue of their role as boundary spanners, marketing is in a good position to identify strategic windows of opportunity for the firm. The marketing-R&D interface is thus important because it brings user needs into the product design and development processes. Through the cooperation of top management, the firm can match skills in marketing, technology, design, and/or production to market opportunities in a timely manner, while the strategic window is open.

Note that the term "strategic window" refers to the match between a firm's distinctive competencies and the factors that are critical to success in a given market (Abell, 1978; Day, 1984). The strategic window is said to be open if there is a good match. As the market

TABLE 4.2   The Integration of Marketing and R&D

Marketing is involved with R&D in
Setting new product goals and priorities
Preparing R&D's budget proposals
Establishing product development schedules
Generating new product ideas
Finding commercial applications of R&D ideas and technologies

Marketing provides information to R&D on
Customer requirements of new products
Regulatory and legal restrictions on product performance and design
Test marketing results
Regular feedback from customers regarding product performance
Competitor strategies

R&D is involved with marketing in
Preparing marketing's budget proposal
Screening new product ideas
Modifying products according to marketing's recommendations
Developing new products according to market need
Designing communication strategies for customers of new products
Designing user and service manuals
Training users of new products
Analyzing customer needs

SOURCE: Gupta, Raj, and Wilemon, 1985a.

evolves, requirements for successful entry can change, and the window can quickly close. Thus, correct market assessment for proper timing is of prime importance. In particular, in an environment marked by uncertainty and rapid change in technologies and markets, strategic planning becomes even more important since the firm will not be able to survive with "trial-and-error" planning (Day, 1984).

The need for integration across the marketing-R&D interface is perhaps best illustrated by briefly examining the controversy over the use (or misuse) of the marketing concept. Some have claimed that by being driven solely by market needs, firms have developed products of questionable value such as the pet rock, deodorants, and new-fangled potato chips (Bennett and Cooper, 1979, 1981; Hayes and Abernathy, 1980). Bold new ideas that lead to products of higher quality and long-term value to customers and to the firm must come from "technology push." Examples such as lasers, xerography, and instant photography are often given here. The implication is that, by misusing the marketing concept, America has managed its way into economic decline (Hayes and Abernathy, 1980). But there is another side of this argument. Without the input of the marketing department, the technology push model can also lead to failure. Examples are Corfam, the picturephone (currently making a comeback), and the Nimslo 3-D camera. These were all technically excellent products, but they failed to satisfy customer needs. Thus they all failed in the marketplace. It has been argued that "the product concept must be subservient to the marketing concept for a product to be successful.... Scientists and engineers may well be the source of new ideas, but they are also the proponents of products that end up serving no market at all" (Schnaars, 1991, p. 11). Thus, it would seem that the call for better coordination of the marketing and technical tasks to guide new product development is reasonable in light of the risks of not coordinating effectively.

Early concept evaluation and screening

Through technology push, market pull, or a combination of these, many product concepts can and will be generated. Many of these product concepts are simply not feasible for a firm: Financial, technological or marketing resources are lacking, the market is viewed as too volatile, or there is a poor match with top management long-run objectives. Concepts thus must be screened. The screening stage is the point where the firm can make some early go/no-go evaluations before very much time or money is invested in the concept. The objectives of the screening stage are several: (1) to decide whether, and how much, R&D resources should be committed, (2) to send some concepts back into concept development for more work, (3) to select standby options to pursue, should difficulties arise with an ongoing project, (4) to assist in record keeping on concept development, (5) to stimulate communication across departments and increase the sensitivity of managers to the needs and desires of other departments, (6) to identify potential political disagreements across departments regarding a proposed concept, and (7) to help a firm assess what its distinctive competencies are and to redefine them if necessary (Crawford, 1991, p. 197; Cooper, 1986, p. 101–113).

Prior to screening, a firm may engage in "prescreening" activities that give a quick and very inexpensive read on a product concept.

Crawford (1991, p. 180) gives Dow Chemical's prescreening of chemical specialty products as an example. Full screening of these products typically costs only about $1000, so prescreening evaluations would cost even less. Prescreening evaluation usually consists of a short meeting of four or fewer experts and costs Dow about $100. A simple concept test among potential customers, using perhaps a picture or short written description of the product, may also be employed at this early stage to eliminate misdirected concepts before the full screening stage.

Methodologies for screening

Many different methodologies are used by firms for screening product concepts. Some firms simply rely on expert judgment. A study of the new product development process at nine large companies (some making industrial goods, some consumer goods) showed scant evidence of formal screening or concept-testing procedures. Screening criteria used by these firms tended to be general, subjective, and unwritten and often resembled gut feel more than anything else (Feldman and Page, 1984). Some firms make what have been termed "single-drive" decisions. This term refers to the fact that only one of the drivers of product innovation (technology push or market pull) is considered by the firm. A consumer goods firm may consider only whether the idea passes a concept test among consumers. A technology-driven firm may look only at whether the R&D department likes the product (Crawford, 1991, p. 198).

More advanced screening methods will rely on a portfolio approach (that is, rank order the best projects according to some acceptable ranking model). Some of these may be quite involved mathematically. They have a danger of leading a firm to committing to the best of a bad lot. Crawford (1991) implies that it is better to use a method that judges each project on its own merits, not whether it is superior to some other less desirable projects.

Many firms use a checklist or a scoring model for product concept screening. Although there are many forms of these models, they essentially are all similar in that each has the same objective: to judge each product concept on a set of factors that the firm considers to be the most important criteria for selection. Checklists are easier to implement but provide less information than other methods. They may require only that the firm identify whether a concept is strong on each of the criteria. A scoring model requires more managerial input, in that a ranking on a numerical scale must be agreed upon for each criterion, for each concept. Agreement must also be reached on how important each criterion is, again on a numerical scale. That is, a scoring model is a weighted, linear compensatory model. Ideally, managers from all the departments involved (marketing, R&D, design, and so on) ought to participate in the ranking and weighting procedures. This will lead to discussions and perhaps disputes over the relative importance of criteria and the attractiveness of different criteria to the firm. But this is a healthy situation. This hashing-out process sensitizes the participating managers to the views and desires of their counterparts elsewhere in the firm and may identify communication or coordination problems needing attention that no one knew were there.

The newer the product technology or the market, the more risk is incurred by the firm. This has implications for management. Some alternative screening models are designed especially to assess the risk of product projects involving revolutionary technology. New product project risk can be assessed by examining four factors: (1) How new is the technology? (2) How new is the application of this technology? (3) How new to the firm are the customers in this application? (4) How innovative is the product in the marketplace? (Abette and Stuart, 1988). New product risk may also be conceptualized in terms of development costs, marketing synergies, development complexity, competitive advantage, and buyer risk (More, 1982).

Advanced screening methods are based on empirical studies in which variables that are critical to past successes or failures are identified and become components of the model. As part of Project NEWPROD (Cooper, 1980), an empirical screening regression model for industrial products was built that explained almost 40 percent of the variance in the observed degree of success or failure. This model expressed the relationship between product success and eight independent variables: product superiority; resource compatibility between project and company; market need, growth, and size; economic advantage of the product; newness to the firm; technological resource compatibility; market competitiveness; and product specialization. A firm could assess each of these variables for each concept via the judgment of an expert (or team of experts). Then, using the regression equation, a "product score" can be obtained that could be used to select high-potential concepts and screen out poorer ones. This model has recently been updated by Cooper, as more data has become available (Cooper, 1985a).

Studies by Calantone and Cooper (1979, 1981) used more sophisticated methodologies for sorting out winning concepts from losers. In their 1979 study, they used cluster analysis to group about 100 new industrial product failures into six groups based on the extent to which the firm conducted critical marketing and technical activities. The names of the clusters indicate the kinds of scenarios to avoid: (1) "The Better Mousetrap Nobody Wanted" (the result of R&D developing products without the guidance of marketing input), (2) "The Me-

Too Product Meeting the Competitive Brick Wall" (competition too firmly entrenched, product not superior), (3) "Competitive One-Upmanship" (competitor has a better product), (4) "Environmental Ignorance" (situation analysis faulty), (5) "The Technical Dog Product" (technical and design flaws sabotaged the product), and (6) "The Price Crunch" (cost escalations led to price spiraling). The 1981 study extended this analysis to include industrial product successes as well as failures.

Strategic implications for management

Armed with information about prior new product successes and failures, managers are in a better position to select concepts that are consonant with their key strengths and abilities and to screen out the others. They can look for early-warning signals that indicate low success potential before too much time and working capital are invested. Balachandra (1984), for example, developed a list of warning signals that indicate "keep out": low likelihood of probability of technical success, lack of guarantee of access to raw materials, and possibility that the market itself will not continue to exist. In addition to these "red-light" signals, he also identified "yellow-light" signals that denote "proceed with caution." Any one of these might not indicate project termination, but too many of these cannot be ignored by management. The yellow-light signals include lack of a project champion, high expected levels of competition, low company profitability, and likelihood of low commercial success.

One important finding of NEWPROD III (Cooper and Kleinschmidt, 1990) was that consistency in execution of both marketing and technical activities is just as important as screening. Firms launching successful products were much more likely to have done a good job in initial product screening and also on several later stages in the product development process. These findings indicate that the more steps performed well, the greater the chances for success. Clearly, superior performance of all technical and marketing activities is an important precursor to consistent product success.

Cooper and Kleinschmidt (1990) also found some interesting shortcomings in screening execution that have substantial implications for management. In 88 percent of the cases they studied, initial screening was reportedly poorly done. And projects that should have been "killed" often were not, once development had begun. In general, a useful set of strategic recommendations for R&D managers in the development of new products was proposed by the European Industrial Research Management Association (1982). These include, among others, establishing relations with legislative bodies and consumer organizations, anticipating changes in consumer and legal trends, anticipating opposition and constraints in developing new product specifications, performing adequate safety tests during new product development, planning for reuse or disposal of used products, ascertaining that process capabilities are adequate, and having R&D take an active role in developing instructions for product use and in analyzing production defects.

Product use testing

By the time the product passes full screening and a working proto-type is developed by the R&D department, a new set of questions emerges: Does the newly developed product really work? Will the targeted market buy it and understand how to use it? Should we make more improvements to the prototype or are we ready to initiate production? These and other questions can be answered through product use testing.

In fact, the firm has a more basic question to answer first: Is it worth it to do any product use testing at all? For one thing, concept tests may have indicated that the product is a can't-miss proposition. Much time and money may have already been devoted to market assessment. Use testing may slow down the development process, perhaps letting a competitor enter first with a similar product. Many firms may have the opinion that they know everything there is to know about their customers already. Crawford (1991, pp. 220–221) suggests that firms making these arguments have forgotten that one of the main reasons for product failure is "not meeting customer needs." In actuality, a trade-off must be made by management between risk reduction and speed of product development. These are conflicting objectives. The firm should choose between these objectives, depending on its goals and the situation it faces. "Doing it fast" (or, at least, faster than usual) was most important to IBM in its personal computer venture; product development time was cut by two-thirds and the PC was able to establish an early stronghold in the market. By contrast, "getting it right" was more important to Boeing in the development and testing of its 767 (Krubasik, 1988).

There are two broad classes of product use testing: "in-house testing" and "market user testing." Sometimes these are referred to as Alpha and Beta testing, respectively. In in-house testing, as the name implies, the new product is put to use by company employees. An accounting software system might, for example, be used by the accounting department of the developing firm. A product that passes this step would be moved to market user, or Beta, testing, where it is actually implemented by customers. While these are widely used product use tests, Crawford points out the potential for their misuse:

...both can be misleading (since they test whether the product "works"), not "that it meets the needs of the customer.".…Here is a typical procedure, used recently in a firm that makes computer hardware. Of the "80 pilot units made…10 units went to the company's field test department for installation at selected customer sites…. While pilot testing was under way, manufacturing completed the pilot run and went into full production." No way did those users have time to judge whether the new equipment met their needs, how cost effective it was, how various employees in their firms adjusted to the new item, and so on. Computer industry manufacturers know this, but they usually go ahead and produce and introduce the product. Beta testing does not meet the developer's real needs (Crawford, 1991, pp. 222–223).

To combat some of these problems, firms may use what is known as "Gamma testing," which tests not only whether the product "works" but also whether it actually solves the adopting customer's problem. Needless to say, Gamma testing can be much more time intensive, and if time pressure is sufficient, a firm may have to rely on the relatively shorter Alpha and Beta tests.

A useful way to categorize product use testing procedures is provided by Crawford (1991, pp. 223–231). In-house testing can be carried out by lab personnel, experts, or employees. Lab personnel usually get involved in the testing of the product's technical performance. Experts (such as food tasters or automobile design specialists) may look at a product in a different way than the average customer but will be especially careful in their estimation and may be much more critical of potential product drawbacks. Employees are the most widely used participants in in-house testing. The inevitable problem of bias toward the company's products or brands can be overcome by hiding product identities and by properly motivating the personnel. For example, the R.J. Reynolds Tobacco Company uses in-house testing on new cigarette brands. Employees taste-test new cigarette blends and may also be asked to sniff or feel cigarettes and provide feedback. Employees with good sensory abilities may be asked to help the chemists and blenders keep the taste of popular brands consistent over time, despite differences in tobacco crops or manufacturing processes (Scheuing, 1989, p. 187). Kodak used employees rather than consumers when it was testing its Instamatic camera to avoid tipping off competitors (Scheuing, 1989, p. 107).

In Beta testing, participants may include customers and users as well as noncustomers and nonusers and users of competitive products. A further decision facing the firm doing Beta testing is whether to conduct the study at the point of use or at a centralized location such as a test kitchen or testing theater. There are obvious problems involved in centralized testing (e.g., the sample is probably not representative or respondents may behave in a way different than they would otherwise because they know they are being watched). But for consumer goods, centralized testing is very popular. Industrial products are primarily tested on site. Paired comparisons are often used to get precise results from consumer testers. For example, if a manufacturer is choosing which of three colors of household cleaner would be preferred by consumers (clear, blue, or amber) and whether glass or plastic containers should be used, six separate consumer groups would have to be set up to compare each possible two-way combination of attributes (Scheuing, 1989). The extra cost involved in such a setup may be more than offset, however, if a clear-cut winner can be identified.

One of the best illustrations of a product use testing system in action is that used by Gillette in its development of Dry Idea Deodorant, described in Fig. 4.1. Two things are notable about this

---

1. Technical lab work in 1975 suggested available technologies to achieve a drier deodorant.

2. A 2000-person concept study (cost: $175,000) determined that "Yes, roll-ons are good, but they go on wet and make you wait to get dressed." A concept was at hand.

3. Laboratory project assigned to scientist: find a replacement for water as the medium for the aluminum-zirconium salts that did the work.

4. A prototype using silicone was developed, and it wasn't wet or sticky. But it did dissolve the ball of the applicator. (In-house lab test.)

5. Next prototype was tested by volunteers from the local South Boston area. It was oily. (Outside research firm employed to test college students in the area. Gillette often used in-house test of employees, too.)

6. By late 1976, a later prototype tested well on women recruited to sweat for hours in a 100° "hot room." (Test of market users in the Boston area who served on a regular panel.) Unfortunately, although it worked well, it eventually turned into a rock-hard gel.

7. By early 1977, another prototype had passed the "hot room" tests and was then sent to company-owned medical evaluation laboratories in Rockville, Maryland. (In-house test on rabbits and rats.) It passed the test.

8. Packaging was being developed and tested by in-house package design engineers. Early packages leaked.

9. However, the package dispensed a product that test subjects felt was too dry going on. (Test of market users.)

10. They then returned to a conventional roll-on bottle, added a special leak-proof gasket, and enlarged the ball so the antiperspirant could be applied in quantities large enough to be felt. Another test of market users confirmed that people did indeed feel drier. This conclusion, when put with the earlier data that the product did have a good antiperspirant effect, was enough to go to market.

Figure 4.1 Illustration of product use testing: Dry Idea Deodorant. (*Source: Crawford, 1991, p. 224.*)

procedure: the number of different kinds of tests employed and that the testing procedure was nonlinear. Several steps in Fig. 4.1 could be classified as "back-to-the-drawing-board" steps, including one at the very end when the researchers were surprised to find out that dryness was not a "motherhood and apple pie" dimension (that is, more dryness is always better). It was possible for a product to feel too dry going on. This was a finding no one in the development of this product had anticipated. The example serves as a good counterpoint to the above discussion, which might mislead the reader into thinking that product use testing proceeds in a progressive fashion from Alpha to Beta to Gamma testing and on to full production.

There are risks involved with product use testing. As with any other kinds of testing, if the test is not representative, the results will be useless or misleading. Lavidge (1984) cataloged several common mistakes made by market researchers that can mislead product planners. Testing under carefully controlled, exacting conditions can lead to problems for makers of products like coffee or cake mixes, where one probable success factor is how well the product performs if the user does not follow instructions to the letter. The firm should not evaluate unrepresentative products (for example, those which are obviously of higher quality than the ones the firm plans to mass produce). Nor should the firm forget to study all the key population segments (as did the dog food manufacturer in the possibly apocryphal story that tested pet owners' reactions but forgot to see if dogs would eat the product). Relying strictly on focus groups for information gathering can also be misleading, as can using mean ratings (half of the market may want mild spaghetti sauce, half may want spicy, but nobody may want medium). The fact that some of these dos and don'ts need to be pointed out at all is a good indication that the practice of product use testing could be improved in many firms.

Market testing

The result of the preceding steps moves the original innovation to a physical product. The next step is to determine what response this newly developed product will have in the market, that is, whether the full-scale launch of this product, in this form, will meet the company's objectives as stated in the marketing plan. This step is called "market testing" and can be defined as "the phase of new product development when the new item and its marketing plan are tested together. A market test simulates the eventual marketing of the product and takes many different forms" (Crawford, 1991, p. 538).

The terms "market testing" and "test marketing" are often confused. Test marketing, broadly speaking, is an attempt by a manufacturer to forecast sales based on monitoring and projecting actual sales in stores limited to a representative geographical area. Thus, test marketing is but one of the techniques available for market testing. There are obvious drawbacks as well as advantages to "full-sale" test marketing: The manufacturer "tips their hand" to competitors, a complete test market can be very costly and time consuming, and so on. These and other drawbacks will be examined in greater detail later. Alternate forms of market testing have been developed to help circumvent some of these problems. In this section, several of these alternate forms will be presented.

Pretest market models

Pretest market models can be used by managers to gain an inexpensive early read on the likely performance of a new brand or product. These models, sometimes known as simulated test marketing procedures, generally follow similar data gathering procedures, although actual practice can vary widely. Usually, the brand name and full label are on the product, and it is packaged in much the same way as is it would be for the nationwide launch. Note how this is unlike the Beta test, where such details may not yet be worked out (Scheuing, 1989, p. 193). Participants in the study are obtained via mall intercepts. Attitudes toward the product category in question, brand purchase behavior, demographics, and other information are gathered by questionnaire; then the respondent may be exposed to advertising for the new brand. A trip to a simulated shopping aisle might follow, where the brand(s) purchased by the respondent will be noted; possibly those respondents who did not choose the brand under study will be given a sample to take home. Later stages of the study include telephone follow-up with all respondents to gain information on whether and how they used the product, how well they liked it, and if they are likely to purchase it again. The respondent may have the opportunity to get more of the brand, and another call may be placed later to gather more information (Shocker and Hall, 1986). A variation of this procedure is used by Procter & Gamble, which operates its own test store in the Cincinnati area rather than contracting out the test to a market research agency that specializes in pretest market analysis (Scheuing, 1989, p. 193).

Pretest marketing can provide several advantages over full-sale test marketing. Data are gathered rather quickly (usually in several weeks, substantial data are available). Full test marketing can take years. Costs are also far lower: While a pretest market study may cost from $15,000 to $100,000, a full-sale test market can easily cost 10 or more times that amount (Shocker and Hall, 1986). The pretest market procedure keeps the product relatively secret from competitors and also may provide the manager with diagnostics on how the marketing mix can be fine-tuned.

Many well-accepted pretest market models exist, including ASSESSOR (Silk and Urban, 1978), TRACKER (Blattberg and Golanty, 1978), NEWS (Pringle, Wilson, and Brody, 1982), COMP (Burger, Gundee, and Lavidge, 1981), LITMUS (Blackburn and Clancy, 1982), LTM (Yankelovich, Skelly and White, Inc., 1981), and others. While these are all different in some respects, most are grounded in an awareness-trial-repeat, or A-T-R, model, which represents the process of diffusion of an innovation through a market. Much of the terminology encountered in pretest marketing pertains to consumer nondurables (repeat purchases are less of a concern with consumer durables and many industrial products), and these models are widely used in a consumer nondurable context. But the models can be appropriate to other situations as well (perhaps suitably modified) since the key process of interest is that of innovation diffusion.

ASSESSOR is one of the earliest and most successful of the pretest market models. It is unique in that it uses input from management, such as figures from the marketing plan and brand positioning objectives, to complement the customer-derived input. The output of ASSESSOR and similar models is a forecast of brand share, as well as draw (gains in market share coming from competing firms' brands) and cannibalization (market share gains coming at the expense of other brands also sold by the firm testing the new product). The firm is undoubtedly interested in knowing not only if its new brand can attract market share but also from where that share will be derived. Ford, for example, would be very concerned about whether a new model will draw from sales of competing auto makers or will simply cannibalize other older Ford models. ASSESSOR can provide useful information to the manager on marketing program options (such as favorable advertising, price, and distribution levels) because it, and other pretest market models, allow the manager to carry out "what-if" analyses very easily.

The LTM, or Laboratory Test Market model (Yankelovich, Skelly and White, Inc., 1981), analyzes laboratory store purchase results in terms of repeat rates, product form novelty, and intensity of promotional effort and has been shown to be very effective in real settings. TRACKER (Blattberg and Golanty, 1978) can also be used in test marketing (as further discussed below).

A study that investigated the performance of ASSESSOR determined that two-thirds of the products that ASSESSOR rated favorably went on to pass full-sale test market procedures with ease (Urban and Katz, 1983). This raises the question of whether a firm should bypass the time and expenses of test marketing altogether if the pretest market provides sufficiently favorable results. Although one could probably argue this in either direction, Urban and Katz (1983) themselves stated that both pretest marketing and test marketing are necessary steps in new product development. Similarly, Shocker and Hall (1986) indicated that significant reductions in costs of development were obtainable through the use of pretest market models in screening out bad concepts. A firm can gain efficiencies in the long run by eliminating test market costs on the losers and concentrating on test marketing only the better products.

While they provide some important advantages over full-sale test marketing procedures, pretest market models have some disadvantages. They may not adequately address problems in implementation of the launch. For example, they provide information about whether the customer will buy the product but little information about the retailer's interest in stocking the product. Some parameters in these models are set by managerial judgment and may thus be in error. Pretest market models have also been criticized for ignoring competitive strategies, and their validity has been questioned. The contrived settings they sometimes use may be unrepresentative of the real sales setting (Shocker and Hall, 1986).

Controlled sale market testing

In addition to pretest markets, controlled sale market testing has been successfully used to provide information on product adoption. This method provides many of the advantages of full-sale test marketing but avoids the problems (such as assuring distribution) that often hinder implementation of the latter in practice. Crawford (1991, pp. 270–275) identifies four major categories of controlled sale market testing procedures: informal selling, direct marketing, minimarkets, and scanner markets.

Informal selling may not be considered by many firms as a market testing procedure, but it can be an important source of information about the market. It is very useful in the testing of industrial products, where firms may have less funds available for more formal techniques and most promotion takes the form of personal selling. In a typical informal selling procedure, sales representatives are provided with the new product and some information on it and simply promote it on their sales calls or at trade shows. According to Crawford (1991, p. 272), consumer goods manufacturers and, in particular, consumer service providers have also found this kind of controlled sale technique to be appropriate for their purposes. A simple informal selling procedure that costs less than $30,000 to implement is described by Scheuing (1989, p. 197). A new brand may be available for a weekend at a given store, surrounded by attention-getting promotions and salespeople. Consumers' phone numbers are taken, and a couple of weeks later, they are called and asked to assess the product. While the results are anything but controlled and generalizable, some quick preliminary information can be obtained in this way.

Direct marketing can also be used as a controlled sale technique. Many firms market products directly to customers, through catalogs, by mail or phone, by van, or over computer or fax linkups. Sales of these products can then be monitored as an indication of the likelihood of product acceptance if they are launched nationwide. The countless catalog firms are, for example, in a perfect situation to test new products on a limited basis by including them in their catalogs. There are many advantages to direct marketing controlled sales, including secrecy, quick feedback, positioning opportunities (several versions of the product's ad can be tested in regional editions of a catalog, for example, and the best can be selected for use nationwide), low cost, and easy adaptation to phone ordering or sales to mailing lists (Crawford, 1991, pp. 272–273).

Minimarkets are sometimes used to gain preliminary information about what makes the customer adopt a product. A couple of desirable retail stores might be singled out for participation in the minimarket test and would be asked to carry the product being tested. The product thus is actually for sale in real stores under relatively realistic circumstances. Usually this takes place in a relatively small city of 50,000 inhabitants or less (Scheuing, 1989, p. 196). Media advertising, of course, is not feasible (only a handful of stores in a city might carry the product), so in-store promotions such as attention-getting displays or live demonstrations might need to be used. Minimarkets offer the testing firm great flexibility in in-store promotion and pricing levels. A big drawback is that they do not provide results that are in any way generalizable to the national level: Too few stores are used, and they by definition have to be stores that agreed to participate and are probably not representative. Nevertheless, key information about product trial (and possibly repeat purchase behavior) can be collected through minimarkets, and several research firms can provide this service to product manufacturers. This information can then be used to make adjustments to the product or the marketing mix before a full-sale test market takes place.

Scanner markets are a result of advancing technology in checkout scanners and in cable television. Information Resources, Inc. (IRI) offers several services including BehaviorScan, one of the first scanner market testing procedures (Crawford, 1991, p. 274). Several cities around the country are BehaviorScan markets. In each city, most of the households have agreed to shop at stores hooked up to BehaviorScan where, whenever they make a purchase, it is recorded via the checkout scanner. The television commercials shown in these cities can be targeted specifically to certain households via the cable system, and print media (magazine purchases, newspaper subscriptions, and so on) are also monitored. This provides for an endless number of possible research studies. For example, half of the city may see version 1 of a commercial for a new product, half may see version 2, and then, through supermarket checkout scanning, it can be determined whether the half that saw version 1 bought more. Or, a manufacturer may want to target a new ad campaign to target households (say, young families) to determine whether the desired target was encouraged to purchase by the commercial.

> The variations and controls stretch the imagination. (A food manufacturer) can find out how many...upscale homes who watched the initial commercial bought some of the product within the next two days. And what they bought on their prior purchase, what they paid, what else they bought at the time, and the like.... Johnson & Johnson's problems when (Tylenol) was poisoned were lessened when its scanner market service showed that the item's share of the market dropped from 47 percent to a bit over 6 percent in the problem city—but began to revive the very next week (Crawford, 1991, pp. 274–275).

Full-sale market testing

Full-sale market testing, or test marketing, involves selling the product being tested in limited geographical areas in real retail stores, at representative prices, supported by representative promotional campaigns—in short, in as realistic a selling environment as possible—and projecting the results to nationwide sales forecasts. Often, a decision may have already been made to launch the product, and test marketing will help the firm decide on the proper level and allocation of promotional funding or the correct price—in other words, not *whether* to market the product but how *best* to market it (Crawford, 1991, p. 276). Most of the models used in test marketing are A-T-R models, where advertising and promotion are viewed as influencing the awareness stage, and awareness and distribution intensity are seen as increasing levels of trial.

Test markets take longer to run than the controlled sale market testing procedures. A consumer goods firm such as Procter & Gamble can keep a new product in test market for up to 2 to 3 years (Urban and Star, 1991). They also have a much more intense informational requirement. Often, data such as purchase intentions and advertising and promotional recall are gathered via questionnaire. One earlier model (Parfitt and Collins, 1968) uses panel data as input. Another early model, the Ayer Model (Claycamp and Liddy, 1969), looks at promotion and distribution levels, as well as packaging, category usage, and customer satisfaction with the product category. Some of the other earlier models, many of which have been updated and modified and are still in use today, include STEAM (Massey, 1969) and SPRINTER (Urban, 1970). Later versions of SPRINTER were built to include many additional variables that might influence the A-T-R process (such as word-of-mouth effects) that had been left out of previous models. Some of the pretest market models previously mentioned (such as NEWS) can also be used at this stage.

TRACKER (Blattberg and Golanty, 1978) is a particularly interesting test market model in that it estimates the number of new triers in a given period using a dynamic model. At any one time, new triers will include some individuals who have just become aware of the product and immediately try it, plus some individuals who were made aware in previous periods but had not tried it yet. Thus, dynamic estimates of the number of newly aware individuals and nontriers are used to drive the model. Other interesting later models include those of Assmus (1975) and Mesak and Mikhail (1988).

The reader wishing to compare test market models has several good comparison and review articles to use as sources. NEWS, STEAM, SPRINTER, the Ayer Model, and several others used in both test marketing and pretest marketing were compared by Larreché and Montgomery (1977) on several factors including adaptability, completeness, and ease of use. While all models scored quite well, NEWS and the Ayer Model were preferred overall in this study. A later study (Mahajan and Muller, 1982) examined and compared one component common to each of these models, the adoption-diffusion submodel, and found SPRINTER to be superior. A third comparison study (Narasimhan and Sen, 1983) investigated the performance of several test market models on a new set of criteria including the amount of complexity they could handle and the likelihood of use by product managers. NEWS and TRACKER were rated highest. Two additional studies (Assmus, 1984; Mahajan and Wind, 1988) each reviewed a large number of available test market models, comparing the strengths and weaknesses of each and examining trends in the development of these models. This has been an area that has attracted the critical eye of quantitative marketing academics.

As stated earlier, there can be many drawbacks to full-sale test marketing. The product will be sold in real stores under realistic conditions, which cause expenses to be very high. Enough product must be made and shipped to the test market areas, advertising and other promotion (e.g., billboards) of sufficient intensity must be bought in local markets, slotting allowances may need to be paid to retailers, and so on. Added to this expense is the cost of doing the market research itself: phone surveys, diary panels, and store audits of sales. A firm conducting a test market may want to test several possible marketing mixes in as many cities, and recent estimates place the cost of a test market in the range of $300,000 to $500,000 per city

(Crawford, 1991). In two recent test markets, Quaker Oats spent over a million dollars on marketing research procedures alone, incurring additional expenses of production, advertising, and distribution (Scheuing, 1989, p. 196). Another real risk facing the test marketing firm is that it tips its hand to competing firms. At the very least, a competitor's sales representative is likely to spot an unusual product on store shelves, surmise a test market is under way, and contact the home office. In the worst-case scenario, the competitor might actually sabotage the test market (e.g., by dropping coupons on its own brand in the test market city or by buying up the product to inflate the sales figures). Timing is also a factor since a full-sale test market may take 2 to 3 years. This is enough time for a competitor to launch a competing brand and preempt the market. The cost, time, and loss of secrecy factors are of great enough concern to make some firms wonder whether it is always worth it to test market their products. Figure 4.2 provides several recent examples of test markets that were sabotaged

Richardson-Vicks took Olay Beauty Bar national in the United States and took Climacel national in Australia, both solely on the strength of positive simulated test market results obtained with ASSESSOR. The company believes ASSESSOR results to be sufficiently accurate so as to bypass full-sale test markets. Both products were successes.[1]

S.C. Johnson tried Agree cream rinse in a simulated test market, in which it succeeded very well. It was moved into test market, but positive response was so overwhelming that the test market was shortened by 6 months and the product was launched nationwide.[1]

The following products recently went from simulated test marketing directly to national launch: Sara Lee meat-filled croissants, General Foods International Coffees, Quaker's Chewy Granola Bars, Pillsbury's Milk Break Bars.[1]

Procter & Gamble has gone directly from simulated test market to national launch with some of its recent products but kept both Cinch dishwasher detergent and Certain bathroom tissue in full-sale test market for 3 years each.[1]

Kellogg tracked the sale of General Foods' Toast-Ems while they were in test market. Noting they were becoming popular, they went national quickly with Pop-Tarts before the General Foods test market was over.[1]

General Foods was test marketing frozen baby food. Sales in the test markets were very high, which was good news until it was discovered that most purchases were made by Gerber's, Libby, and Heinz. Test market results were therefore inflated.[2]

While Procter & Gamble had their "soft" chocolate chip cookies in a lengthy test market, Nabisco and Keebler both rolled out their own "soft" cookies nationwide, beating P&G to the market.[3]

General Foods invented freeze-dried coffee and was in the process of test marketing its brand in this category (Maxim) when Nestle beat it to the market with its own brand, Taster's Choice.[4]

Figure 4.2 To test market or not to test market? (*Sources: 1. Crawford, 1991, 2. Guiltinan and Paul, 1991, 3. Urban and Star, 1991, 4. Scheuing, 1989.*)

Factors Favoring Test Marketing

1. Acceptance of the product concept is very uncertain.

2. Sales potential is difficult to estimate.

3. Cost of developing consumer awareness and trial is difficult to estimate.

4. A major investment is required to produce the product at full scale (relative to the cost of test marketing).

5. Alternative prices, packages, or promotional appeals are under consideration.

Reasons for Not Test Marketing

1. The risk of failure is low relative to test marketing costs.

2. The product will have a brief life cycle.

3. Beating competition to the market is important because the product is easily imitated.

4. Basic price, package, and promotional appeals are well established.

Figure 4.3 Considerations in deciding whether to test market. (*Source: Guiltinan and Paul, 1991, p. 187.*)

by competing firms, as well as several examples of products that were extensively test marketed and ended up being very successful. It is instructive to note that Procter & Gamble, long the prime example of how a firm should conduct a careful test market, is now bypassing test markets for some products. While the decision to test market or not may not be a clear-cut one, a pragmatic guideline to use is provided by Guiltinan and Paul (1991), as shown in Fig. 4.3.

Today, many firms are opting for a "rollout introduction" as an alternative to test marketing. In a rollout introduction, "target markets are divided into several geographical areas and the new product is initially introduced in only one or a very few of these areas. If the product is successful in the small initial effort, additional areas are opened until all target geographic markets are being served" (Guiltinan and Paul, 1991). This differs from test marketing in that the initial effort is not in representative markets but rather in markets that the firm is seeking to enter and stay in and, as the name implies, from which it can rollout to other regions.

One of the reasons rollouts are becoming more prevalent is the availability of scanner data. IRI, the firm behind the BehaviorScan service, also provides another service known as InfoScan. This is a market measurement service that provides complete data on all sales in food stores in 60 cities in an extremely timely manner (data are provided with only a 1-week delay). Thus, a firm can, in a sense, be test marketing a product as it rolls it out, constantly monitoring its progress (Crawford, 1991, p. 282).

Customer input to value analysis

As a final and relatively overlooked way in which user needs can be incorporated into the product design and development process, consider value analysis. A product's *value* is "determined by the relationship of worth to cost [conforming] to the customer's wants and resources in a given situation" (Fallon, 1971). In other words, value is defined in terms of worth as perceived by the customer. *Value analysis* can be thought of as "an organized effort to obtain optimum value in a product...or service by providing the necessary function at the lowest cost" (American Society of Tool and Manufacturing Engineers, 1967). A step-by-step value analysis program is outlined in Table 4.3. Benefits of a successful value analysis program can include increased sales and product performance, improved delivery, better use of scarce resources, greater standardization, and more innovation by employees and suppliers, as well as cost reduction (Reuter, 1986).

Value analysis should ideally be a team effort with participation from all involved functional areas in the firm. It would seem that marketing has an important role to play in value analysis, namely discovering what customers really value in products and transmitting this information to production or product design so that they can make adjustments to improve service to customers. In practice, however, the active participants in value analysis programs have been engineering and production departments. Marketing departments are often less than enthusiastic about such programs (Gordon, Calantone, and di Benedetto, 1990). Perhaps because of greater participation by production and engineering, value analysis has tended to focus on decreasing costs of materials and production. While traditional incremental cost-cutting steps (such as using more off-the-shelf

---

TABLE 4.3   Steps in a Value Analysis Program

---

1. Information is collected.

2. The source of information is examined.

3. The validity of the information is evaluated.

4. Requirements for additional information are determined.

5. The function of the product is defined.

6. Appropriate costs for providing each function are determined.

7. A search for better ways to satisfy the need is undertaken.

8. The proposed options are evaluated as to cost and do-ability.

9. The accepted options are implemented.

SOURCE: Fallon, 1971.

parts) and backward engineering of competitors' products are an important aspect of a value analysis program (Morris, 1991), they should not be the only components of the program. Value analysis should be more than simply "cost" analysis. By formally integrating marketing into the value analysis process, more relevant information about user needs can be collected and relayed to R&D and production. In other words, value analysis can be a means of improving the interface between marketing and R&D. Customers can be more formally integrated into value analysis through involvement in value analysis training programs sponsored by the company by providing design suggestions through the sales force or being asked for input via periodic questionnaires. Value analysis, then, can benefit all involved parties. R&D and production receive more complete and more timely information about their markets. Marketing improves its customer data reporting mechanisms and, ultimately, customer relationships. And customers have the opportunity to provide input into the design and production of products they will ultimately be using (Gordon, Calantone, and di Benedetto, 1990). Value analysis that incorporates user needs into product design and development is thus consistent with the current tendencies toward Total Quality Management and other quality programs among manufacturing firms. Quality Function Development (QFD) is one currently popular method that some firms employ to take customer requirements and translate them into engineering characteristics and production and process specifications (Hauser and Clausing, 1988).

User-Development Matching

It is clear from the work of Cooper (1980, 1986), Crawford (1984, 1991), Gupta, Raj, and Wilemon (1985a, b), and others that, in order to improve its new product development process, a firm must first conduct an honest self-assessment of its skills in several critical areas: marketing, technical/R&D, product design, and manufacturing. As previous sections of this chapter have indicated, all of these areas play a vital role in the development of products that truly satisfy user needs and requirements.

What, then, is the best way for the developer to attempt to address user needs? Is it better to "stick to the knitting," that is, develop new products having great marketing, production, and manufacturing synergies with the current product line or to take more chances and work on riskier new product projects in uncharted territory? As one might expect, both paths have merit; the idea of two paths to successful new products somewhat parallels the idea of two sources of ideas (technology push versus market pull). Some good evidence for the viability of using two paths (which might be called the "risky" and "conservative" paths to success) was provided by Calantone and Cooper (1981). They grouped about 200 industrial products (half successes, half failures) into nine categories, using cluster analysis on self-reported answers to questions about activities, skills, environmental conditions, and so on. Three clusters were shown to have the greatest proportion of successes: the synergistic "close-to-home" product (72 percent successes), the innovative superior product with no synergy (70 percent successes), and the old but simple money saver (70 percent successes). The first and third of these are variations of the conservative path to success. By staying close to home (that is, by seeking new product projects with great marketing, manufacturing, and technical synergies) and focusing on satisfying the basic customer need of reducing costs, firms can be very successful. The second cluster is noticeably different and is characteristic of the riskier path to success. In it, as long as the product is truly innovative (possibly innovative in a revolutionary sense) and superior to competing product forms, the firm has a good chance to succeed almost in spite of poor synergy. Technology-push-based innovations that proceed without much or any guidance from the marketing department may frequently fall into this category.

Later empirical work by Cooper (1985b) supported this dichotomy of risky versus conservative paths to success. In this study, 19 strategy factors were examined, including technical sophistication, technical synergy, production synergy, market potential, and market newness. Cooper found that the strategies that combined technological aggressiveness with a market orientation (risky strategy) tended to lead more often to success but that developing products with maximum fit and focus with the current product line (conservative strategy) can also lead to substantial success. Other empirical studies by Cooper (1984, 1985c) complemented these findings. In these studies, firms were found to have five different patterns or scenarios of new product strategies: (1) technology-driven, (2) balanced, (3) defensive-focused, technology-deficient, (4) low-budget, conservative, and (5) high-budget, diverse. According to all measures of success employed by Cooper, the balanced strategy firms were superior. In general, these firms could be characterized as being technologically sophisticated, yet possessing a strong marketing orientation. These firms tended to choose high-potential, large, growing, and noncompetitive markets. Knowing that this balanced strategy tends to lead to more success with new products, managers can use this strategy as a benchmark against which they can compare their own firms' strategies (Cooper, 1984).

It is important to recognize that having a balanced strategic focus is not a guarantee of success. Indeed, there are no guarantees. In the

TABLE 4.4   New-Product-Specific Technical, Marketing, and Launch Activities

Technical activities

Initial screening of product concept or idea

Preliminary engineering, technical, and manufacturing assessment

Product development (engineering, design, R&D)

Prototype or sample testing (Alpha)

Sample testing (Beta)

Pilot production or trial production

Start-up of full production and ramp-up

Marketing activities

Preliminary market assessment

Detailed market potential and consumer behavior research

Financial viability study of demand

Sample testing with customers

Test marketing or trial selling

Gathering intelligence on competitors

Gathering market intelligence on customers

Launch activities

Selling

Promotion activities

Distribution activities

SOURCE: Adapted from Calantone and di Benedetto, 1988.

Calantone and di Benedetto (1988) study of new industrial products mentioned above, it was shown that new-product-specific activities (technical, marketing, and launch activities), together with product quality, were the most direct and important precursors to new product success. Simply having the marketing or technical resources or skills is not enough; having the resources and skills only enables the firm to do a *better* job on the activities. By examining the variety of new-product-specific activities as shown in Table 4.4, one can see the importance of the user's input throughout the product development process, from preliminary market assessment and sample testing to promotion and distribution activities at product launch. An important corollary of this observation is that should the firm be deficient in any of the skills or resources (for example, a computer software developer with high R&D skills but little marketing experience), it can and should seek ways to compensate. For example, the firm may be able to obtain marketing or planning support from a consulting firm. It is also of interest to note that, in the original study, the direct precursor to product success with the strongest effect was product quality. Thus, even with excellent execution of marketing, technical, and launch activities, if the product lacks high enough quality to satisfy user needs, it runs a large risk of failure. All of this evidence of the importance of identifying and matching user needs is consistent with a marketing orientation on the part of the developing firm. Understand customer needs and wants, and then set out to satisfy them in a way that provides the firm with a long-term competitive advantage—this is the path to success.

Summary

Several topics have been discussed in this chapter. These include factors leading to successful product development, customer willingness and ability to buy, concept screening, and product testing of various forms. Product tests of all kinds are available to the practicing manager: product use testing (in-house and market user testing), market testing and pretest marketing, controlled sale market testing (informal selling, direct marketing, and so on), and full-sale market testing. The advances made in recent years in scanner technology have made available product purchase and consumption data in seemingly limitless quantities, presenting opportunities as well as challenges to analysts. In addition, customer input into value analysis, effective interfacing between marketing and technical departments, and the need for solid execution of new-product-specific marketing and technical activities are all growing areas of concern for developers of new products for industrial or consumer markets. All of these topics are related. They all point to the prime importance of understanding user needs *and* of incorporating this information into the new product development process.

The final section of this chapter highlights the importance of identifying user needs and matching them to firm capabilities, with the long-run goal of establishing a sustainable competitive advantage through quality commitment and customer satisfaction. Some of the newer total-quality-management techniques, such as House of Quality or Quality Function Deployment (Hauser and Clausing, 1988) are designed to accomplish just this. They are oriented toward improving the level of interfacing and cooperation among departments involved in the new product development process and building customer or user needs and requirements directly into product engineering.

References

Abell, D. F. "Strategic Windows," *Journal of Marketing*, 42, July 1978, 21–26.

Abernathy, W. J., and Utterback, J. M. "Patterns of Industrial Innovations," *Technology Review*, 80, 1978, 2–9.

Abette, P. A., and Stuart, R. W. "Evaluating New Product Risk," *Research Technology Management*, 31 May–June 1988, 40–43.

American Society of Tool and Manufacturing Engineers. *Value Engineering in Manufacturing*, Englewood Cliffs, N.J.: Prentice-Hall, 1967.

Anderson C. R. and Zeithaml, C. P. "Stage of the Product Life Cycle, Business Strategy, and Business Performance," *Academy of Management Journal*, 27: 1, 1984, 5–25

Assmus, G. "NEWPROD: The Design and Implementation of a New Product Model," *Journal of Marketing*, 39: 1, 1975, 16–23.

———. "New Product Forecasts," *Journal of Forecasting*, 3: 2, 1984, 121–138.

Avlonitis, G. J. "Revitalizing Weak Industrial Products," *Industrial Marketing Management*, 14, 1985, 93–105.

Balachandra, R. "Critical Signals for Making Go/No Go Decisions in New Product Development," *Journal of Product Innovation Management*, 1: 2, 1984, 92–100.

Bennett, R. C., and Cooper, R. G. "Beyond the Marketing Concept," *Business Horizons*, 22: 3, 1979, 76–84.

——— and ———. "The Misuse of Marketing: An American Tragedy," *Business Horizons*, 24: 6, 1981, 51–61.

Berenson, C., and Schnaars, S. P. "Growth-Market Forecasting Revisited," *California Management Review*, 28, 1986, 10–15.

Blackburn, J. D., and Clancy, K. J. "Litmus: A new Product Planning Model," *TIMS Studies in Management Science*, 18, 1982, 43–46.

Blattberg, R., and Golanty, J. "TRACKER: An Early Test Market Forecasting and Diagnostic Model for New Product Planning," *Journal of Marketing Research*, 15: 5, 1978, 192–202.

Bremmer, B. "Two Big Macs, Large Fries—And a Pepperoni Pizza, Please," *Business Week*, August 7, 1989, 33.

Burger, P. C., Gundee, H., and Lavidge, R. "COMP: A Comprehensive System for the Evaluation of New Products," in *New Product Forecasting: Models and Applications*, Lexington, Mass.: Lexington Books, 1981.

Buskirk, B. D. "Industrial Market Behavior and the Technological Life Cycle," *IMDS*, 12, November–December 1986, 8–12.

Calantone, R. J., and Cooper, R. G. "A Discriminant Model for Identifying Scenarios of New Product Failure," *Journal of the Academy of Marketing Science*, 7, Summer 1979, 163–183.

——— and ——— "New Product Scenarios: Prospects for Success," *Journal of Marketing*, 45: 2, 1981, 48–60.

——— and di Benedetto, C. A. "An Integrative Model of the New Product Development Process: An Empirical Validation," *Journal of Product Innovation Management*, 5: 3, 1988, 201–215.

——— and ———. "Canonical Correlation Analysis of Unobserved Relationships in the New Product Process," *R&D Management*, 20: 1, 1990, 3–23.

———, ———, and Gordon, G. L. "A Conceptual Integration of Innovation and Diffusion Drivers of Industrial New Product Success," *Proceedings, American Marketing Association 1991 Winter Educators Conference*, Chicago: American Marketing Association, 1991a, 102–110.

———, ———, and ———. "Using Product and Technological Life Cycles as Guides in Strategic Opportunity Identification," *Proceedings, Product Development and Management Association 1991 International Conference*, Indianapolis, Indiana:

Product Development and Management Association, 1991*b*, 60–68.

——, ——, and Meloche, M. S. "Strategies of Product and Process Innovation: A Loglinear Analysis," *R&D Management,* 18: 1, 1988, 13–21.

Claycamp, H. J., and Liddy, L. E. "Prediction of New Product Performance: An Analytical Approach," *Journal of Marketing Research,* 6: 11, 1969, 414–420.

Cooper, A. C., and Schendel, D. "Strategic Responses to Technological Threats," *Business Horizons,* 19: 2, 1976, 61–69.

Cooper, R. G. "The Dimensions of Industrial New Product Success and Failure," *Journal of Marketing,* 43: 3, 1979, 93–103

——. *Project Newprod: What Makes a New Product a Winner?* Montreal: Centre Québecois d'Innovation Industrielle, 1980.

——. "New Product Strategies: What Distinguishes the Top Performers?" *Journal of Product Innovation Management,* 1: 3, 1984, 151–164.

——. "Selecting Winning New Product Projects: Using the NEWPROD System," *Journal of Product Innovation Management,* 2: 1, 1985*a*, 34–44.

——. "Industrial Firms' New Product Strategies," *Journal of Business Research,* 13, 1985*b*, 107–121.

——. "Overall Corporate Strategies for New Product Programs," *Industrial Marketing Management,* 14, 1985*c*, 179–194.

——. *Winning at New Products,* Reading, Mass.: Addison-Wesley, 1986.

—— and de Brentani, U. "Criteria for Screening New Industrial Products," *Industrial Marketing Management,* 13, August 1984, 149–156.

—— and Kleinschmidt, E. J. "New Products: What Separates Winners from Losers?" *Journal of Product Innovation Management,* 4: 3, 1987*a*, 169–184.

—— and ——. "Success Factors in Product Innovation," *Industrial Marketing Management,* 16, 1987*b*, 215–224.

—— and ——. *New Products: The Key Factors in Success,* Chicago, Ill.: American Marketing Association, 1990.

Crawford, C. M. "Protocol: New Tool for Product Innovation," *Journal of Product Innovation Management,* 1: 1, 1984, 85–91.

—— *New Products Management,* 3d ed. Homewood, Ill: Irwin, 1991.

—— and Tellis, G. J. "The Technological Innovation Controversy," *Business Horizons,* 24: 1, 1981, 18–23.

Davis, J. S. "New Product Success and Failure: Three Case Studies," *Industrial Marketing Management,* 17, May 1988, 103–110.

Day, G. S. *Strategic Market Planning: The Pursuit of Competitive Advantage,* St. Paul, Minn.: West, 1984.

——. *Analysis for Strategic Marketing Decisions,* St. Paul, Minn.: West, 1986.

De Bresson, C., and Townsend, J. "Multivariate Models for Innovation: Looking at the Abernathy-Utterback Model with Other Data," *Omega,* 9: 4, 1981, 429–436.

Dhalla, N., and Yuspeh, S. "Forget the Product Life Cycle Concept!" *Harvard Business Review,* 54: 1, 1976, 102–112.

Di Benedetto, C. A., and Calantone, R. J. "Effective Management of the R&D-Marketing Link for Improving New Product Success Rates," *Journal of Managerial Issues,* 2: 1, 1990, 75–90.

Dickinson, R. A., Ferguson, C. R., and Herbst, A. F. "Evaluating the Capital Purchase of a New Technology," *Industrial Marketing Management,* 13, 1984, 21–24.

Ettlie, J. E., Bridges, W. P., and O'Keefe, R. D. "Organization Strategy and Structural Differences for Radical Versus Incremental Innovation," *Management Science,* 30: 6, 1984, 682–695.

—— and Rubenstein, A. "Firm Size and Product Innovation," Journal of Product Innovation Management, 4: 2, 1987, 89–108.

European Industrial Research Management Association. "Ten Recommendations for R&D Managers," *Research Management,* 15: 4, 1982, 13–15.

Fallon, C. *Value Analysis to Improve Productivity,* New York: Wiley, 1971.

Feldman, L. P., and Page, A. L. "Principles Versus Practice in New Product Planning," *Journal of Product Innovation Management,* 1: 1, 1984, 43–55.

Fernelius, W. C., and Waldo, W. H. "Role of Basic Research in Industrial Research," *Industrial Marketing Management,* 7, 1982, 128–132.

Foxall, G., and Johnston, B. "Strategies of User Initiated Product Innovation," *Technovation*, 6, 1987, 77–102.

Golden, S. L., Huerta, J. M., and Spivak, R. B. "New Product Marketing: A Team Approach," *Proceedings, Simulation Profession Conference*, 1985, 17–19.

Gordon, G. L., Calantone, R. J., and di Benedetto, C. A. "Marketing: The Missing Piece to the Value Analysis Puzzle," *Proceedings, Product Development and Management Association 1990 International Conference*, Indianapolis, Indiana: Product Development and Management Association, 1990, 64–73.

———, ———, and ———, "Mature Markets and Revitalization Strategies: An American Fable," *Business Horizons*, 34: 3, 1991, 39–49.

Guiltinan, J. P., and Paul, G. W. *Marketing Management: Strategies and Programs*, 4th ed., New York: McGraw-Hill, 1991.

Gupta, A. K., Raj, S. P., and Wilemon, D. L. "R&D and Marketing Dialogue in High-Tech Firms," *Industrial Marketing Management*, 14, 1985a, 289–300.

———, ———, and ———, "The R&D-Marketing Interface in High-Technology Firms," *Journal of Product Innovation Management*, 2: 1, 1985b, 12–24.

———, ———, and ———. "Managing the R&D-Marketing Interface," *Research Management*, 30: 2, 1987, 38–43.

Gupta, A. K., and Wilemon, D. L. "The Credibility-Cooperation Connection at the R&D-Marketing Interface," *Journal of Product Innovation Management*, 5: 1, 1988, 20–31.

Hauser, J. R., and Clausing, D. "The House of Quality," *Harvard Business Review*, 66, May–June 1988, 63–73.

Hayes, R. H., and Abernathy, W. J. "Managing Our Way to Economic Decline," *Harvard Business Review*, 58, 1980, 67–77.

Hisrich, R. D., and Peters, M. P. *Marketing Decisions for New and Mature Products*, 2d ed., New York: Macmillan, 1991.

Hlavacek, J. D. "Toward More Successful Venture Management," *Journal of Marketing*, 38: 4, 1974, 56–60.

Hopkins, D. S. "New Product Winners and Losers," *Research Management*, 24: 3, 1981, 12–17.

——— and Bailey, E. L. "New Product Pressures," *The Conference Board Record*, New York: National Industrial Conference Board, 1971.

Krubasik, E. G. "Customize Your Product Development," *Harvard Business Review*, 66: 6, 1988, 46–52.

Kulvik, H. *Factors Underlying the Success or Failure of New Products*, Report no. 29, Helsinki: University of Technology, 1977.

Larreché, J. D., and Montgomery, D. B. "A Framework for the Comparison of Marketing Models: A Delphi Study," *Journal of Marketing Research*, 19: 11, 1977, 487–498.

Lavidge, R. J. "Nine Tested Ways to Mislead Product Planners," *Journal of Product Innovation Management*, 1: 2, 1984, 101–105.

Link, P. L. "Keys to New Product Success and Failure," *Industrial Marketing Management*, 16, 1987, 109–119.

McIntyre, S. H. "Obstacles to Corporate Innovation," *Business Horizons*, 25: 1, 1982, 23–28.

Mahajan, V., and Muller, E. "Advertising Pulsing Strategies for Generating Awareness for New Products," *Marketing Science*, 5: 2, 1982, 89–106.

——— and Wind, Y. "Business Synergy Does Not Always Pay Off," *Long Range Planning*, 21: 1, 1988, 59–65.

Maidique, M. A., and Hayes, R. H. "The Art of High-Technology Management, " *Sloan Management Review*, Winter, 1984, 18–31

———, and Zirger, B. J. "The New Product Learning Cycle," *Research Policy*, December, 1984, 18–31.

Mansfield, E. and Rapoport, J. "The Cost of Industrial Product Innovation," *Management Science*, 21, 1975, 1380–1386.

Massey, W. F. "Forecasting the Demand for New Convenience Products," *Journal of Marketing Research*, 6: 11, 1969, 405–412.

Mesak, H. I., and Mikhail, W. M. "Prelaunch Sales Forecasting of a New Industrial

Product," *Omega*, 16: 1, 1988, 41–51.

Moore, W. L., and Tushman, M. L. "Managing Innovation Over the Product Life Cycle." In M. L. Tushman and W. L. Moore (eds.), *Readings in the Management of Innovation*, Marshfield, Mass.: Pitman, 1982, 131–150.

More, R. A. "Risk Factors in Accepted and Rejected New Industrial Products," *Industrial Marketing Management*, 11, 1982, 9–16.

Morris, M. H. *Industrial and Organizational Marketing*, 2d ed., New York: Macmillan, 1991.

Narasimhan, C., and Sen, S. K. "New Product Models for Test Market Data," *Journal of Marketing*, 47, Winter 1983, 11–24.

National Industrial Conference Board. "Why New Products Fail," *The Conference Board Record*, New York: National Industrial Conference Board, 1964.

Parfitt, J. H., and Collins, B. J. K. "Use of Consumer Panels for Brandshare Prediction," *Journal of Marketing Research*, 5: 5, 1968, 131–145.

Peck, M. J., and Goto, A. "Technology and Economic Growth: The Case of Japan," *Research Policy*, 10, 1981, 222–243.

Penfield, P. L., Jr. "Small is Big: The Microeconomic Challenge." In *The Innovative Process: Evolution Versus Revolution, Proceedings of a Symposium for Senior Executives*, Cambridge, Mass., Massachusetts Institute of Technology, 1982.

Pringle, L. H., Wilson, R. D., and Brody, E. I. "NEWS: A Decision-Oriented Model for New Product Analysis and Forecasting," *Marketing Science*, 1: 1, 1982, 1–29.

Quinn, J. B. "Managing Innovation: Controlled Chaos," *Harvard Business Review*, 63: 3, 1985, 73–84.

Reuter, V. G. "What Good are Value Analysis Programs?" *Harvard Business Review*, 64: 2, 1986, 73–79.

Rink, D. R., and Swan, J. E. "Product Life Cycle Research: A Literature Review," *Journal of Business Research*, 7, September 1979, 219–242.

Rothwell, R. "Factors for Success in Industrial Innovation," *Project SAPPHO: A Comparative Study of Success and Failure in Industrial Innovation*, Brighton, U.K.: Science Policy Research Unit, University of Sussex, 1972.

———— "Some Problems of Technology Transfer into Industry: Examples from the Textile Machinery Sector," *IEEE Transactions on Engineering Management*, EM–25.1, 1974, 15–20.

Scherer, F. M. *Industrial Market Structure and Economic Performance*, 2d ed., Chicago, Ill.: Rand McNally, 1980.

Scheuing, E. E. *New Product Management*, Columbus, Ohio: Bell & Howell, 1989.

Schnaars, S. P. *Marketing Strategy: A Customer-Driven Approach*, New York: Free Press, 1991.

Schnee, J. E. "Government Programs and the Growth of High-Technology Industries," *Research Policy*, 7, 1978, 2–24.

Shocker, A. D., and Hall, W. G. "Pretest-Market Models: A Critical Evaluation," *Journal of Product Innovation Management*, 3: 2, 1986, 86–107.

Silk, A. J., and Urban, G. L. "Pretest-Market Evaluation of New Packaged Goods: A Model and Measurement Methodology," *Journal of Marketing Research*, 15: 5, 1978, 171–191.

Souder, W. E. "Promoting an Effective R&D-Marketing Interface," *Research Management*, 23: 4, 1980, 10–15.

————. "Disharmony Between R&D and Marketing," *Industrial Marketing Management*, 10, 1981, 67–73.

————. and Chakrabarti, A. K. "Managing the Coordination of Marketing and R&D in the Innovation Process," *TIMS Studies in the Management Sciences*, 15, 1980, 133–150.

Stern, M. E. *Marketing Planning: A Systems Approach*, New York: McGraw-Hill, 1966.

Urban, G. L. "Sprinter Mod III: A Model for the Analysis of New Frequently Purchased Consumer Products." *Operations Research*, 18, September–October 1970, 805–854.

———. and Katz, G. M. "Pretest-Market Models: Validation and Managerial Implications," *Journal of Marketing Research,* 20: 3, 1983, 221–234.

———., and Star, S. H. *Advanced Marketing Strategy,* Englewood Cliffs, N.J.: Prentice-Hall, 1991.

Utterback, J. M. "Innovation in Industry and the Diffusion of Technology," *Science,* 183, 1974, 620–626.

———. "The Innovative Process: Evolution Versus Revolution." In *The Innovative Process: Evolution Versus Revolution, Proceedings of a Symposium for Senior Executives,* Cambridge, Mass., Massachusetts Institute of Technology, 1982.

Utterback, J. M., and Abernathy, W. J. "A Dynamic Model of Product and Process Innovation," *Omega,* 3: 6, 1975, 639–656.

Von Hippel, E. "Successful and Failing Internal Corporate Ventures: An Empirical Analysis," *Industrial Marketing Management,* 6, 1977, 163–174.

———. "The Sources of Innovation." *McKinsey Quarterly,* Winter 1988, 72–79.

Wind, Y., and Claycamp, H. "Planning Product Line Strategy: A Matrix Approach," *Journal of Marketing,* 40, January 1976, 2–9.

Yankelovich, Skelly and White, Inc. "LTM Estimating Procedures." In *New Product Forecasting: Models and Applications,* Lexington, Mass.: Lexington Books, 1981.

Zaltman, G., Duncan, R., and Holbek, J. *Innovations and Organizations,* New York: Wiley, 1973.

Decision Implementation: Managing R&D, Product Development, Manufacturing, and Launch

Part 2 of this book focuses on the implementation of the critical front-end decisions (from Part 1). These activities involve managing R&D, product development, manufacturing, and product launch.

In Chap. 5 Professors Mariann Jelinek and Joseph Litterer discuss organization structures and related management issues which affect technological innovation. They articulate how traditional assumptions and management practices can actually inhibit the efficient development of a technology. As a result of intensified global competition, greater speed in new product design, increasing variety in products and services, higher-quality standards, and shorter delivery cycles are required today. To meet these challenges, Jelinek and Litterer characterize "the real-time organization." This characterization includes decentralization, greater use of cross-functional product development teams, employee empowerment, shared strategic vision, and a culture which supports continued process improvement. When coupled with the use of sophisticated management information system technologies, technology development performance is maximized.

In Chap. 6, Professors Dorothy Leonard-Barton and Warren Smith explore how the manufacturing function can be more completely integrated with R&D in the new product development process. In addressing this issue, Leonard-Barton and Smith show how organizational, human resource, and R&D issues affect the modern manufacturing function. In addition to noting the importance of continuous process improvement, they emphasize the importance of continuous and rapid learning. This includes not only learning from internal sources but from customers, vendors, and competitors in the external environment. The modes of interaction between process developers and users are discussed, in addition to the differences between contracting out process development and developing new manufacturing processes internally. Critical issues like key roles, resource allocations, and reward systems are discussed as they relate to new process development.

Following Leonard-Barton and Smith's discussion of manufacturing issues for new products, Chap. 7 focuses on product launch and follow-on marketing activities. In this chapter Professors Roger Calantone and Mitzi Montoya outline the specific steps in a process that begins with the development of a marketing plan keyed to potential customers, competitors, key opportunities or threats, and market-share forecasts. Calantone and Montoya describe the issues involved in developing the marketing strategy, including decisions on promotion, pricing, and distribution. This strategy also includes financial planning and controls, product design finalization, market testing, and launch timing. The final stages discussed by Calantone and Montoya involve the development of channels of distribution and major issues associated with channel selection and promotional strategy decisions.

The final chapter in Part 2 reviews the development cycle and introduces the reader to an emerging paradigm of new product/technology development. In Chap. 8, Professors David Wilemon and Murray Millson contrast the traditional paradigm with the emerging paradigm and present evidence to support a fundamental paradigmatic shift. The authors' characterization of this change focuses on flatter organizational structures, increasing numbers of joint ventures, new product development which crosses international boundaries, accelerated development cycles, total quality management, and the use of new information technologies to speed new product development. Other factors *which Wilemon and Millson observe are the organization cultures which emphasize continuous learning and continuous improvement, the development of prospector cultures which facilitate the creation of new market opportunities, and the development of more integrated or holistic organizational approaches to new product development. These observations summarize the traditional and emerging trends in new product / technology development and serve as both a close-out to Part 2 and a lead-in to Part 3 of this book.*

5

Organizing for Technology and Innovation

Mariann Jelinek
College of William and Mary

Joseph A. Litterer
University of Massachusetts

During the concept development and market definition steps (Chaps. 3 and 4), serious organizational decisions must be made to prepare for the ramp-up of subsequent activities (e.g., product R&D, manufacturing, launch, and follow-on; see Fig. 1.1). Organizing for effective technological innovation is a major challenge to any new product process. In this chapter, the authors show how traditional ways of organizing can severely handicap modern new product developments. How can these handicaps be overcome? How can the new product development process be organized to achieve high-speed development, high product quality, product variety, and on-time delivery? How can today's organizations adapt to the challenges of fast-paced markets, intense global competition, and rapidly changing technologies? These are some of the important questions that are resolved in this chapter.

WM. E. SOUDER AND J. DANIEL SHERMAN

Introduction

Technology has made major contributions to today's intense global competition. Developments in computers, telecommunications, and management information systems (MIS) have created the technological potential for rapid, flexible response. However, traditional organizational forms have proven inadequate to support effective competition in this environment. Traditional firms are fundamentally incapable of fully exploiting available technological capabilities without radical change in their organizational paradigm and its underlying assumptions. As typically constituted, traditional firms cannot make full use of available technological and human resources to meet more strenuous contemporary competitive demands. An alternative more suited to today's conditions, the real-time organization offers more effective means to enable organization members to tap their technological potentials.

This chapter describes how the form of organization common in much of U.S. industry handicaps and limits firms in their efforts to innovate and to be responsive to customer needs and desires. We continue with an examination of the conditions and practices successful firms have found necessary to foster innovation and then describe a newly emerged organization form that satisfies these conditions to permit responsive, flexible, cost-efficient operations that deliver quality products or services. A central feature of this new organization is the effective use of computers and telecommunication systems. The resulting real-time organization draws on a much larger array of human talents and capabilities than any previous form, making the new organization both powerful in the competitive marketplace and highly satisfying as a workplace for its members.

The First Technological Paradox

U.S. technology performance presents a curious anomaly, a paradox of potential versus performance. While they have enormous areas of technological strength in basic research and discovery, U.S. firms generally have turned in a far weaker performance both in the marketplace and in the utilization of technology within organizations than would be anticipated from their strengths in basic technology. Evidence for the strength of basic technology in the United States is visible in the size and productivity of the U.S. research establishment: universities and private research laboratories, the training of doctoral students (including many foreign nationals), to say nothing of the almost-science-fiction quality of Desert Storm military hardware, and the generation of numerous fundamental discoveries that have launched entire industries. A host of examples would include xerography, television, semiconductor electronics, computers, and peripheral devices.

Evidence for the weakness of U.S. firms' marketplace performance is equally obvious, and in many cases it includes some of the same areas of former U.S. dominance. Industries of former U.S. strength now under attack include commodity semiconductor components, consumer electronics (televisions, tape recorders, radios), plain paper copiers, automobiles, personal computers, and machine tools, among others. And the performance shortfall—or the technological paradox—is not all that new.

When American business was first shocked by strong global competition in the early 1970s, the cry went up that low foreign wages, favorable foreign bank rates, government subsidies, tariffs, and other arrangements were the cause (Hayes and Abernathy, 1980; Thurow, 1985; Cohen and Zysman, 1987). In this context, the pressure to make better use of apparent technological strength to meet contemporary competitive challenges became even greater. Yet it was already clear that U.S. technological capabilities were missing something important (Hofheinz and Calder, 1982; Kang, 1989; Pascale and Athos, 1981; Peters and Waterman, 1982). The consequences of the first technology paradox can be readily summarized in loss of market share, decline of formerly dominant U.S. firms vis-à-vis typically foreign competitors, an embarrassing and persistent trade imbalance, and an apparent inability to compete with a growing number of international competitors who even relocated to the United States.

What was the outcome of these circumstances? U.S. firms employed more assets—physical, financial, and human—for less output than did their more successful competitors. As an illustration, consider the dilemma in automobile production, as shown in Fig. 5.1.

Some claimed the problem was merely obsolete "rust belt industries" coming to their logical demise. If only this could be true. During the 1980s, U.S. firms in computers, semiconductors, and airframes struggled as visibly as had the auto industry. U.S. semiconductor firms, among the highest of the high techs, together lost over $2 billion in a harrowing series of downturns. Worse still, U.S. merchant makers' share of the U.S. semiconductor market fell from 90 to 67 percent, and from 80 to less than 40 percent of the global market. In a number of widely used products like commodity DRAMs (dynamic random access memories, a key computer component), Japanese market share is widely reported to be well above 90 percent worldwide (Prestowitz, 1988).

Similarly, in computers, U.S. firms' share of the American market fell from 94 to 66 percent, while in the airframe market, long dominated by U.S. manufacturers, the European consortium's market share went from 0 to 18.8 percent of narrow-body jets and from 25.2 to 49 percent of wide-bodies (Prestowitz et al., 1991). In industry after industry, although U.S. firms might create basic insights and breakthroughs—VCRs, microwave ovens, miniature radios, calculators, computers, plain paper copiers, and automobiles are among a host of examples—other firms, often from other countries, are reaping the marketplace benefits.

	Toyota #9 Engine	Chrysler Trenton	Ford Dearborn
Products	2.4 L 4-cylinder 2.0 L 4-cylinder	2.2 L 4-cylinder including turbo	1.6 L 4-cylinder HO; turbo; EFI
Plant size	$310,000^2$ ft	2.2 million^2 ft	2.2 million^2 ft
Hourly workers	180	2250	1360
Production rate	1500/day	3200/day	1960/day
Worker hours/engine	0.96	5.6	5.55
Shifts	2		1 assembly, 2 machining
Inventory (average)	4–5 hours	2.5–5 days	9.3 days
Absenteeism	5% (scheduled)	3% (AWOL)	N.A.
Wages	$11.35/hours, excluding fringes	*	*
Robots	None	5	N.A.

*U.S. auto industry average is $12.50 per hour excluding fringes. Chart is adapted from "Quality Goes In," *Automotive Industries,* November 1984, 51–52.

Figure 5.1 1984 manufacturing comparisons.

It's tempting to define the problem as "poor American competitiveness in the face of Japanese success." Yet American firms aren't the only ones experiencing difficulty, nor are all American firms in trouble. Equally, not all Japanese firms are succeeding—nor all German, French, or Italian firms, although some firms in each category are. Industry per se is not a good indicator of difficulty, either: Some firms in mature industries seem to prosper, while others in "growth" industries struggle or fail. Neither reputation nor market leadership nor past technical dominance seem a match for newer methods. Some firms, in virtually any industry or country, are in trouble; others in those same industries and countries do well.

More curious still, the successful firms seem to completely outclass their competition of whatever national origin. Successful firms attain levels of performance that seem unreal in the context of traditional arrangements, "impossible" by comparison with norms of not very long ago. A few examples will suffice: Inventory turns of twice a year were considered adequate in prior decades; contemporary standards have risen to 20, 50, 100 turns per year—or even "negative inventory," products bought and paid for with no inventory holdings at all. Similarly, 5 percent failure rates were once thought acceptable, while today Federal Express strives for "100 percent customer satisfaction," and many manufacturers track single-digit errors per million, or better. In this context, too, technology seems to hold great promise, but few firms achieve the goal.

The first technology paradox can be restated more succinctly: Despite enormous technological resources, many traditional firms seem unable to innovate effectively, to adapt to marketplace changes, or to comprehend what their difficulty seems to be. What's the problem? Numerous partial explanations—the growth of worldwide technological capabilities, increasing competition, foreign government industry subsidies, the decline of the work ethic—fail to embrace the full width and depth of the technological paradox. They hardly explain why firms with so much richer a technology base to draw upon have done so poorly. A closer look reveals that the real cause is organizational. The answer is that rich technological capabilities are not being as effectively utilized by firms in difficulty as by their more successful rivals.

In a larger perspective, the environment has changed, requiring different organizations more appropriate to the new circumstances. To really understand the nature of the difficulty, we must take a closer look at traditional organization modes, structures, and practices and how they interact in the contemporary environment. A look backward, to the last technological high-water mark of U.S. industry, is instructive.

Organization Is Standing in the Way of Technology

Twenty-five years ago, when many in Europe feared that American firms were going to take over business on the Continent, Servan Schreiber (1969) argued that *the American challenge* rested on organizational and managerial superiority. Rather than using legal barriers to keep American firms out, Servan Schreiber advised European firms to adopt American management practices and organization forms.

Despite differences among firms and industries, many commonalities gave American firms a distinct character and contributed to their successes. These firms embodied the development of decades, first in perfecting techniques of mass production, then, in close parallel, the techniques of mass marketing and distribution. Let us describe the type of firm to which Servan Schreiber was referring, which we will call the scale-based organization.

They were large companies, which by exploiting the economies of scale, produced large volumes of a limited variety of products, in large, specialized facilities, using long production runs on specialized equipment to attain low cost. Deliveries were typically from finished goods inventory, giving fairly prompt delivery to customers except in the case of stock-outs. When finished goods inventories were exhausted, delivery could be uncertain and lengthy. They were oriented toward large markets, aiming their standardized products at substantial market opportunities, and were willing to let smaller markets go. Exploiting economies of scale, they created large plants using specialized equipment and facilities to get costs down. The high cost of new plants effectively kept new competitors out. The few large competitors in an industry competed on price and/or offering features to differentiate their products. New products were rarely introduced, but upgrades of extant products were introduced often, as with automobiles, on a yearly basis.

The underlying premise of these organizations was efficiency (Weber, 1964), implemented in practice by a heavy emphasis on cost and cost cutting. Because of the basic emphasis on low costs, production scheduling was driven by the need for long stable production runs, not by market demand (Sloan, 1963). It was the task of marketing to dispose of what was made—a practice of market push (Womack et al., 1990; Lacey, 1986)—rather than for manufacturing to produce what was wanted. A common belief held that quality was costly and that therefore a trade-off had to be made between quality and cost. The standard was to have "adequate" quality in order to achieve acceptably low costs.

The characteristics of scale-based firms

Scale-based firms had many thousands of employees, arranged in many hierarchical levels, with many specialties both at the management and the worker level (Holden, Pederson, and Germane, 1968; Janger, 1973). Internally, scale-based firms were centralized, with most decisions made at the top of the organization (Kuhn, 1986). Even in decentralized firms, decentralization seldom went far below the division headquarters (Wolff, 1964). Top decision-making elites assisted by cadres of specialists (Holden, Pederson, and Germane, 1968) made all major and even relatively minor decisions. Decision-making throughout the organization was carefully controlled over many organization levels and across functional specialties. Nonmanagement members often had almost no decision-making authority, while lower-level managers enjoyed little more.

Specialization was carried to considerable lengths across both functions and hierarchical levels. This often resulted in individual positions or units operating very efficiently but made integration of individual and unit outputs difficult (Lawrence and Lorsch, 1967). Coordination was difficult, slow, and inefficient. There was an implicit distrust of organizational members, particularly those at the lower levels of management and especially nonmanagers. Structures were established to restrict what could be done by precise job definitions, based on detailed rules, limited spending authority, signatures required to approve an action or decision, and audit review of performance by an independent party (Holden, Pederson, and Germane, 1968). Members at all levels were expected to follow directions, do their job as directed, and produce what was specified. Communication was channeled, mostly along hierarchical lines and by specified channels to functional specialists.

To permit the specialized machines and tasks to operate most effectively, work was stabilized as much as possible. Cost cutting was the norm. Firms buffered internal activities from one another to prevent "disturbance." Thus, manufacturing activities were buffered from market and supply demands as much as possible (Thompson, 1967). Research and engineering were kept separate from the problems of operations and sales in order to promote creativity and to get designs done with the minimum of distraction. The same underlying stability also made major change in product or production methods very costly and therefore undesirable. In automobiles, while new models might be brought out annually, new automobiles would be introduced only every 6 years or so (Womack, Jones, and Roos, 1990).

When the effects of specialization were combined with the effects of buffering, departments, units, and even individual positions became increasingly separated from one another by social barriers, making coordination and integration increasingly hard to accomplish. Members of specialized units tended to understand the importance of their own work and the conditions needed to perform it well. They frequently developed different sets of values as to what was important, different codes of behavior, and even different ideas about time (Lawrence and Lorsch, 1967). They were all too often indifferent to, and ignorant of, the needs of other specialities within the firm. Different specialties pursued their own subgoals, often with little regard for the overall goals of the firm. Chronic conflicts and misunderstandings developed not only between major functional areas of marketing, manufacturing, and engineering (Hower and Orth, 1963) but also between units within these areas, for instance between manufacturing operations and production scheduling, advanced engineering and product design, and sales and advertising. One thing almost all shared was an indifference to, and sometimes a hostility toward, customers. These factors created other, less desirable characteristics in addition to those mentioned earlier. Firms could be slow and unreliable on product delivery, product development cycles could be lengthy, and variety was restricted to support long production runs and reduce changeover costs. All too often the relentless pressure to cut costs allowed quality to erode.

Externally, scale-based firms became increasingly decoupled from customers, vendors, and economic trends. They operated in a protected world on their own time base, using lead time and inventory to buffer operations, enabling the firm to continue operating as if wholly independent of the environment (Thompson, 1967). The decoupling continued within the firm. Specialists too moved to their own internal clocks, worked to meet their subgoals with diminished attention to overall company goals or customer needs, and evolved their own set of values which grew from the needs and traditions of their specialty (Lawrence and Lorsch, 1967). Highly specialized, repetitive, and unchallenging work coupled with ubiquitous controls and their implied mistrust frequently produced alienation among nonmanagerial personnel and even among lower managerial personnel (Walker and Guest, 1957; Roethlisberger, 1961).

Against smaller competitors organized along the earlier lines Servan Schreiber had observed in Europe, scale-based firms enjoyed distinct efficiency advantages. Competing against other traditionally organized firms, they were successful as long as they skillfully utilized the capabilities of their paradigm, as a few large firms in industries like automobiles, steel, consumer goods, and chemicals did for decades. With all their difficulties, scale-based firms prospered. But when firms with more effective characteristics began to appear, their less desirable traits made them seriously vulnerable.

A brief comparison of these performance characteristics with contemporary competitive requirements is instructive. For large sectors of the U.S. economy, both markets and competition have become global. Many competitors are using variety as a competitive weapon. As a result, consumers are getting more sophisticated and demanding. They want a variety of products to choose from, they want their choice promptly and reliably delivered, and they expect excellent quality. Industrial customers, operating under comparable competitive pressures, Just-in-Time (JIT), and quality constraints, are no less demanding. Firms that cannot meet these requirements are in trouble because aggressive competitors stand ready to displace them. Unfortunately, as Fig. 5.2 summarizes, the scale-based organization is not able to meet these conditions.

Scale-based organizational characteristics that impede successful response include the following:

- Scale-based organizations are not well linked to their environment; they tend to be inwardly oriented and inattentive to external change.

- Scale-based organizations are designed for stability in products, processes, and markets, with the archetype being the mass-production and high-volume manufacturer.

- Because scale-based organizations resist early, incremental change, they must endure the alternative of reconfiguring them-

	Traditional firm performance	Contemporary requirements
Orientation	Market push	Customer pull
Basis for strategy	Minimum cost, price commodity features	Price, uniqueness, variety, delivery, quality
Production run	Long	Short
Delivery cycle	Long	Short
Delivery reliability	Moderate to poor	High
Product change cycle	Infrequent	Frequent
Customer reaction	Slow, focus on commonality	Fast, address uniqueness
Quality	"Adequate"	High

Figure 5.2 Traditional firm performance characteristics and contemporary requirements compared.

selves in "big bites," a far more risky task typically carried out much later on, under more pressure.

These scale logic responses are quite simply inadequate to the challenges facing firms today. They make no headway whatsoever in resolving the technological paradox. None of these responses in any way assists in effective use of technology. The traditional response to adversity has been that survival requires cutting where cuts can be made, but such "belt-tightening" assumes that eventually business will return to "normal" so that belts can again be loosened. This time, it's not at all clear that our old notions of "normal" will ever return because it is highly unlikely that the older conditions will return. The era of minimal technological challenge and limited competitive rivalry or contest for markets because of pent-up demand and the easy predominance of North American methods, tastes, and technology are gone. Moreover, the nature of the challenges has also changed: Manufacturing or servicing in the same old way, but doing it cheaper, simply isn't enough.

Where scale-based organization fails to support technology utilization

How is it that scale-based organizations fail to adequately support effective technology use? Traditional scale-based organizations are weakest in the very capabilities that are most important for meeting contemporary challenges. These new capabilities have displaced older competitive advantages based on standardized products, high volume, and low cost.

Speed in new product design or product update. In automobiles, telecommunications, and consumer electronics, to name a few industries under pressure, speedy new product development has opened a discernible technology performance gap between scale-based firms and contemporary firms capable of more rapid response. Where American and European automobile firms took 3 million engineering hours and 60 months to produce so-called "clean sheet" new products, new Japanese cars required 1.7 million engineering hours and 46 months (Clark and Fujimoto, 1991; Womack et al., 1990). AT&T was taking 36 months to design and market new telephone products, while Pacific Rim competitors averaged 18 months or even 12.

In gross terms, such differentials mean that the slower firms have one- to two-thirds the number of opportunities their competitors have to improve their product, add new features, substitute better performing components or materials, improve production processes, upgrade design, and the like. Slower development raises costs substantially (Foster, 1986). Rapid new product development enables a firm to exploit new technology when it is newer and can command a premium in the marketplace, while slower firms in subsequently more competitive circumstances cannot. Worse still, should a slower firm succeed in designing a genuine advance, competitors' rapid development skills enable them to respond to it, reducing the innovator's potential to recoup its higher (because longer) development costs (Birnbaum-More, 1990; Ungson, 1990; Walleck, 1985). Cutbacks and layoffs do nothing to speed new product development, and of course "speed" in the abstract, without producing an appropriate, competitive product, is useless. Nevertheless, all else equal, the speedier-to-market innovator appears to have an important advantage.

Increasing variety in products and services. Strong market pressures for cost cutting would argue for decreasing the number of SKUs (stock keeping units: different models, products, or varieties). In scale-based firms, variety is expensive, and the trade-off between cost and variety is an article of faith. Today, even stronger market pressures demand increasing variety in response to market niches, custom opportunities, and local needs. At the same time that markets are becoming global, markets are simultaneously fragmenting. Customers are more sophisticated, more demanding that products fit their specific needs (Toffler, 1970, 1981). Often, they are willing to pay a premium for such products—especially if delivered quickly and reliably (Stalk and Hout, 1990).

Highest-quality standards. High-quality standards are rapidly becoming a commodity feature in many product areas. Some authorities speak of "world class standards" (Schonberger, 1982, 1986), while electronics manufacturers routinely describe product quality in terms of millions of hours of mean time between failures. Quality norms have increased to the point where having excellent quality is not enough to set a firm or a product apart. Having less than excellent quality does—with disastrous results. Much higher quality is the minimum price for entry. Unceasing work to maintain and to increase quality is no longer an attractive alternative or luxury to differentiate the firm but is ostensibly effort required for survival. The minimum quality requirements of contemporary products are, frequently, well beyond those considered "premium" or "laboratory" quality not long ago. Examples abound: Commodity chemicals, reagents, and silicon substrates used in semiconductor manufacture are available today from commercial suppliers off the shelf in quality attainable only in advanced research laboratories a decade or so ago.

Shorter delivery cycles. The move to JIT manufacturing has drama-
tized a more general trend: Customers want things now or at least
far sooner than just a few years ago. Federal Express's success in
overnight delivery (AMA, 1991) stands as testimony to this desire.
So, too, does the emergence of others in a vigorously competitive
fast-delivery industry virtually nonexistent 25 years ago. Rapidly
rising demand for fax services corroborate customers' desire for
"hurry-up" service, as do such innovations as drive-in oil changes
guaranteed in 30 minutes or Citibank's "15-minute mortgage."
Inventory, including inventory in a delivery pipeline, constitutes
wasteful slack that must be financed and owned, with all the atten-
dant risks for loss or obsolescence.

Smaller order quantities and utterly reliable delivery dates. Customers
want to place smaller orders but receive more frequent delivery to
cut down on floor inventory and therefore inventory cost. Inventory
holdings and turns are one measure of such trends. Where
American automobile manufacturers regularly held months or even
years of steel inventory (as a safeguard against steel strikes) in past
decades, today their holdings are more apt to be measured in days;
the Japanese measure in *minutes* of production time (Stalk and
Hout, 1990; Womack et al., 1990). Inventory turns for the best contemporary firms are no longer low single digits but high double digits or better; some firms speak of "negative inventory" sold before it
is ever delivered. But lower inventory raises the ante for missed
delivery, so utterly dependable delivery throughout the channel is
essential. At the same time, because the product pipeline is shorter
(and less product is in it), there is less inertia and greater potential
for still more change in customer demands for product volumes and
for product features.

None of these demands is facilitated by scale-based organizations
or is amenable to traditional responses. Indeed, on the contrary,
these demands seem quite incompatible with traditional assumptions of control, rigidity, hierarchy, and continuity. Organizations
designed to deliver rigid control as a mode of quality assurance, or
long production runs as a mode for cost reduction, for instance, are
fundamentally unable to deliver flexibility and variety along with
high quality. Scale-based firms have traded off speed for cost, quality for variety, responsiveness for control—a devil's bargain that has
rendered large-scale production possible but has rendered scale-
based firms utterly vulnerable to lean, flexible rivals (Goldhar and
Jelinek, 1983; Womack et al., 1990). In a competitive environment of
rapid change, the stable, conservative strengths of scale-based organizations turn to fatal weaknesses, frailties ripe for exploitation by

faster-moving firms that can change, change, and change again. Thus are long-held assumptions about competition, markets, and realities called into question, while new standards and survival criteria emerge.

Examples could be proliferated, but the point has been made. Poorly organized and poorly managed by contemporary performance criteria, scale-based organizations have limited capability for using technology effectively. Conflict was common, the result of isolated units operating in competition with one another. Coordination was achieved with difficulty, if at all, and only through elaborate, expensive effort. As a result, technical work was slower and often of lesser quality than needed, and it frequently failed to address crucial needs—of other departments in the firm, of customers, of cost constraints, of organizational strategy, or of external developments.

The Second Technological Paradox

U.S. firms' technology difficulties in global competition pose a second compelling paradox with regard to computers. American technological dominance in computers and software is definitive, yet U.S. firms have not benefited from this strength. Indeed, enormous expenditures for computer systems have not improved white collar productivity (*Business Week*, 1990a). Computers have been heralded for 30 years as the "technological fix" to reduce labor costs definitively by automating a wide range of routine activities, as indeed they have. Insurance companies alone would probably be able to eliminate present U.S. unemployment were they to carry on their present level of operations without the aid of computers. Significant decreases in middle management staffing, also long predicted (e.g., Leavitt and Whisler, 1958; Simon, 1960), seem to be occurring—but again without the anticipated benefits.

Expectations and reality match in the vastly increased penetration of computers, in their potential for eliminating jobs, and in their simplification of other jobs to mindless rote activity. Repetitive motion syndrome injuries for grocery check-out clerks or data-entry personnel are another adverse consequence. Modest changes in decision-making support, such as credit, charge approvals, and insurance claim processing suggest broader potentials that have gone largely unrealized.

Computers seem all too often to have been used to mechanize the hand labor motions and reinforce the traditional control-oriented organizing assumptions of scale-based bureaucratic organizations. The possibilities computers offer of redefining what work needs doing and how it should be done have generally not been realized. The possibilities computers offer for meeting the competitive challenges of the global economy have not been generally exploited, even though contemporary conditions seem to cry out for the very capabilities that computers promise.

In place of the scale economies, low wages, and low-cost capital advantages of the past, many of which favored sheer size, a much more sophisticated blend of sheer expertise, flexibility, and rapid response characterizes the most successful firms today (Goldhar and Jelinek, 1983; Foster, 1986), and computer technology seems highly apropos. In the new conditions, a "logic of industrial success" remains (Chandler, 1990), but the contemporary logic emphasizes close attention to rapidly changing market conditions, not simply established position or size, as in the past. Speed (Stalk and Hout, 1990), creative knowledge, rapid new product development (Bower and Hout, 1989; Clark and Fujimoto, 1991; Wheelwright and Sasser, 1989), continued improvement and innovation (Jelinek and Schoonhoven, 1990a; Schonberger, 1986; Womack et al., 1990)—all performance characteristics that seem especially amenable to computer assistance—offer the basis for rapid response. Information technology should provide the means for speeding response, broadly sharing information to support effective and coordinated decisions and participation, and improving the quality of organizational coordination. It rarely does.

The second technological paradox is that despite abundant and long-standing computer expertise, scale-based firms have not achieved the promise of computer technology. The underlying reason is that computers have all too often simply supported and extended scale-based organizational forms and methods. Scale-based firms are poor platforms for information technology development, innovation, or implementation. Put differently, the possibilities for designing and coordinating work which information technology holds out cannot be realized within the scale-based organization.

Information technology possibilities

Because "technology" is a very broad term indeed, we shall focus principally on the application of computers to illustrate our discussion, construing organizations as information-processing engines—artifacts designed to facilitate human use of information to make effective decisions, take effective action, and create effective new products and services (Galbraith, 1973). Computers have long promised support to just such an organizational mission, but pundits have differed radically on whether the advent of computers would support human decision making or supplant it (Leavitt and Whisler, 1958; Simon, 1960). In light of contemporary competitive conditions, it seems clear that centralized, routine-based, and control-oriented use of computers, while feasible and even (as some have argued) prevalent (Child, 1987), is not the way to reap the potential benefits. Such computer usage essentially automates the inadequate practices of the past, offering no real answer to contemporary competitive challenges.

We have argued that scale-based firm characteristics are especially deficient at extending responsiveness because they favor control, stability, and replication over effectiveness, innovation, and responsiveness to changing circumstances. These same characteristics render scale-based organizations virtually incapable of exploiting the technical capabilities that computers offer that might provide the basis for effective competitive response.

Four broad categories of computer applications can be defined to help summarize the technological possibilities organizations might exploit to meet the competitive challenges we have been outlining (Broderick and Boudreau, 1992; Jelinek, 1987). The simplest and most conspicuous application is widespread: *Monitoring, or tracking, routine activities,* particularly machine operations, is both obvious and appropriate. Data can be collected, ordered, stored, and used to ensure that activities are properly carried out and any errors can be detected for rapid correction. Computers never tire, daydream, or take coffee breaks—so their data records are more complete and correct than human records are likely to be.

Such information can be used to control human operatives and check up on them, but this is perhaps the least productive application of it. Better uses include scheduling machine maintenance, tracking machine utilization and thus capacity needs, providing the basis for statistical process control and the like, training human operatives and honing their skills on exceptions, and predicting future demand patterns. Data collection in routine monitoring and tracking applications is typically automatic, as is simple statistical analysis of the data; no great computer expertise is required on the part of operating-level personnel for either data collection or access and use.

A second category of information technology use is closely related: *Integration by means of computers* improves possibilities for coordination of diverse organizational activities. Basic integration of organizational activities arises from simply making information from one activity—sales order entries or bill of materials for a design or inventory information, for instance—available to multiple users in other activities. This may require only minimal computer skills from operating personnel, who can use any information they can access through the computer system. It does, however, require an organization that will authorize sharing information and foster integration based upon it.

Networking enables wide collection and distribution of data, which increases its potential value in two ways. Collected data about many activities can be brought together for planning and coordinating and monitoring performance of positions and units. This has already happened on a centralized basis, which essentially degrades networking to a tracking and monitoring function (Child, 1987). However, networking can also distribute data to multiple and widely separated parties. Both individuals and organizational units can receive information on their own performance and that of others to enable comparisons and error correction. Such feedback has long been recognized as highly effective quality improvements (Feeney, 1973), but traditional organizations have made little use of it, preferring instead to continue to rely on far less effective, infrequent supervisory performance reviews.

Networking offers further benefits. Wide distribution of data increases information's potential value because it increases its potential for reuse by multiple interested parties. Information about operations, shared broadly, permits many people to make decisions dealing with an overall effort. Because all can have timely and accurate access to the decisions of others, they can more readily coordinate their efforts and collaboratively improve their decision processes over time. Although some work has been done in this area, this promise is largely unrealized in large part because it, too, is contrary to the tendency for hierarchical centralized decision making in scale-based organizations. To benefit fully from networking requires that the organization enfranchise many decision makers and decision sites.

Further development of such integration, which constitutes a third category, can be achieved by the *use of expert systems to ensure more competent decisions with better information at remote, local sites.* For instance, credit card approvals within broad parameters are now made by computer, thus improving speed of response to predictable problem or decision situations. Decisions beyond the routine parameters are referred to experienced human decision makers, whose training was expedited by their use of the computer database of "good past decisions." Similarly, complex computer systems configurations, loan applications, geological data interpretation, medical diagnosis, and other tasks have been supported by expert systems.

Such decision support systems can be used to improve the quality of all decisions by making information and decision criteria widely available and also to improve system performance incrementally by using information gathered on performance over time. What works best can be systematically determined and incorporated into "best contemporary practice" in the updated expert system for the use of all on the network. Such applications require sophisticated programming to set them in place but only moderate expertise on the part of users. More important than computer expertise per se is appreciation of what the provided information means, how to make good use of it, and how to interpret exceptions when they arise. The potential for improving the quality of decisions on routine matters and allocating valuable experienced human decision resources to extraordinary decisions has been demonstrated repeatedly for decades (e.g., Simon, 1960). Such potential cannot be realized by traditional organizations, which limit and constrain information access, decision power, and action patterns.

A fourth category, *decision support for creative work, strategic planning, or other nonstructured problem formulation and solution situations,* is the most difficult and complex computer application. Genuinely helpful computer programs have only recently begun to appear, and because their operation is necessarily unstructured and context-free, they demand the greatest computer capability on the part of their users (Broderick and Boudreau, 1992). Of course, such programs do not provide solutions or "answers." Instead, they support human decision processes. The human decision makers must be deeply knowledgeable and technically sophisticated enough not to be misled by the apparent authority of a computer response. In short, hardware, software, and users all require development before fourth category applications begin to make sense. Much developmental work remains, but nevertheless, fourth category applications are under way.

Integrated systems

These task categories can be adapted to technical, managerial, and strategic uses of information, with integration and coordination of organizational activities as an overriding aim for information systems. Within the context of a manufacturing organization, for instance, these broad uses can be illustrated and highlighted against typical functional needs and information flows. Product design, process design, process control, scheduling, materials control, order entry, payroll, and warranty are all capable of support via monitoring and tracking, for instance. Computer-aided design (CAD) and computer-aided engineering (CAE) systems offer examples of more sophisticated support, combining monitoring of such routine matters as arithmetic accuracy or preset design parameters with known strengths of materials, for instance. Links between product design and manufacturing, as well as the scheduling and information coordination that links order entry with inventory, provide examples of coordination supported and facilitated by computer networks. Expert systems for CAD and CAE, for routine maintenance, inventory management, and process control, offer high potential for performance improvement, as does "on the fly" production scheduling in response to order priority or materials availability.

Figure 5.3 illustrates the key categories and their relationship to managerial tasks in an archetypal manufacturing organization. At virtually every step, computers can assist and support human activity. Order entry can be wholly automated or remotely sited at customer locations or can be assisted by data fields that alert the order taker to inappropriate entries (typographical errors, improper model numbers, unavailable items, incomplete addresses, telephone numbers with missing or inappropriate area codes, and the like). CAD and CAE can perform routine calculations to check a designer's use of materials, layout of circuits, or mathematics automatically, leaving the designer freer to concentrate on the creative aspects of the work. Simulation tests can prove out circuit design, three-dimensional modeling can ensure that tightly packed components actually fit (and can be manufactured with existing equipment), and so on. Creativity support programs can be utilized both in obvious applications like planning or design and also in a multitude of other activities throughout the firm.

Two nodes of information flow are identified, the technical side and the managerial. Within each, the sequence tends to flow from top to bottom—that is, from product concept to tool point on the technical

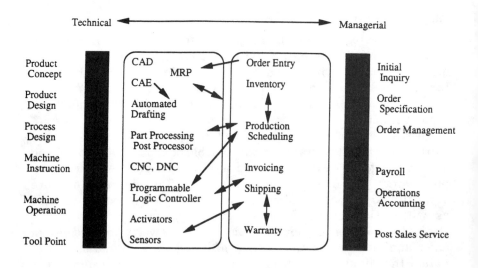

side, with information accumulating, elaborating, and being reused as the sequence proceeds. Similar uses exist with managerial information. However, while the nodes appear to be separate, they necessarily carry important implications for one another and should be linked to maximize information use and benefit. Networks can speed information transfer and use; ensure common versions of designs, process sequences, or inventory records; and create other positive "ripple effects" throughout the organization by simply making accurate information commonly available. For instance, a customer's initial inquiry may require "custom" design, a new design may have inventory implications, production activities (whether successful or not) will have implication for later scheduling, warranty experience ought to feed into the next design cycle, and so on.

Networks permit wide distribution of data which, coupled with expert systems and decision support, supports genuine dispersion of decision making that nevertheless reflects best available data and empowers real distributed flexibility. Rather than merely replicating centralized perspectives (as bureaucratic rules might), such a system creates new potential for effective, distributed organizational response with commensurate assurance that decisions will be made with good data at hand.

Computer applications can be used to control and limit discretion, to constrain human potential, and thus to make work limited and boring—or they can support effective discretion, decision making, and response by enriching the information accessible to the decision maker. Technology permits vastly increased information productivity. Potential information capture, use, sharing, transformation, transfer, access, use, and repeated reuse of information with variations suggest vastly increased organization responsiveness, efficiency, and innovation simultaneously.

From a more general perspective, these capabilities show how information technology can improve organizational performance when coupled with genuine organizational change. It is through information productivity that computers may finally deliver their long-promised benefits, because the information that they collect, store, and transmit can be so effectively reutilized, either directly or with modifications. Adaptation of "custom" designs is estimated to involve typically only 20 percent truly new information, with about 80 percent capable of reuse, for instance—so the second "custom" design will cost a good deal less than the first if information "already paid for" can be reutilized. Engineers have long reused information in just such a fashion, but the concept of information as an asset is far more general. However, clearly this will transpire only if employees are empowered to do more than merely replicate past practice. True information productivity will reflect new uses, not merely old ones.

Considered as an asset, information has great potential for improving human and capital asset productivity against criteria of the organization's strategic goals. The analog appears to hold for a multitude of situations ranging from word processing (templates for form letters and "boilerplate" in law offices, for example) through spreadsheets (tax programs or accounting programs), discrete parts, and integrated circuit design. Figure 5.4 illustrates the information productivity concept.

All of these technological capabilities for increasing the productivity of information run the risk of stultifying organizational response by rigidifying it—unless appropriate human creativity, judgment, and adaptation are invoked, rather than supplanted by scale-based hierarchical, centralizing assumptions. All of the information technology possibilities for integration and improved organizational response involve automating routine decisions in order to explicitly delegate the nonroutine. Better informed human judgment and innovation are made possible but are not guaranteed by the technology. The objective of technology applications is to enable organization members' capabilities to be more effectively utilized, not, as has so often in the past been the goal, to enable equipment to be fully occupied or to enable the organization to check up on members' performance.

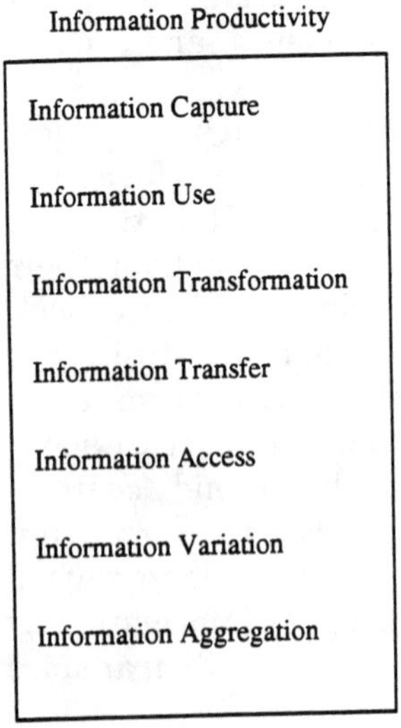

Information Productivity

Information Capture

Information Use

Information Transformation

Information Transfer

Information Access

Information Variation

Information Aggregation

Figure 5.4 Information productivity.

Active participation in real decision making on the part of a broad array of organizational members and commensurate amounts of trust, training, and responsibility are involved to ensure effective exploitation of the technological possibilities. But none of this comes automatically; it must be the result of genuine commitment on the part of management, built into the structure and operating norms of the organization. As we have argued, these are precisely the blind spots in scale-based organizations, the responses most contrary to traditional organizations' hierarchical, centralized, control-oriented paradigm.

Technology-Based Challenges

The competitive developments we have been describing raise three underlying challenges for today's organizations:

1. *How can organizations make their products more responsive to changing customer needs?* More product variety, more rapid response to emerging customer needs and technological possibilities, and improved new product development seem closely intertwined. The faster the effective new product development cycle, the easier a company should find it to incorporate frequent changes in product design or features, incorporate new technology, increment design improvements, or respond to clarifying customer needs. Technology, and more specifically computer applications to speed information generation, checking, and sharing, offer potentials both for speeding new product development and for improving its quality by facilitating checking, use of expert input, and a variety of perspectives in a timely, cost-effective fashion.

2. *How can organizations adapt their operations (manufacturing or service and distribution) so that process technology pays no quality penalty for the speedier deliveries, small order quantities, and reliable delivery schedules?* Since high demands for quality are not going to dissipate, organizations must deliver "both/and"—both quality and new products; quality *and* rapid, assured, frequent delivery of short quantities; and "custom" products and low inventory carrying costs. Operations must be highly effective, as well as highly efficient, and they must meet these formerly traded-off criteria simultaneously. Technology suggests that here, too, information sharing and analysis, together with automated decisions for routine activities and computer-supported delegated decisions for nonroutine decisions, could enormously speed, improve, and coordinate decision making—if traditional barriers between departments and levels can be dissolved and decision responsibility genuinely shared.

3. *How can organizations use the technological potential of computers and telecommunications to speed response yet not fall into massive disorganization?* Old approaches based on paper systems, "line-of-sight" supervision, or elaborate rules are clearly inadequate, and computers seem to offer instantaneous interconnect between different organizational departments, units, or activities. Simply "automating" the old systems has not proven successful, however. Effective instantaneous response and rapid capture of cue data indicating change suggests enormous response power diffused throughout the organization, which must somehow be coordinated in service of strategy. It won't help if organization members are busy marching off in all directions or making serious errors unchecked.

Overarching these specific issues and implicit in all of them is a major difference in the utilization of technology. Until recently, while technology has been acknowledged as important, its evolution was assumed to be intermittent, with perhaps many years intervening between major changes. Often, when changes were made, they were made at the time and in the degree chosen by management. GM's revolutionary introduction of annual car model changes in the 1920s masterfully exploited apparent novelty with underlying stability (Sloan, 1963). New automobile styles were introduced yearly—but underlying major components and their technology changed only rarely, generally at management-determined rates. Moreover, technology change was considered not to imply change in the organization itself or in its underlying assumptions.

Indeed, even earlier attempts to automate had the perverse effect of slowing production technology change (Abernathy, 1978), with disastrous long-term impact upon the U.S. automakers' capabilities to respond to technological change. New clothing fashion styles were introduced in major seasonal offerings at set times of the year, and no competitor behaved differently—until Bennetton and Liz Claiborne moved to multiple offerings per season. In some industries, new products were not introduced until development costs of existing products were completely written off—until increasing new product introductions cost the stable firms dearly. Without competitive rivals, such delays bore no discernible consequences. Firms dominant in their markets were able to set their own calendars for change, working on what might almost be called an internal, artifactual clock.

The increasingly interconnected, increasingly competitive global environment firms face today demands a radically different focus: on changing customer preferences, emerging technical possibilities, potential new variants or product families competitors are introducing, and shifts in competitors. Costs must constantly decline while quality increases. Both product and process change occur much more frequently than in the past, as firms march to the more rapid pace of externally moderated change. Today change is externally oriented and its impact is pervasive.

Organizational Requirements for Effective Technology Use

While information technology offers promise for responding to change and radically improving organizational performance, we clearly need a different set of organizational premises than those of traditional firms. Dispersed decision making that can draw upon local knowledge and unique expertise; technological decision making that includes knowledge about customer needs, competitive conditions, production capabilities, and the strategic intent of the firm; widespread participation in "management" and strategy; widespread problem-finding and problem-solving participation; concurrent product and process design—-these are all characteristics that fit with the potentials technology makes feasible.

The traditionally utilized bases for structure—time, technology, function and physical location, and client served (Jelinek, 1981)—are no longer so obviously immutable or fundamental. Time is less relevant when electronic networking can link shifts and integrate their activities. Geographical distance, even to the point of locations around the planet, is no longer a barrier to efficient and timely communication or collaboration when firms like GE-Fanuc can employ three "shifts" of engineers (one in the United States, one in Europe, and one in Japan) to collaborate on round-the-clock development work. Functional specialization makes less sense when many former specialist tasks, ranging from calculations, monitoring, and checking to the application of sophisticated judgment, can be more reliably implemented by computer programs and expert systems. Clients' needs may no longer be so dissimilar. When computers permit the electronic analysis of what it takes to satisfy them, the effective reuse of information generated in one client situation with modifications becomes a new "custom product."

Effective use of available technology can be described by looking at some of the successful practices visible in contemporary firms and contrasting them with the responses of the past. The scale-based firm sought advantage from mass production and market dominance. Such a firm might be very large, with many thousands of employees, arrayed in many levels of the organization, with many specialties both at the management and worker levels. By contrast with traditional firms, rapid-response firms are often much smaller. For decades, Toyota has had a much smaller workforce producing more per employee, on a plant-by-plant or organizational basis, than Ford or GM, for instance (Miles and Snow, 1978). Figure 5.1 illustrates such differences in 1984. Smaller firms are easier to coordinate and avoid the diseconomies of scale (Hayes and Wheelwright, 1984).

To permit their specialized employees, machines, and tasks to operate most beneficially, bureaucratic traditional organizations stabilized work procedures and conditions as much as possible. Unusual requests from customers in such settings quickly become a nuisance rather than an opportunity. A new idea for a product, a production process improvement, or a new machine meant slowing or stopping operations to make a change. Even creating an off-line model shop or pilot plant took extra resources and was therefore likely to be resisted—especially in competitive times. Manufacturing was protected from market and supply disturbances as much as possible by inventory build up, queuing, or rationing. Research and engineering were kept separate from the disturbances of operations and sales in order to promote creativity and to get design done with the minimum of distraction. Specialist engineers, designers, marketers, manufacturers all spoke special languages, often mutually unintelligibly, and pursued their own agendas. Design is a segmented, sequential task to be completed and "thrown over the wall" for manufacture and "push" by the sales force.

By contrast, successful contemporary organizations integrate with a vengeance. Designers do their work in cross-departmental teams including members from manufacturing and engineering, marketing and finance, and perhaps even production workers. Ford Motor Company credited just such cross-functional teaming for creation of the highly successful Taurus design (Mishne, 1988), while a similar practice has been credited as the source of rapid-design advantages in Japanese automobile firms (Clark and Fujimoto, 1991; Stalk and Hout, 1990; Womack et al., 1990) and Korean consumer electronics firms (Magaziner and Patkin, 1989), as well as American high-technology firms (Jelinek and Schoonhoven, 1990a; Kearns and Nadler, 1992).

Yet such activities are difficult to accomplish, even for technically sophisticated firms. Three cases which follow examine how well-known firms, rich in technology, struggled to implement effective, technologically based responses to competitive challenge. IBM is currently restructuring its businesses, reducing unit size, and pressing for improved response through greater autonomy and responsibility; the jury is still out on whether or not the effort will succeed. Xerox has apparently succeeded in "reinventing" itself after a massive struggle to counter the organizational ills that made it vulnerable to more nimble Japanese competitors. Ford developed a completely new car, using a radically different management approach, beating both Japanese and domestic competitors with its popular Taurus. Each case has lessons for us. Let us now examine these examples.

IBM, 1992: Restructuring again

International Business Machines, a bellwether firm long lionized as a technological leader and surely equipped with abundant resources—human, financial, software, hardware, and otherwise—offers a cautionary tale. IBM had successfully weathered numerous major shifts in its business and technology, from mechanical machines to electrical devices. IBM typewriters and collators ran many offices and dominated business equipment for decades before its computers did. Twenty years ago, IBM stood unchallenged as the preeminent firm. "In all the world one corporation dominates the shape of the future," began *Think,* the William Rogers book on the company and its founding family (1974). More recently, IBM also epitomized the technological paradoxes of faltering performance despite mighty technological resources.

By 1992, the firm's shortcomings in delayed or failed innovation efforts, lost market opportunities, and service distress were the subject of a spate of articles detailing the substance behind Chairman John Akers's assertion that his business was in crisis. IBM remained big—but rivals outperformed it. Specialization, size, centralization, and concomitant difficulties with coordinating a business grown vast and complex on its past successes were all part of IBM's troubles. IBM had grown distant from customers and market realities as well. Shortcomings with one customer alone, Wal-Mart, cost IBM service contracts at 400 of the 1900 Wal-Mart stores nationwide and $2.3 million in revenue (Lohr, 1992). "Big and bloated," IBM was characterized by a "culture of complacency" that produced a "follow the leader" stance in evolving computer businesses (Kirkpatrick, 1992).

IBM has frequently restructured in the past to respond to change (Jeffery, 1986), often with great success. But improving technology has so whittled the elements of the mainframe business where IBM was strong, while generating so much competition in markets with different dynamics, that structural change alone will be insufficient to remedy IBM's ills. The challenge is pervasive. A look at the general trends in computers tells why. In the 1960s, mainframe computers dominated computing, and "desktop systems" meant adding or calculating machines. Today, much smaller computer systems deliver the power that formerly required whole rooms of space, millions of dollars, and corporate connections. Desktop computer systems and peripherals to serve them are smaller-ticket items, vastly different in sales dynamics from the mainframes where many of IBM's current managers earned their spurs. Many of "the ways we do things around here," which were eminently suitable for mainframe sales and service, especially in an era of little competitive challenge, spell disaster in markets dominated by competition and the need for compatibility with many vendors' equipment, software, and components and increasingly reliant on smaller, more powerful personal computers and workstations that may be networked.

In five key new markets that successively undercut the mainframe business to redefine the industry, IBM trailed more innovative, nimble competitors who profited by Big Blue's ponderous response. Digital's highly popular PDP-8 minicomputer beat IBM's first mini by 11 years—allowing DEC plenty of time to establish an enduring niche, while substantially eroding IBM's market. In personal computers, the Apple II enjoyed 4 years of market penetration before IBM's initial anemic PC response. Apollo's DN100 engineering workstation enjoyed 5 years of nonresponse from IBM, as did Toshiba's T-100 laptop computer. Sun Microsystems launched the Sun 4 RISC workstation 3 years before IBM could reply. Today, minicomputers exceed the capabilities of the mainframes of a few years ago, networked personal computers substitute for mainframes in many instances, and desktop-sized workstations are nearly as powerful as mainframes. In each instance, competitors defined the market away from IBM's strength, carving out their segments of IBM's market unopposed.

IBM's current "restructuring" does address more than structure per se. Thirteen new entities have been designated, some former businesses, some former functions, others now wholly owned subsidiaries comprising manufacturing and development of hardware and software as well as marketing and service. Where formerly "The whole IBM business model has been based on centralization—in mainframe computers and in corporate organization" (Kirkpatrick, 1992, p. 45), the 13 new entities stress autonomy. Rules, procedures, and constraints—for everything from budgeted sales, pricing, and expense approvals to programming rules and dress codes—are being jettisoned in an effort to speed innovation and bring responsive, informed decision making nearer the scene of action. Business managers are being assessed and rewarded on their line of business's performance.

And some public, symbolic actions, like removing nonperforming executives, have shaken the "business as usual" attitude (Kirkpatrick,
1992; Lohr, 1992).

But explicit attention to "remaking the culture of complacency" gets
no more than a D grade from *Fortune* so far. The company is said to
remain "extraordinarily mainframe-centered," a serious flaw in a
marketplace moving inexorably toward distributed computing and
desktop systems. There are signs of change, including a host of new
ventures quite at odds with IBM's past culture—risk-sharing partnerships with former competitors and suppliers, plus others. These
efforts have brought IBM into new markets, exposed it to the expertise of nimble partners, and reduced the risk of such efforts while
potentially opening the company's thinking to the new partners' radically different visions of the world. While morale is said to be suffer-
ing in the midst of the changes, decades of IBM commitment to its
employees has created a potent sense of mutual obligation (Mills,
1988). The potential for change, like the technological muscle, is
apparent at IBM, but the jury remains out on the final results, with
1991 chalking up the company's first-ever net loss.

Reinventing Xerox

Xerox, founded on the strength of a superb technological achievement, fell into a pattern of traditional response, the victim of its own
successes (Kearns and Nadler, 1992). Unopposed and dominant for
decades, by the 1980s Xerox was slipping and badly out of touch. The
company was not even close on its estimates of competitors' costs or
capabilities, customer needs, prevailing quality, or price standards in
its own industry. The company's vaunted technical strengths simply
failed to materialize into comparable marketplace results. The unusually frank account of Kearns and Nadler (1992) details both the ease
of losing touch with customers, markets, and competitors' capabilities
and the difficulty of "reinventing" a firm to change the long-calcified
habits of traditional organization.

Xerox's difficulties highlight the dilemma of good, formerly successful firms organized on traditional lines. The functional and departmental distinctions that once made sense and helped people to comprehend the business can easily become barriers to successful operation. Necessary information, even crucial information, simply fails to
get through these barriers because the hierarchical structure implicitly assumes and asserts that "higher ups know best." To his credit,
Kearns reported his own chagrin when a factory floor worker challenged him: "Why didn't you ask us? We knew that product would

never work." While Kearns vowed that such input would never again be cut off, it's an uphill battle in a scale-based firm. So much in the structure, information flow, communication patterns and assumptions, authority, culture, incentives, and mores of traditional organizations encourages just the shortfalls Kearns experienced.

Kearns and Nadler (1992) identify four necessary elements for regaining competitiveness and reversing decline. First, a basic requirement for competitiveness is viable competitive strategy. Clear targets, specific objectives, and sustained efforts to attain the objectives are all important. However, the process by which strategy is developed is critical: It must be broadly inclusive and pervasive throughout the organization. From our perspectives, this is excellent advice because without a genuine commitment on the part of senior management to widespread strategic thinking, the necessary pervasive understanding, ownership, and enthusiasm will not emerge. "Qualitatively different and better strategic thinking and execution" begin with obvious commitment to a shared vision—public, so that all within the firm know of it; mutually designed, to draw upon the essential critical information that will support implementation at every stage; and participative, so that all can help test it, hone it, and understand their responsibility for its execution.

Second, quality is a critical success factor. Quality standards, ranging from "compliance to specifications" in products to genuine interest in customer service, cannot be achieved without wholehearted commitment on the part of all organization members to that goal. Nor can it be achieved without real tools for local analysis and improvement of operating efforts. In short, quality cannot be imposed from above. It must be generated on a grass-roots level throughout the organization. But equally, especially in a traditional organization, quality will not happen without genuine, persistent urgency and conviction from above.

Third, organization design is a critical success factor. Competitive success requires innovation and speed to sustain effective and dynamic linkage to the marketplace. "New architectures of work and organization," say Kearns and Nadler (1992, p. 305), must be evolved to encourage boldness and the "ability to think outside the boxes and lines that define organizations today." A key element of the required organization design, we believe, is that it will explicitly support both highly effective integration and reconfiguration to achieve new sorts of integration in support of innovative products and services. Neither one alone is enough to ensure ongoing success in a dynamic environment.

Fourth, organizational learning is essential. Developing and utilizing a capability to reflect upon unavoidable mistakes and to correct them quickly enables a firm to turn experience into insight. As Kearns and Nadler (1992) note, "productive failure" can provide a key ingredient to later success if people learn from the failure. But virtually everything in traditional organizations stymies this requirement because of the pervasive emphasis on control and conformity. Hiding problems or ignoring them seems far safer for the individual—but it is disastrous for the firm. Finding effective solutions, identifying emerging problems, and thinking about the longer-term effectiveness of the organization all require empowering lower management, according to Kearns and Nadler (1992). We would add that empowering all organization members is ultimately required.

Team Taurus

One well-publicized illustration of new product development helps to identify some of the organizational capability necessary to permit technology to be used effectively. A brief description of the Team Taurus approach will show how Ford benefited from an approach very different from traditional practices. Taurus was a car that began with a "clean sheet of paper" (Mishne, 1988). Beneath its aerodynamic exterior, the Taurus was a totally new car, with more than 4000 newly designed components and all new major systems (transmission, engine, and suspension). The new design required much greater precision and higher quality levels in manufacture than prior designs.

Team Taurus was assigned the task of designing both a creative, "new departure" car product and the production process necessary for its manufacture to exacting quality standards. Ford had already invested several years and much effort in improving quality and had discovered that close linkage between product design demands and manufacturing capabilities was essential. Thus from early stages of the design effort, manufacturing engineers were involved, and development was a team effort—a sharp departure from traditional norms. As a result, Ford's manufacturing engineers had precious months' advance time in which to develop key skills, such as those needed to ensure the much closer fit between the door and car body necessitated by the Taurus's advanced "soft-shouldered" profile. Equally important, the design was shared with the factory workers who were to produce it. Engineering drawings were posted on the factory wall for workers' comments and suggestions. They recommended several changes that were incorporated to improve the car's ultimate manufacturability and thus its quality. Taurus included a host of suggestions—gleaned from benchmarking competing products and luxury vehicles far beyond Taurus's intended market, as well as from customer surveys.

The design and the approach were both such departures that they were seen as risky. "Ford bet the family jewels on Taurus," according to the team leader, who added that he wasn't sure the company would have risked such change if its circumstances had been less dire. The bet was fully covered: Senior Ford management repeatedly asked the design team, "Are you sure you've reached far enough?" They also wanted to know how the team planned to make the new car world class in quality and customer satisfaction.

Acknowledged from the outset as critical to Ford's survival, the Taurus project generated widespread ownership, commitment, and acceptance. The project was effective—Taurus came to market more swiftly and with better quality than previous new products, despite its unique design. Taurus utilized technology effectively, not just the product or the production process but into the design effort as well. In order to do so, Team Taurus broke a number of older traditional norms. Key features visible in the Team Taurus approach included the following aspects:

1. *Explicit integration across functions and departments.* Early involvement of engineers from manufacturing as well as designers and marketers broke down "chimneys" within which functional specialists in the auto industry often spend their entire careers. Worker input was another new departure. Willingness to iterate the design process in order to get it right in terms of all functional needs as well as design terms is a further indicator of real integration. Computers facilitated design sharing and iteration, but organization and management cemented integration.

2. *Strategic overview to guide team efforts.* "Everybody" knew that Taurus was intended to sharply differentiate Ford from its competitors and to do so not merely by external looks but by the quality of the design and its execution. Both coordination efforts and a host of local decisions for implementation were harmonized by this strategic perspective. Top management emphasized the importance of quality in the design and its strategic role, reinforcing the overview for organization members. Top management also validated the risk of excellence, insisting that the quality differential was strategic.

3. *A culture of innovation was created.* Organizational functions and departments were prepared to interact differently by a long-standing effort at quality improvement that validated adaptation. The best ideas were sought out, not the familiar. Management support for Team Taurus's autonomy and repeated encouragement to produce a really excellent, really creative design were embodi-

ments of this culture as well. "Innovation" included quality in execution and envisioned future improvement.

4. *Marketplace links were explicitly forged.* Rather than turning inward, Ford undertook a major benchmarking activity, exhaustively examining "the best of the best" in automobiles to learn what "best" meant to the marketplace. Customer surveys and serious attention to "niceties" from the consumer's perspective were also included. Team Taurus exhibited vastly more sensitivity to environmental cues—on quality, size, style, conveniences, and other factors—than other U.S. automobiles.

5. *New roles for leadership are visible.* Senior Ford management cheered the Team Taurus effort and held it to high standards of excellence, repeatedly urging genuine creativity, real breakthroughs in design rather than incremental tweaking. This encouragement to take risks and management's cheerleading role were important departures from traditional roles. So, too, was the supervisory role of the team leader: As team captain, rather than "boss," his role differed sharply from the traditional project head function in scale-based organizations. Boundary spanning, interface with senior management, and creative flair—all capabilities long advised for team leaders but too rarely found—are clearly visible.

Each of these features distinguishes Ford's effort from traditional organizational practices. Together, they describe a radically untraditional approach. "Simultaneous engineering," cross-functional teams, and iterative reconsideration of designs typify a fluid, interactive mode of organizing that is very different from the rigid departmental boundaries of bureaucratic firms. Shared common goals, acceptance of change, and close contact with customers are almost diametrically opposite the identifying properties of the scale-based organization. Traditional firms' complexity and their unchanging procedures encourage an inward focus that risks becoming increasingly decoupled from customers, vendors, and economic trends over time. They operate in a protected and imaginary world, on their own time base and without external reference. Firms may even pride themselves on never having visited a competing factory. After all, particularly if the firm has been highly successful in the past, it is easy to believe that rivals have nothing to teach.

In these circumstances, technology implementation suffers. People in sales, finance, distribution, or other areas leave technical matters to engineers and thus have a difficult time understanding the needs, possibilities, and limitations in product development, new process capabilities, or technology use within the firm. But equally, design engineers out of touch with other areas fail to take into account the problems of manufacturing or the needs of customers. Advance engineering sees its job as done when it shows a product to be technically feasible. In these conditions, products and technology often fall short of their promise.

The Real-Time Organization

The traditional scale-based organization is unable to meet either the conditions of the contemporary global competition or the requirements to more successfully employ technology. Is there an alternative that is not limited in the same ways? The success of firms like Motorola, Intel, Federal Express, Square D, Nucor, Digital, Delta Airlines, and others suggest that the answer is Yes. Are these firms always successful? No. Digital's first entry into the PC market was "the right product at the wrong time." Federal Express has found it more difficult and costly to enter the European market than anticipated. Delta Airlines has been buffeted by the storm of confusing price changes and economic adversity that has swept the airline industry. Nobody bats 1.000. But perfection is less the point than differential success, better capability in the face of contemporary challenges than rivals, and repeated realization of effective new products, improved methods, and better service. Moreover, these firms are not just performing better than their rivals; they are different.

The above stories of IBM, Xerox, and Team Taurus suggest some highly interesting approaches to doing business that overturn the assumptions built into traditional scale-based organizations and their hierarchical structures, their control-oriented communication and incentives, and their limited participation operating modes. The difficulties encountered by Ford, Xerox, and IBM, and their diagnoses of the problems, underline the systemic and pervasive roots of the technological paradoxes we have been examining. They also point toward the outlines of an effective response.

Successful firms offer partial descriptions of successful real-time response, elements that can be put together into an ideal type, which we call the real-time organization (RTO). Because it is organized to respond "here and now" rather than buffering change in hopes that it will go away, the RTO's defining characteristics are oriented toward providing added capability to respond.

An organization so oriented is deceptively similar to traditional firms in many ways. Because its visible structure and indeed many of its elements seem similar, real differences are often unnoticed and unrecognized. There is no short list of new elements or practices that differentiate the RTO, which makes it difficult to communicate just how different RTOs are. Even relatively new concepts or practices like JIT or autonomous work groups, while common in RTOs, are not defining elements. Instead, RTOs differ from traditional firms in fundamental paradigm, philosophy, and operating mode.

Partial descriptions of the RTO have been appearing for years (Ouchi, 1981; Pascal and Athos, 1981; Peters and Waterman, 1982; Abernathy and Utterback, 1978; Jelinek, 1979; Jelinek and Schoonhoven, 1990*a*; Stalk and Hout, 1990; Kanter, 1983), and even announcements of "new" organizations are not unique. All too frequently what has been described as "new" has been a single, particular practice, like JIT, autonomous work groups, quality improvement, or a style of managing. These are important topics, but all too often they were only part of what is going on.

Describing the RTO is difficult because the usual ways of describing organizations are inadequate. Practicing managers, consultants, and scholars have been struggling with inadequate ways of describing organizations for decades. Initially what was described was the formal structure of an organization: its authority structure, formal procedures, throughput sequences, and rules. This seemed especially important for rules-based, bureaucratic organizations. Yet this perspective was long ago recognized as inadequate, so organizations' social structure, their informal organization, and culture have also attracted attention, often at the cost of ignoring formal structure and throughput processes. Decision-making perspectives offered some capability of integrating formal and informal organization and culture but focus largely on individual decisions, providing little help on how the many decisions taken by organization members might be integrated. Regrettably, none of these partial approaches takes into account the goals of the organization and the plans it has for achieving them, although abundant evidence suggests that meaning, goals, and purposes are central to human interaction.

Many basic elements well beyond structure or authority or communication patterns—the typical elements of organization analysis—are included in RTOs. Few of the insights they utilize are new. Indeed, virtually all of their characteristics are to be found—perhaps in isolation, certainly in repeated admonitions from academics over the decades, and increasingly in accounts of successful turnaround efforts or successful firms. What is different is RTOs explicitly attend to all of these elements essentially as interactive tools for achieving their managerial mode.

To transcend these limits, we have found it necessary to use six basic perspectives or dimensions to describe the RTO, moving toward an interactive perspective. This has us using as descriptors features and properties not typically included as part of an organization. There may be more dimensions; we feel quite sure that we cannot adequately describe and understand the RTO with fewer. In the description that follows, our intention is to relate the dimensions interactively, to highlight their mutual impact upon one another in producing the RTO.

1. *Member commitment.* The behavior of members of the RTOs differs from that which is typical in traditional firms. It is not just that members work hard but what they work hard at and how they make sense of what they are working on that marks the difference. Tracy Kidder's (1981) *The Soul of a New Machine* described how work can be pervasive and time consuming, not because people are driven to it but because they have become absorbed in it, because it has become very important to them personally. RTO members see their individual jobs as part of a larger effort directed toward a specific organizationwide goal. For them work is not just intrinsically satisfying, although it usually is. They find it rewarding because their work contributes to a larger effort of which they are proud to be a part. The firm and what it is doing are important for them. Thus, a special shipment, the solution to a particular problem, an effort to get a particular product into the market to hold off a competitor are all seen, not as ends in themselves, but as part of broad organizational intentions.

RTO members usually work with a minimum of direct supervision. In fact they have considerable autonomy, often deciding for themselves what to do, as well as when and how to do it. Their autonomy is used in working to meet goals and deadlines established with a superior. When they encounter a problem, RTO members tend to go directly to a source of help, wherever that might be in the organization, rather than referring the problem to a superior. When asked for help by others, they will frequently decide on their own whether and how to respond, consulting with their superior only when some change in schedule, major resource allocation, or other prior commitment becomes necessary. "After the fact," they are quite likely to brief superiors or others, ensuring that interested or affected parties are kept informed.

Members of the RTO frequently spot problems or opportunities for improvement and initiate action themselves to take care of the problems or exploit the opportunities. They see themselves as responsible for problem solution and for improvement as well, autonomously committed to both "routine operations" and to problem finding and solution, all in service of the larger organization goals.

2. *Shared strategic vision.* Implicit in this description is member knowledge of the larger purpose their work has. Organization members know and understand the strategy of the firm and work hard to fit their efforts to it. While higher levels of management may know more details of the strategy or may have a broader purview of it, RTO members know the basic goals and design of the strategy: who customers, competitors, and suppliers are; how each is to be addressed; the characteristics of the key products or services; and how all these components contribute to the firm's functioning. They also know how those factors relate to their jobs, their daily activities, and the appropriate criteria for assessing their activities against firm strategy.

Not only do organization members share an understanding of the overall strategy of the firm, but they usually also have an understanding of the strategy of their plant, department, or unit and how it fits into the strategy of the firm. In short they have an interlocking set of visions of how what they are individually doing fits into the work of their unit and how what the unit is doing fits into the work of the organization of the firm.

A key part of any strategy is the customer and what the firm intends to do for the customer. Customers, when RTO members discuss them, are not abstractions but persons with specific needs. Examples of such RTO characteristics are legion in firms renowned for their clear understanding of their customers. In Hewlett-Packard's early days, "the customer" was envisioned as an engineer who needed some state-of-the-art capability to handle a technical problem. To ensure successful sales, HP products were expected to meet the "next bench" test. An HP engineer with an idea had to be able to convince "the engineer on the next bench" to make room on that crowded surface for the proposed product. Today, HP makes a variety of computers, peripherals, and instruments, still servicing technical problems—but often for customers who are far less technically self-sufficient. Keen awareness of just who the customer is has guided HP's evolution and its contemporary successes.

For years, J.C. Penney's customer was a housewife who would spend the family's hard-earned money only for good, reliable products at fair prices (Penney, 1948). Today, with the overwhelming majority of "housewives" themselves employed, the customer definition must emphasize ease of shopping and convenience more than formerly, while competing with Wal-Mart and K-Mart on price. Neiman-Marcus aimed to serve the newly affluent, upwardly mobile person who wanted to buy an attractive, upscale item but also be assured that their purchase was tasteful and of fine quality, "a BUY for the customer" as much as "a SALE for Neiman-Marcus" (Marcus, 1974). "Being close to the customer" certainly isn't new, but successfully structuring an organization, its systems, and operations to accomplish this task is sufficiently rare to be the source of great rewards (e.g., Marcus, 1974; Penney, 1948; Peters and Waterman, 1982; Trimble, 1990).

3. *Shared, differentiated management.* Hierarchy is far less obvious in RTOs than in traditional firms, and the basis for status distinction is also different. In a given situation, it may be difficult to tell who is the boss and who is the subordinate: A junior person whose technical expertise is relevant may simply take charge of a meeting; a subordinate may instruct a superior, even to the extent of informing the superior that lessons will be held (Jelinek and Schoonhoven, 1990a). Nominal nonmanagers make many decisions reserved as "managerial" in traditional firms. Shop floor workers will seek out help from other departments without authorization by their superior. Superiors attend meetings called by subordinates. Subordinates revise schedules and tell their superiors later.

Such behaviors make more sense if we first take into account two characteristics of RTOs: Management is a responsibility shared by all members of the organization, while relationships between superior and subordinate positions and organizational units are less directive and authoritarian than in scale-based organizations. Instead, the entire enterprise is more participative, collaborative, and cooperative. There are two underlying premises: All organization members are responsible for and committed to seeing that the organization is managed well, not just the nominal managers, and management is a collaborative or team effort which basically involves getting input and commitment from everyone involved with an objective, problem, or project.

Successful RTOs generate widespread responsibility for managing the operations of the company. As "management" is not just the work of managers, but of organization members at all levels, so too, strategic management is not just the concern of senior managers. People at many different levels have important roles to play (AMA, 1991; Jelinek and Litterer, 1992). However, this does not mean that all have equal concern or responsibility. Differences both in the area of managerial responsibility and level or magnitude of responsibility are recognized. It is clearly understood that selecting strategy is the primary work of senior managers, but it is equally clear that people at many different levels have important roles to play in strategy formulation and implementation (AMA, 1991; Jelinek and Litterer, 1992) and that many essential nitty-gritty details depended on "local knowledge" (Geertz, 1973; Baba, 1990) that only "nonmanagers" lower down in the organization can supply. That such knowledge

exists has long been known. What is different here is that this is recognized and its use deliberately fostered to contribute explicitly to
real-time response capability (Jelinek and Litterer, 1992), thereby
building the organization around that capability. Team Taurus again
offers an example: Those who would be affected by the change, or
who had essential information, were involved in the project from the
outset. Xerox's new product design efforts were similarly supported
by input from a broad array of people who brought their insights to
bear on customer information, surveys, intelligence about the market
place, and benchmarking.

4. *Structure.*   Ask a member of a traditional firm to describe its
organization, and they will almost surely draw a typical pyramidal
authority structure. Ask how the structure operates, and members
will specify that instructions and feedback on performance come from
higher authority, while questions and requests go up. Similar questions in a real-time firm get a different response. Many members will
describe their organization as a chain beginning with the initiation of
a new product or service, going through manufacturing, sales, and
distribution, and ending with the customer. Or the chain will begin
with the customer and flow through a series of departments and functions, looping back again to end with the customer. In any case, what
is described is an output system, with directions and feedback about
its output coming from the customer.

To outsiders, firms with the RTO form can appear to have little
structure or a fuzzy one. Yet successful firms are not "ad hoc," even in
"high-velocity" environments like electronics (Jelinek and
Schoonhoven, 1990a; Bourgeois and Eisenhardt, 1988). What structure do RTOs have, then? One very different from what traditional
experience prepares us to see. Jobs are usually quite specific—but
expandable and changeable. Meanwhile, hierarchical relations are
loose. Problems and project assignments to which a person or unit
will direct attention are firm but can be quickly revised. Direction on
what to do and when to do it comes more typically from users or the
situation—customers, other departments or positions further down
the work flow who are treated as "internal customers," or problems
that need to be solved.

What matters is less one's assigned "turf" than the organizational
goal and the outcomes of customer satisfaction, quality operation, and
problem solving. Authority derives in part from position, more from
expertise, and most from embracing responsibility on a project or a
problem. This is a highly pragmatic logic. A production worker may
directly call a design engineer to address difficulty in an assembly
thought to be a result of the design. A salesperson will contact a pro-

duction unit to describe how some aspect of manufacturing is not meeting the customer's needs, seeking information to transmit directly back to the customer—and perhaps to design engineers as well. While the situational specifics may vary, the underlying message is, "I have this responsibility for the firm, and to carry out that responsibility I need your help in this way." In effect, authority comes more from responsibility than through formal position or delegation. Structure cues people as to where the expertise might be found to solve problems, and it helps signal who is responsible for particular areas of activity. But above and beyond all this, all members are responsible for the success of the firm as a whole.

In many RTOs, when the term "structure" is mentioned, it carries a negative connotation. Some senior executives will strongly state their opposition to structure and describe how hard they work to minimize it. When speaking of structure, they usually mean structure as it is used in traditional, scale-based organizations—as arrangements to limit and restrict people and constrain their behavior. Nevertheless, there is considerable structure in RTOs, although it is very different from the traditional structure.

Allocation of resources is one basic structure. It involves what types of skills will be hired and more directly how a member's time will be used: how much for training, how much for which projects, how much is "free" to be used at the member's discretion. These allocations are usually reached through a dialogue between superior and subordinate, and they are frequently changed to meet new conditions. Norms also affect how "free time" will be used: Expectations prevalent in an organization's culture can communicate that people should be "110 percent busy" as in some high-technology firms (Jelinek and Schoonhoven, 1990a)—placing a very explicit decision responsibility on members for allocating a portion of their own time.

Specified relationships exist in RTOs, but except for the relationship with one's superior, these are usually specified to address a temporary situation and will change as the conditions or projects change. Organization members in RTOs have significant responsibility for initiating and maintaining such relationships and for ending them as well. Procedures exist in RTOs, but they are used less to tell people what they must do and more to tell them how to get done the things they want to do. For example, personnel procedures will tell people how to get money for a trip but will rarely include the names or titles of higher authorities whose approval is needed to get the money because such approval is often not required. Another example is the "Wild Hare" program at Texas Instruments, where anyone can apply for funds to explore an idea (Jelinek, 1979), and multiple channels exist for getting a Yes.

Communication structures, both formal and informal, distribute vast amounts of information in RTOs. This includes information on costs, technical developments, schedules, problems, competitors' products, new products being considered, and so on. RTOs expend much effort to keep as many people as possible informed about the present state of affairs facing the firm through reports, announcements, meetings, company newspapers, and internal mail systems, as well as by managers constantly moving throughout the organization, no less than the electronic means described above. In addition, RTOs share a widely held norm that individuals will pass on information to those who may be affected by it.

Whether the real-time loop or the customer chain is described, the difference from traditional perspectives of structure is striking, especially in how organization members see their roles and how they see the organization fitting together. In the RTO, work is not seen in isolation, and the department is not the limit of the member's purview. Nor is work intended to satisfy a superior. Rather, tasks are part of a larger effort that is not finished until the right output is delivered. For example, engineers in these firms have a mental framework in which their work is seen in part as a response to customers which must go though several other functions before its success can be judged. Throughout the organization, managers and other members share a common view of the customer and the marketplace as ultimate judges of products, services, and accomplishment (Jelinek and Schoonhoven, 1990a).

5. *Leadership.* One thing characteristic of RTOs is the familiarity organization members seem to feel with key executives, especially the CEO. They speak as if they personally know him or her, although few have ever had close face-to-face conversations with the CEO. Yet the CEOs and other senior executives are known, important figures in the lives of most organization members. They are the ones who articulate what the firm is about, where it is going, how it is doing. They are often the embodiment of the values and ideals that are held to be important. In short, they fill a charismatic role, pointing to overall company goals and directions and at the same time letting each member feel that he or she is personally important (Bass, 1985; Conger, 1989; Kotter, 1988; Kouzes and Posner, 1987; Walton, 1972, 1977).

The senior executives also put considerable time into building the organization and assessing its health. They lead, not by directing but by pointing to goals and challenges, not by controlling but by getting commitment, not by acquiring or preserving power but by sharing it, not by delimiting what people can do but by facilitating what they do.

This transformation style of leadership, then, becomes the style throughout the organization. Managers throughout the organization lead by example; by setting goals and challenges, they support and facilitate others as they collaborate on getting work done.

6. *Culture.* RTOs have strong, dense cultures oriented toward effective response, not to perpetuating the past (Kotter and Heskitt, 1992; Jelinek and Schoonhoven, 1990a, 1990b; Wheelwright and Clark, 1992). Clear values, explicit beliefs and articulated codes of behavior, and a broadly shared vision of purpose link with understandings of how things work and how things came to be as they are. Such cultures include conviction that the firm and its work are important, even noble. Statements of intent and mission help to focus and reiterate core values. Some examples: "We help engineers by providing the most advanced measuring instruments available." "We provide absolutely reliable overnight delivery." "Always the low price. Always." What seems different in RTOs is that people take such statements seriously, build them into a shared internal language, and operate the firm to accomplish them.

RTOs commonly share a belief that "people count." Many firms say this, of course, but in RTOs this value finds substance in a multitude of organizational arrangements. RTOs behave as if people really matter, in observable ways. People are trusted. Initiative is expected from them, and their ideas are sought out, listened to, and often acted on. Potential new hires are carefully screened not only or even mostly for their background or training but for their potential fit with the organization's culture and members. Educational programs for continued training and development, both inside the firm and in more academic settings, are widely available. Management conveys its respect for people by keeping them "in the loop," acknowledging their importance by treating them as valuable resources.

Another widely shared value in RTOs that illustrates this situation is the expectation that every member will know about the firm's technology. In successful high-technology organizations, Jelinek and Schoonhoven (1990a) found that even nominally nontechnical personnel far from the production floor—secretaries, for instance—were conversant with the major technology tasks, challenges, and current difficulties and with many of the nuances as well. Senior executives, even those far from direct technical responsibilities on "the bench," were nevertheless deeply sophisticated about the firm's technology, its possibilities and potential linkages to customer needs. Such deep familiarity with the firm's basic technology is clearly essential for senior executives who fill important information-gathering responsibilities. Without effective knowledge, they would be unable to recognize important information when they encountered it or to appreciate its implications for the firm. Yet such sophistication is reasonably rare in many *Fortune* 500 firms, where senior executives may rise through their careers without any technology experience (Jelinek and Schoonhoven, 1990b; Kearns and Nadler, 1992).

In RTOs, design engineers are expected to be conversant with production technology, while those in production are expected to be informed about the technology and its broad capabilities, not merely focused on current practices. Much of this cross-functional familiarity arises from frequent interaction across functional and departmental boundaries—which are frequently changed anyway—expected in the normal course of both operational problem solving and new product design. As a result, RTO members share a common knowledge about product and process technology or about the service and how it is delivered. The depth and scope of knowledge vary, but enough sharing exists to support and benefit from broad dialogue which serves a common goal of improvement.

Some key facets of these norms are of special interest. While the existence and value of local knowledge have long been recognized (Geertz, 1973; Baba, 1990), what is unique is that RTOs have moved to build their organizations around such local knowledge, using it to support a real-time response capability (Jelinek and Litterer, 1992). Organization members at all levels and in all areas are consulted, and they are expected to contribute their insights. Moreover, their contributions have both weight and consequence: The firm runs on such inputs, so individuals have genuine responsibility. People know that they matter.

One thing that follows from this is the belief and expectation that individuals will initiate action when they see it is necessary rather than merely reporting problems to others or waiting for direction. Then, when they take action, members are expected to get the help and support they need, to form teams to address the issue, to design and carry out solutions. For example, in one firm a decision was made at the senior level to move the production of a product from the plant where the product had been developed to a plant 2000 miles away which had excess capacity. Handling the move was assigned to the two plant managers rather than to a member of higher management or some staff unit. The two managers pulled together a team of other plant managers who had been through previous moves: the managers in charge of the product, members from the logistics department, personnel, and engineering. The team saw itself as responsible for successfully carrying out the move rather than being advisory to the two plant managers.

At other times changes can start at the bottom. In one instance an organization development specialist recognized that she was dealing with a large number of problems of conflict and stress coming from problems of inventory. She found, for example, that the firm had a 4- to 5-year stock of some parts for products that had life cycles of only 2 to 3 years. She pulled together various product managers, production schedulers, and plant managers and observed that they could keep on fighting the same problems or they could correct the underlying problem. From this came a task force which analyzed the problem and presented a plan of action to higher management for financing. Only then did higher management get involved, which they did by authorizing a substantial budget for the group to continue its work.

Finally, RTO cultures share a universal conviction that "there has got to be a better way," a better product, a better design (Jelinek and Schoonhoven, 1990a; Peters and Waterman, 1982; AMA, 1991). Change and improvement are not only spoken of with approval but are expected and desired. Indeed, the absence of change is often cause for concern—"too much stability" means that improvement has slowed. Commitment to change is frequently prescribed as appropriate to the continual incremental improvement of production processes, or Total Quality programs (Crosby, 1979; Schonberger, 1982, 1986; Walton, 1986). What is different is that successful organizations are pervaded with this commitment to constant improvement.

Comparing the RTO with the traditional scale-based organization on the same six dimensions sharply distinguishes the two (see Fig. 5.5). Since these six dimensions address both the competitive environment's demands and the necessities for effective technology use, Figure 5.5 highlights the RTO's strengths for both tasks.

The RTO as a system

We have identified six core dimensions which need to be part of any adequate discussion of the RTO: member commitment, shared strategic vision, leadership, structure, shared differentiated management, and culture. None is new, and abundant research and experience validate them. What is not common is that to describe an RTO all six must be taken into account.

The second defining characteristic of the RTO is that it is a system and these six dimensions are interactive. Looking at one element or even a few of the elements in isolation will miss what is different. To look only at structure, autonomous groups, empowerment, or quality improvement as a separate, independent item is to miss the fact that to succeed, each element will have to be a functioning part of the organization that contains it, and that the actions of the element will influence the way other elements function. A look backward, to an early effort to implement some of these elements, where the interac-

	Traditional organization	RTO
Strategy	Guarded closely, known only to a few	Widely shared, disseminated to all organization members
Structure	Hierarchical, authority-based, rigid and specialized	Output-based, systemic, flexible, shared responsibility
Leadership	Specialized, segmented, hierarchical, functional, compliance-oriented	Core-technology or task-based, transformational, charismatic, seeks commitment
Culture	Thin, fragmented	Thick, homogeneous
Member commitment	To task, boss, discipline, specialty	To firm, customer, output, problem at hand
Management	Specified, exclusive to managers	Shared and differentiated

Figure 5.5 Characteristics of traditional and real-time organizations.

tive nature of these dimensions was ignored, will illustrate the need to employ a systems perspective.

Over 20 years ago, the Topeka Pet Foods plant of General Foods (Walton, 1972) received considerable attention as a demonstration of effective autonomous work groups. Work groups in the plant decided on work schedules and methods, interviewed and hired new team members, decided on wage increases, and handled performance assessment, training, rotation, and maintenance. Productivity and employee satisfaction under this regime increased. Six years after this experiment began, however, it was in disfavor and slated to be abolished. While there were problems with the way the teams were functioning, these could have been worked out.

Higher management did not support the way the plant was operating, while other plants were uncomfortable with the comparison and its implied criticism of traditional methods (Walton, 1977). Autonomous groups "failed," not because of any inherent flaw or because they were ineffective. Autonomous groups arguably succeeded in Topeka but failed within General Foods because they did not fit with the management system and culture of the rest of the firm.

Their failure is more accurately described as transplant failure—they were rejected as alien. And indeed so they were. Without a supportive culture and consistent management methods, incentives, and broad comprehension, rejection was predictable. So different a mode of operation, rooted in philosophy and operating assumptions so radically different from those of the traditional firm, is alien indeed to a scale-based, traditional, mass production organization and to its culture and assumptions.

Integrating the elements of real-time response

Success, then, rests upon how organizational elements fit together and on how well they integrate into a dynamic entity capable of effective response to competition and opportunity. If a rejection like that at Topeka can be predicted, however, so too can the necessary elements that seem to support such efforts. We need not only specify six dimensions of an organization, but we must ensure that the elements of these dimensions fit together and support each other. Let us examine some illustrations of how this occurs in the RTO.

Guiding autonomous decision making. In the RTO, members have considerable autonomy. Higher managers are willing to lay aside detailed direction, close supervision, narrowly defined jobs, and elaborate procedures to keep individuals from doing the wrong thing and to allow them to decide for themselves about how work will be done so that they may do the right thing. How can they risk this? It is partially a matter of trust. But even when people are trusted, there is the question of how one can be sure that large numbers of individuals, often widely scattered, will make decisions that will converge and support each other in achieving what the firm wants to accomplish.

In RTOs, much is in place to ensure that the individual member, acting autonomously, will make decisions that support the firm. It begins with the understanding members have that when a problem or a need for action arises, they share responsibility for management and action. Considerable information and assistance are available to analyze the issue and generate alternatives. But they also have knowledge of the operating structure of the firm, in particular the operating structure of which they are part. They know how things get done, who can help them, and whom they need to inform about what is going on because their work will be affected.

Above all, members have acquaintance with both the firm's strategy and their unit's strategy as part of a dynamic portrait to help them define appropriate objectives and guide the selection of alternatives. Members also have the values, codes of behavior, and mission of the organization and their own job within the organization's culture to help guide in selecting trade-offs and filling in blank spaces in the strategic vision. Pervasive pride in membership and commitment to the firm's larger values and long-term interests also help ensure members' appropriate actions. Altogether, a rich array of interactive insights will shape RTO members' decision making, and they act knowing that the elements of their array are shared by higher managers. It is not too surprising that, while the specific details may differ, the decisions individuals make typically converge and integrate to support overall intentions of the firm and thereby meet with higher-management approval.

Evolution of structure. Structure in RTOs is continually changing and evolving, although it may be local details which change the most while the large configurations remain stable. Cultural norms favor constant effort at improvement, to shorten throughput time and eliminate quality reducers or cost increasers. Members' managerial responsibility to take action requires continual examination of work flows so that members learn them and stay abreast of changes. Similarly, changes in organization strategy lead members to reexamine their own unit strategy, making changes if necessary. This in turn often requires a change in the web of work and information flows to accomplish strategy.

Organization members become intimately familiar with the positions and departments that feed them work and information or provide support for their work. They learn who is dependent upon them, and what and how things are done. In making decisions with others about how their system will work or can be improved, they learn what trade-offs are to be made, what work is to be optimized and which suboptimized, and why. They learn what variances develop in the system and what their effects are.

In short, organization members of necessity become very well informed of how work in their area of the firm gets done and what is involved with managing the work. They know who will make decisions that will influence their work and who will be influenced by actions and decisions they take.

The unique role of leadership. Leadership fills the role of caretaker for the other five dimensions. Senior leaders craft the vision of the future, which is the foundation for the organization's strategy— though often through broad participation that enlists both shared support and understanding. It is often their beliefs and values which become the beliefs and values of the organization—though with important input from others that holds them accountable for the consistency and integrity of those values. Leaders' struggles and experiences become the heroic myths that members recount to help identify who they are as a firm, how they are to behave, what they value. Yet those struggles and experiences become "heroic" in the value and validation that members accord them because they are seen to be consistently harmonious with the values of the firm. It is senior managers' belief in the value of individuals that creates widespread trust, a willingness to share management, and a willingness to risk allowing individuals to initiate action.

Leadership obviously involves more than just having these ideas, values, visions. They must be transmitted to others and accepted by them to attain currency in the culture. Senior managers in RTOs work hard to have this happen. For instance, CEOs and presidents work to make themselves known and to be available. They tour plants, offices, labs; attend meetings; and watch demonstrations, consciously lending importance to activities by their presence. They also use in-house television networks, often with two-way voice and video capabilities. Firms like GE-Fanuc, Texas Instruments, Hewlett-Packard, and Federal Express make heavy use of telecommunication devices such as satellite hook-ups, global computer networks, and data links.

Such efforts are more than just "walking around." Touring senior executives are also reviewing the condition of the firm, gaining intelligence of the market and what competitors are doing from first-line members, and sharing their plans, strategy, hopes, and fears—and hearing those of members. They review new technical developments, noteworthy successes by individuals or departments, arising problems or solutions, corporate culture, and more.

Top managers also listen, look, and acknowledge what they have learned. The incident mentioned earlier by Kearns who, while he was explaining that an important product was failing in the marketplace, was told by a factory worker, "We knew that, why didn't you ask us?" is a good illustration. The comment was not heard as a rebuff but as important information about where to go for information about products. In their visits, senior executives also carry information about what is happening elsewhere in the firm, often in the form of tales of battles fought and successes won. And all the time they hold out the vision of the firm, what it stands for, its goals, ideals, and values, while they model appropriate behavior.

Such accessibility and rapid response to issues or questions does more than merely provide a channel for rapid communications. Spreading the word about strategy and maintaining it at the forefront of attention is an important leadership role. Management's action itself serves as a continuous model of how everyone is expected to behave: focusing on the firm and its strategy, observing, thinking, reporting, helping, and responding rapidly.

The illustrations could be extended, but the point has been made. What is clear is that the active elements are several, that they must be reasonably consistent and mutually supportive, and that conflicting elements must be harmonized. It is also clear that the elements that make up real-time response capability go well beyond what is traditionally recognized as "organization structure." They go far beyond "management style" or authority or culture—although each of these is involved. Finally, the elements interact, mutually enabling, sustaining, and requiring one another. These six dimensions create a powerful synergy that gives the RTO its flexibility and power and make it the form of organization to fit contemporary conditions.

Real-time technology capabilities

Against the context of the contemporary competitive environment, RTO technology capabilities are of special interest. Technology development during the past several decades has moved toward increasing precision and control of complex processes, increasing speed, more integration, and increasing flexibility. All these outcomes, which are simultaneous and which interact, are arguably the result of a thoroughgoing penetration of computers into virtually every phase of business and organizational activity. A few examples will highlight the direction of the trends.

Increasing precision and control of production.   From tool point sensors to point-of-sale registers, data collection is rapid and generally highly accurate. Voluminous amounts of information are now available that were not to be had at any cost not long ago. This data, appropriately analyzed, has revealed much about manufacturing (including such old processes as wine making and metal cutting). With such information, it becomes technically feasible to control production processes precisely, to shift from one production "recipe" to another quickly and with great accuracy. As Figs. 5.3 and 5.4 suggest, integration and widely dispersed decision making in a sizable organization may demand information systems.

Current information.   Point-of-sale registers can generate highly accurate data about sales (who is buying what, where, and for how much). Analysis of this information can permit retailers to "tune" store offerings, purchasing, and inventory to point-of-sale information. Analysis of sale patterns can help project and predict needs for such perishable items as fresh baked goods, ensuring maximum sales and minimum waste. Information technology can make this information available for informed local response and thus support meaningful, effective autonomy.

Available status information. Information on the status of an account, the movement of a package, or the progress of a host of other activities can be made instantly available via computer networks. Such information availability can technically support high levels of customer service—informing a customer about whether or not a check has cleared or a deposit been credited, locating a package and estimating its time of arrival, ensuring the timely delivery of ordered goods.

Active delegation and decision support. Enabling many local decision makers to actively use the information to run the organization will vastly improve its response capability simply by making more effective attention available. However, for these technical capabilities to bear fruit, traditional organizational norms and practices must change. Traditional organizations, because they are set up to facilitate control, stability, and replication, are not particularly friendly to the sorts of changes we are describing. Because information is so central to effective organizational operation and because the application of computer systems has driven much of the technological change in recent decades, we shall use information technology as an exemplary technology to focus our further discussion.

The Third Technological Paradox: Technology and "People Power"

Contemporary information technology poses yet a third paradox: Its effective use requires a thoroughly humanized organization that is pervaded by technology. Throughout human history, technology has substituted for human effort, enhancing capability and extracting a price. Often, since industrial times, the benefits have flowed to the firm and the costs have been borne by individuals displaced by technological advances, their skills rendered obsolete. That, too, is visible today as, for instance, manual drafting skills are replaced by computer programs or advanced clerical and organizing tasks are performed far more quickly by computers.

Technological advances, unmatched by organizational skills, can merely "deskill" jobs (and ultimately, people), eliminating easily accessible job categories. Today many once-numerous high-paying entry-level factory jobs have either disappeared entirely or have been replaced by vastly different jobs with different skill requirements, training and experience paths, and wage structures. The jobs that remain can be mindless "machine tending" jobs—but that is not the only option.

The advent of computers has also created more highly skilled professional jobs, sales jobs, design, engineering, planning, and creative jobs as well. Both individual jobs and the working context as a whole can become more exciting and interesting, more demanding and satisfying, as a result of computers, information systems, networking, and decision support. Available computer technology provides the possibility for effective response to the demanding challenges of global markets, rapid technological evolution, and intensifying competition, but as we have argued, traditional organizational approaches do not support either effective response to competitive challenge or effective deployment of technology.

The core difficulty is that effective use of technology demands spreading information systems, computer support, and access throughout the organization and using that technology to truly empower organization members. Effective use of technology to meet competitive challenges demands an organization oriented in precisely that direction—you cannot exploit the technology without empowering the people. It also demands technical competence in many more organizational members, not merely specialists.

For effective use of technology, especially the computer and information systems technologies, we need to change organizational arrangements that limit and restrict information sharing and decision delegation. In their place we need organizational arrangements that aim to support widely effective decision responsibility with appropriate information support and genuine, meaningful decentralization. Organizations will need to take training seriously to ensure continued skills growth and development of members—a very different perspective from "labor as a cost to be reduced" so prevalent in scale-based organizations.

The more pervasive the technological penetration, the greater the potential for all organization members to be meaningfully involved in running the business—making its fundamental decisions, actively engaging in its problem-finding and -solving activities, interacting with its customers. "Workers" in such a context are far more responsible members of the firm than those of traditional organizations. The workers in RTOs are hired for all of their capabilities, not merely to follow orders or mindlessly repeat some pattern of action. Instead, they can autonomously ensure that the organization's goals are achieved, and provide the local knowledge and attention to sustain on-going innovation and dynamic fit with the marketplace and emerging trends (including technology trends). They can provide the vastly increased response capability needed to remain competitive in the demanding circumstances of today's global markets.

Such possibilities impose costs and risks as well, however. To begin with, as we have argued, organizations and their management must share much that has been considered "management prerogative" since Taylor's time: problem-finding and -solution generation, independent action and responsibility, and more. This shift constitutes a first cost. Organizations will be explicitly and observably more dependent on more actively engaged members. It is predictable that "employees" in such circumstances will feel themselves far more as owners than as hired hands—and will expect to be consulted, taken into account, and remunerated in light of those convictions. Fewer, more skilled, and better paid employees can be anticipated in manufacturing activities. More but also skilled employees may be needed elsewhere.

Virtually all organization members will be networked and involved with the information technology as well. Far beyond technical specialists, sales, support, and marketing personnel and other administrative activities will all require linking into the integrated activities of the firm. Figure 5.3 shows some possible linkages in an integrated manufacturing firm, but the possibilities are, if anything, far greater in service businesses. Insurance services offer an already linked model in the most effective companies. Managed medical care, financial and brokerage services, and shopping all suggest instances for similar cross-function, cross-product linkages (from the provider's point of view) that offer the customer "one-stop shopping" that does integrate currently segregated, yet related needs.

Such distributed power entails genuine risk for organizational disaster if organization members are not effectively trained, technologically literate, and kept abreast of relevant strategies. This is no place for the technically unaware, the illiterate, or the "cowboy individualist." Not only must organization members be technically trained; they must be effectively imbued with real commitment to the organization's mission and must have genuine understanding of how they fit into it. Their skills must encompass both action and understanding.

However powerful the information systems and technological capability of the equipment, its results are only so good as those who run it. Pervasive penetration of information systems and computer applications throughout the firm—a virtual certainty in most organizations—can produce its promised benefits only in the supportive embrace of a highly humanized, empowered organization.

Summary

Our examples show that firms which have innovated and introduced new products successfully have had both rich technological capabilities and an organization that enhanced their integration. Both were needed to bring new products promptly to market. Our attention in this chapter has been on the organizational capability through which firms achieve the potential of technological capabilities. Clearly, such capability goes well beyond what is typically described as a firm's "organization." As clearly, the effort produces an array of strengths closely matched to contemporary competitive conditions and to innovation and new product development.

A brief summary of these organizational capabilities highlights the benefits for innovation and new product development. Such firms:

- Encourage individuals and groups to think "outside the boxes and lines" that (temporarily) define organization structure. This makes it possible for new activities to be evaluated on their own, rather than constraining member thinking to "the way we have always done things."

- Facilitate learning, an obvious benefit for both innovation and new product development since any "new" activity will require some change in perception or activity.

- Ensure that people throughout the organization understand the firm's strategy. Any innovation should either be consistent with the firm's strategy, or, if sufficiently valuable despite its inconsistency, should initiate a reconsideration of the strategy. Both consistency and reconsideration as required will be assisted by widespread familiarity with strategic intent.

- Provide integration across functions, departments, and products. New products often require new processes (as Taurus and other examples show), calling into question routine paths for information transfer, cooperation, and the like. Joint effort is essential for success in cross-boundary endeavors; it appears crucial for speedy innovation.

- Foster a culture of innovation and quality. Contemporary conditions as well as predicted future requirements all trend toward constant innovation and improving quality.

- Explicitly hone marketplace links. In a changing environment, constantly monitored market linkages will provide the distant early warning that can initiate innovations, whether they are new products, new processes, new channels of distribution, or other novelties. Those competitors that lack awareness of trends and developments beyond their own efforts will surely be incapable of survival over the long term, for others will outpace them in implementing new technical options, developing new products, or otherwise responding to challenges and opportunities in the environment.

- Develop strong senior leadership that creates visions, builds commitment to strategic goals, and provides encouragement. Many of the disasters of failed innovation, poor product design, quality, or implementation can be attributed to failures of senior leadership vision, failures to build commitment, and discouragement of innovators. Strong leaders clearly create a climate hospitable to innovation and new product development, as leaders who fail at these tasks do not.

All these are features of the RTO. Leadership that creates vision stimulates enthusiasm, while shared understanding of the firm's strategy bolsters commitment to the firm and its goals. A structure that provides a systems orientation continually highlights links needed to accomplish tasks, while the cultural values of satisfying the customer sustain strong market links. The structure of the RTO, coupled with the shared management style that encourages everyone to connect with those who need to be informed or whose support is needed, assists integration across departmental, functional, or hierarchical boundaries. Better information flow and integration translate to better decisions, better exploitation of innovation and of technical possibilities. Cultural values of continual improvement, speed, and customer satisfaction produce continual innovation and help to ensure quality. With these properties, the RTO is clearly a sound platform for innovation and product development.

References

Abernathy, William J. *The Productivity Dilemma: Roadblock to Innovation in the Automobile Industry,* Baltimore: Johns Hopkins University Press, 1978.
———. and Utterback, James. "Patterns of Industrial Innovation in Industry," *Technology Review,* June–July 1978, 40–47.
AMA Management Briefing. *Blueprints for Service Quality: The Federal Express Approach,* New York: American Management Association, 1991.
Baba, Marietta. "The Local Knowledge Content of Technology-Based Firms: Rethinking Informal Organization." In L. R. Gomez-Mejia and M. W. Lawless (eds.), *Organizational Issues in High Technology Management* (vol. 1), Research Series on Managing in the High Technology Firm, Greenwich, CT: JAI Press, 1990.
Bass, B. M. *Leadership and Performance Beyond Expectations,* New York: Free Press, 1985.
Birnbaum-More, Philip H. "Competing with Cycle Time in the Worldwide DRAM

Market," unpublished paper presented at the *Second International Conference on Managing the High Technology Firm,* University of Colorado at Boulder, January 10–12, 1990.

Bourgeois, L. J., and Eisenhardt, K. M. "Strategic Decision Processes in High Velocity Environments: Four Cases in the Microcomputer Industry," *Management Science,* 34: 7, July 1988, 816–835.

Bower, J., and Hout, T., "Fast-Cycle Capability for Competitive Power," *Harvard Business Review,* 66: 6, 1989, 110–119.

Broderick, Renae, and Boudreau, John W. "Human Resource Management, Information Technology and the Competitive Edge," *The Academy of Management Executive,* VI: 2, May 1992, 7–17.

Business Week. "The Hollow Corporation," March 3, 1986, 56–60.

Business Week. "The Future of Silicon Valley," February 5, 1990*a*, 54–60.

Business Week. "Has High-Tech America Passed its High-Water Mark?" February 5, 1990*b*, 18.

Chandler, A. D. "The Enduring Logic of Industrial Success," *Harvard Business Review,* 68: 2, 1990, 130–140.

Child, John. "Managerial Strategies, New Technology, and the Labor Process." In Johannes M. Pennings and Arend Buitendam, (eds.), *New Technology as Organizational Innovation: The Development and Diffusion of Microelectronics,* Cambridge, MA: Ballinger, 1987.

Clark, Kim B., and Fujimoto, Takahiro. *Product Development Performance,* Boston, MA: Harvard Business School Press, 1991.

Cohen, Stephen S., and Zysman, John. *Manufacturing Matters,* New York: Basic Books, 1987.

Conger, Jay. *The Charismatic Leader,* San Francisco: Jossey-Bass, 1989.

Crosby, Philip B. *Quality is Free: The Art of Making Quality Certain,* New York: New American Library, 1979.

Feeney, Edward. "At Emery Air Freight: Positive Reinforcement Boosts Performance," *Organizational Dynamics,* Winter 1973, 41–50.

Foster, Richard. *Innovation: The Attacker's Advantage,* New York: Summit Books, 1986.

Galbraith, J. D. *Designing Complex Organizations,* Reading, MA: Addison-Wesley, 1973.

Geertz, Clifford. *The Interpretation of Cultures,* New York: Basic Books, 1973.

Goldhar, Joel D., and Jelinek, Mariann. "Prepare for Economies of Scope," *Harvard Business Review,* 61: 6, November–December 1983, 141–148.

Hayes, Robert H., and Abernathy, William J. "Managing Our Way to Economic Decline," *Harvard Business Review,* 58: 4, July–August 1980, 67–77.

Hayes, R. H., and Wheelwright, S. *Regaining Our Competitive Edge,* New York: Wiley, 1984.

Hofheinz, Roy, Jr., and Calder, Kent E. *The East Asia Edge,* New York: Basic Books, 1982.

Holden, Paul E., Pederson, Carlton A., and Germane, Gayton, E. *Top Management,* New York, McGraw-Hill, 1968.

Hower, Ralph M., and Orth, Charles D. *Managers and Scientists,* Boston: Graduate School of Business, Harvard University, 1963.

Janger, Allen R. *Corporate Organization Structures: Manufacturing,* New York: The Conference Board, 1973.

Jeffery, Brian. "IBM's Protean Ways," *Datamation,* 32: 1, January 1, 1986, 62–68.

Jelinek, Mariann. *Institutionalizing Innovation,* New York: Praeger, 1979.

———. "Organization Structure: The Basic Conformations." In Mariann Jelinek, Joseph A. Litterer, and Raymond E. Miles (eds.), *Organizations by Design,* Plano, TX: Business Publications, Inc., 1981, pp. 253–265.

———. "Production Innovation and Economies of Scope: Beyond the 'Technological Fix,'" *Engineering Costs and Production Economics,* 12, 1987, 315–326.

——— and Litterer, Joseph A. "New Formats for Strategic Advantage." In Luis R. Gomez-Mejia and Michael W. Lawless, (eds.), *Managing the High Technology Firm,* vol. 2, Greenwich, CT: JAI Press, 1992.

—— and Schoonhoven, Claudia Bird. *The Innovation Marathon,* Oxford and Cambridge, MA: Basil Blackwell, 1990*a.*

—— and ——. "Managing Innovation in High Technology Firms: Challenges to Organization Theory." In L. R. Gomez-Mejia and M. W. Lawless (eds.), *Organizational Issues in High Technology Management,* vol. 1, Research Series on Managing in the High Technology Firm, Greenwich CT: JAI Press, 1990*b*, 3–17.

Kang, T. W. *Is Korea the Next Japan?* New York: Free Press, 1989.

Kanter, Rosabeth Moss. *The Change Masters,* New York: Simon & Schuster, 1983.

Kearns, David T., and Nadler, David A. *Prophets in the Dark: How Xerox Reinvented Itself and Beat Back the Japanese,* New York: Harper, 1992.

Kidder, Tracy. *Soul of a New Machine,* Boston: Little, Brown, 1981.

Kirkpatrick, David. "Breaking Up IBM," *Fortune,* 126: 2, July 27, 1992, 44–58.

Kotter, John P., *The Leadership Factor,* New York, Free Press, 1988.

—— and Heskett, James L. *Corporate Culture and Performance,* New York: Free Press, 1992.

Kouzez, J. M., and Posner, B. Z. *The Leadership Challenge: How to Get Extraordinary Things Done in Organizations,* San Francisco: Jossey-Bass, 1987.

Kuhn, Arthur J. *GM Passes Ford: 1918–1938,* University Park, PA: Pennsylvania State University Press, 1986.

Lacey, Robert. *Ford: The Men and the Machine,* New York: Ballantine, 1986.

Lawrence, Paul R., and Lorsch, Jay W. *Organization and Environment,* Boston: Graduate Business School, Harvard University, 1967.

Leavitt, Harold J., and Whisler, Thomas L. "Management in the 1980's," *Harvard Business Review,* 36: 6, November–December 1958, 41–48.

Lohr, S. "Pulling One's Weight at the New IBM," *New York Times,* Business Section 3, Sunday, July 5, 1992, 1, 6.

Magaziner, Ira C., and Patinkin, Mark. "Fast Heat: How Korea Won the Microwave War," *Harvard Business Review,* 67: 1, January–February 1989, 83–93.

Marcus, Stanley. *Minding the Store,* Boston: Little, Brown, 1974.

Miles, R. E., and Snow, C. C. *Organizational Strategy, Structure and Process,* New York: McGraw-Hill, 1978.

Mills, D. Quinn. *The IBM Lesson: The Profitable Art of Full Employment,* New York: Times Books, 1988.

Mishne, Patricia P. "A Passion for Perfection," *Manufacturing Engineering,* 104: 5, November 1988, 46–58.

Ouchi, William E. *Theory Z,* Reading, MA: Addison-Wesley, 1981.

Pascale, Richard T., and Athos, Anthony G. *The Art of Japanese Management,* New York: Simon & Schuster, 1981.

Pennings, Johannes M., and Buitendam, Arend, (eds.). *New Technology as Organizational Innovation: The Development and Diffusion of Microelectronics,* Cambridge, MA: Ballinger, 1987.

Penney, J. C. *Main Street Merchant,* New York: Whittlesey House, 1948.

Peters, Thomas J., and Waterman, Robert H., Jr. *In Search of Excellence,* New York: Harper & Row, 1982.

Prestowitz, Clyde V., Jr. *Trading Places,* New York: Basic Books, 1988.

——, Tonelson, A, and Jerome, R. W. "The Last Gasp of GATTism," *Harvard Business Review,* 69: 2, 1991, 130–141.

"Quality Goes In," *Automotive Industries,* November 1984, 51–52.

Roethlisberger, F. J. "The Foreman: Master and Victim of Double Talk." In Dubin, R., *Human Relations in Administration,* Englewood Cliffs, NJ: Prentice-Hall, 1961, 209–217.

Rogers, William. *Think: A Biography of the Watsons and IBM,* rev. ed., New York: New American Library, 1974.

Sakaiya, Taichi. *The Knowledge-Value Revolution,* George Fields and William Marsh (trans.), Tokyo, New York, and London: Kodansha International, 1991. Originally published in Japanese in 1985.

Schonberger, Richard. *Japanese Manufacturing Techniques: Nine Hidden Lessons in Simplicity,* New York: Free Press, 1982.

————. *World Class Manufacturing: The Lessons of Simplicity Applied,* New York: Free Press, 1986.

Servan Schreiber, Jean-Jacques. *The American Challenge,* Ronald Steel (trans.), New York: Atheneum, 1969. Originally published in French as *Le Defi Americain,* Paris: Editions Denoel, 1967.

Simon, Herbert A. "The Corporation: Will It Be Managed by Machines?" In M. L. Anshen and G. L. Bach (eds.), *Management and Corporations—1985,* New York: McGraw-Hill, 1960, 17–55. Reprinted in Harold J. Leavitt and Louis R. Pondy (eds.), *Readings in Managerial Psychology,* Chicago: University of Chicago Press, 1964.

Skinner, Wickham. "Operations Technology: Blind Spot in Strategic Management," *Interfaces,* January–February 1984, 116–125.

Sloan, Alfred P., Jr. *My Years with General Motors,* John McDonald (ed.) with Catharine Stevens, Garden City, NY: Doubleday, 1963.

Stalk, George, Jr., and Hout, Thomas M. *Competing Against Time,* New York: Free Press, 1990.

Thompson, J. D. *Organizations in Action,* New York: McGraw-Hill, 1967.

Thurow, Lester C. *The Zero-Sum Solution: Building a World-Class American Economy,* New York: Simon & Schuster, 1985.

Tichy, Noel, and Devanna, MaryAnn. *Transformational Leadership,* New York: Wiley, 1986.

Toffler, Alvin. *Future Shock,* New York: Random House, 1970.

————. *The Third Wave,* New York: Bantam Books, 1981.

Trimble, Vance H. *Sam Walton.* New York: Dutton, 1990.

Ungson, Gerardo. "Perspectives on Interfirm, Global, and Institutional Strategy and International Competition in High Technology Firms." In Michael W. Lawless and Luis R. Gomez-Mejia (eds.), *Strategic Management in High Technology Firms,* vol. 2, Research Series on Managing in the High Technology Firm, Greenwich, CT: JAI Press, 1990.

Walker, C. R., and Guest, R. H. *Man on the Assembly Line,* New Haven: Yale University Press, 1957

Walleck, Steve. "Strategic Manufacturing for Competitive Advantage," *Electronic Business,* April 1, 1985, 93–98.

Walton, Mary. *The Deming Management Method,* New York: Dodd, Mead, 1986.

Walton, Richard E. "How to Counter Alienation in the Plant," *Harvard Business Review,* 50: 6, November–December 1972, 70–81.

————. "Work Innovations at Topeka: After Six Years," *Journal of Applied Behavioral Science* 13: 7, 1977, 422–433.

Weber, Max. "Wirtschaft und Gesellschaft," 1925; translated as *Max Weber: The Theory of Social and Economic Organization* by A. M. Henderson and Talcott Parsons, London: Oxford University Press, 1947. Reprinted New York: Free Press, 1964.

Wheelwright, Steven C., and Clark, Kim B. "Creating Project Plans to Focus Product Development," *Harvard Business Review,* 70: 2, March–April 1992, 70–82.

———— and Sasser, W. Earl, Jr. "The New Product Development Map," *Harvard Business Review,* 67: 3, May–June 1989, 112–127.

Wolff, H. A. "The Great GM Mystery," *Harvard Business Review,* 42: 5, September–October 1964, 164–166.

Womack, James P., Jones, Daniel T., and Roos, Daniel. *The Machine that Changed the World,* New York: Rawson Associates, 1990.

6

Transforming R&D and Manufacturing Capabilities

Dorothy Leonard-Barton

Warren Smith
Harvard Business School

Product R&D and manufacturing are the next steps in the new technology development process (refer to Fig. 1.1). The authors of this chapter treat these two steps together, consistent with the convention of referring to them collectively as the "operations" phase of the new technology development process. Though R&D remains glamorous, manufacturing continues to be viewed as a low-tech smokestack operation. This chapter quickly dispels that myth. Manufacturing is an important high-technology, knowledge-intensive competitive weapon for new technology developers. How can the manufacturing function be more completely integrated wtih the new product development process? How can manufacturing become a more essential part of the new product business plan? How can new manufacturing processes be expeditiously developed for new products? These are some of the questions answered in this chapter. In answering these questions, the authors show how organizational, human resource, and product R&D issues affect the modern manufacturing function. They quite properly take an integrated view of R&D, manufacturing, organization, and human resource issues.

WM. E. SOUDER AND J. DANIEL SHERMAN

Introduction

Operations have traditionally attracted both talent and corporate resources in continuous process industries. Paper makers, process design engineers in semiconductor manufacturing, and chemical engineer film "designers" in photographic supplies, for example, have all long been recognized as sources of critically important skills. However, only recently have operations in firms in fabrication and service been recognized as a potential source of competitive advantage. Initially stimulated by superior Japanese manufacturing capabilities in automobiles, consumer electronics, and semiconductors, the "total quality" movement has focused enormous attention on process improvements in all industrial settings over the past decade.

Even more recently, in a natural extension of attention to process capabilities, American manufacturers have rediscovered the importance of competing on the basis of "core" competencies.[1] Competencies are considered "core" if they are built up over time; are not easily imitated, transferred, or redirected on short notice; and provide a strategic advantage over competitors (Hamel, Doz, and Prahalad, 1989; Leonard-Barton, 1992a). Some firms have created competitive advantages through developing capabilities in production. In many industries, however, world-class operations are but the most basic requirement to remain in business—an absolute prerequisite to survival. In such industries, excellent operations do not distinguish one firm from another; only inefficient operations do. Whether operations are competitively comparable but nondifferentiating or whether they contribute to a "core" competence in the sense that they are better than those of the competition, managers are increasingly aware of the need to manage them as an essential organizational resource (Hayes, Wheelwright, and Clark, 1988; Itami, 1987).

One of the benefits of the renewed interest in resource-based competition and the resulting attention to identifying core competencies is that this focus requires managers to take a long-range, knowledge-based view of the firm. In this view, the corporation is a mechanism for the creation and control of the interdependent sets of knowledge that constitute a competitive advantage. In assessing competencies, managers are led to view investments in skills, managerial processes, and physical plant in terms of their value as repositories of organizational knowledge. Equipment, procedures, and people are all seen as knowledge assets of the firm (Itami, 1987) and are valued not only for their present benefits but for their potential contribution. The criteria for recruitment of personnel, selection of technical and physical systems, and creation of managerial procedures extend beyond cost alone to include adaptability, flexibility, and potential for growth.

Implicit in this knowledge-based view of the firm is also a recognition that a firm's ability to learn quickly—from the environment, from customers, from vendors, from competitors, from its own members—is also a competitive advantage. If the theme for operations over the past decade has been continuous improvement, the theme for the next is likely to be continuous learning (see Leonard-Barton, 1992a). Therefore, like other chapters in this book, this chapter will emphasize management practices that foster rapid learning.

In the next section of this chapter, the management of new process development projects is discussed, with prescriptions for effective management drawn both from the literature and from ongoing field research by ourselves and others. The role of project guiding visions, heavyweight project leadership, and cross-project learning are highlighted. The third section draws on the senior author's fieldwork to describe the interactions between process developers and users of that process, discussing the differences between contracting out process development and developing processes inside the firm and describing four basic modes of interactions between developers and users. The fourth section of this chapter discusses the "Five R's" of successful new process implementation, namely communicating a business *Reason*, *Reengineering* the process, staffing key *Roles*, marshalling the needed *Resources*, and structuring appropriate *Rewards*. Finally, the last section describes learning as a process capability, using one of the most successful mini-mills in the world as an example.

Throughout this chapter the term "process capability" is used in a sense far broader than the physiochemical transformation of raw materials into finished goods (e.g., manufacturing). Because process routines are established through explicit design or through tradition, the processes embody knowledge accumulated over time through the exercise of best practices. Thus the term "process" (as used here) refers to the actual routines and procedures performed (both physical and mental, e.g., R&D), as well as to the process knowledge embedded in the equipment and other resources used in transforming any set of inputs to outputs. Moreover, operational "processes" include the whole range of transformational systems in the firm, including data handling, managerial, engineering, process improvement processes, etc.

Managing New Process Development Projects

New operations and process capabilities build up over time through incremental adjustments, improvements, and innovations which, though minor, aggregate to a tremendous effect. However, large leaps in competence come through deliberate process innovation projects. The ability to manage such projects well is, in itself, an important competitive advantage; excellent project management results not only in effective resource utilization and in efficient and effective production, but it also increases the firm's capacity to conduct such projects, hence contributing to faster organizational learning. Entire books have been devoted to managing development projects (e.g., Souder, 1987; Clark and Fujimoto, 1991; Wheelwright and Clark, 1992). This section addresses only three elements recently identified as significant contributors to success: project guiding visions, project managers' roles, and project audits for the purpose of cross-project learning.

Project guiding visions

A knowledge-based view of the firm includes the recognition that knowledge assets are diffused throughout the hierarchy—not exclusively concentrated at the top. While this recognition does not necessarily require individual autonomy or consensual decision making, it does imply a greater acknowledgment of the capacity for decision making at lower levels in the organization than has been allowed traditionally and hence indicates the opportunity for more delegation of authority. The paradox of delegation is that the more decision making is to be delegated, the clearer must be the goal to be reached. In the knowledge-based firm, coordination comes through shared goals and objectives rather than edicts. Project teams are often given considerable authority and responsibility.

However, one problem that often plagues both new product and new process development projects is that the project team members have little sense of a shared overall objective and only a vague idea of how their daily design decisions relate to a corporate strategy. Team members may know the market being addressed, or the quality goals set for their function, but such knowledge does not always provide specific guidance for making trade-offs between types of costs implied by their designs. That is, project members often lack knowledge about what organizational capabilities they are trying to build, enhance, or alter. Yet they must make dozens, perhaps hundreds, of apparently minor decisions that in aggregate can significantly affect future capabilities. Is it more important to use a part from an already qualified vendor or to build a relationship with a new source? Would it be better to accept a "quick and dirty" software program to run a particular piece of equipment to hit a production or market window or to require a more elegant solution? Should a user interface be simplified or an operator trained to use a more complex one?

Thus process development engineers or manufacturing representatives on a new product development team potentially have two roles. The first is responsibility for the output of the current project—meeting its goals and objectives. The second role is responsibility for that particular project's contribution to the overall process capabilities of the firm. This latter role is difficult to fulfill if the decision maker is concerned with local optima or is not fully aware of what capabilities are desired.

For instance, in 1984, process engineers at the WORAL aluminum mill conducted an apparently quite localized experiment. A new electromechanical pump was installed to stir recycled aluminum can stock down into the molten bath of metal in the furnace. The pump had been used successfully in over a dozen steel mills and in at least one other aluminum foundry. However, unanticipated effects from its installation reverberated throughout the plant, as the new pump suffered from multiple, interdependent deficiencies. The nose cone material on the pump flaked and cracked under the high temperatures of the molten steel, but the brittle state was at least partially caused by the extreme thermal cycling occasioned when the pump was extracted for frequent repairs, changeovers, and process modifications. Furthermore, the operators did not alter their practices to accommodate the new pump. Accustomed to charging the furnace with large ingots relatively infrequently, they balked at having to feed in a steady flow of lighter-weight recycled aluminum can stock. When the workers force-fed the furnace with too much stock at once, the new pump choked. In response, the workers reverted to their prior method of stirring scrap down into the furnace bath by hooking a fork truck up to a metal rake and manually agitating the molten metal. This relatively violent action broke the pump hoist. Moreover, even when the pump was operating, the increased output strained the capacity of the trucks used to transfer the molten metal to the next work process station. Finally, downstream operations considered recycled aluminum a contaminant and were unenthusiastic about using it. Process engineers faced a dilemma: Was it worth continuing this highly unsuccessful and frustrating experiment? They could make their judgment based only on the local disruptive effects of the experiment on their operations, with no idea as to how important this pump and the ability to recycle aluminum were to corporate strategy.[2] Their decision, not surprisingly, was to terminate the experiment.

Process engineers at a circuit board factory in a major electronics firm were similarly unaware of corporate strategy when they redesigned their work-in-process inventory-handling system. Besides ordering a new physical handling system, they commissioned a software system to monitor and report the whereabouts of every lot of boards going through the factory. The objective was better information about work in process. However a bare 2 months after the system was up and running successfully, the engineers discovered that the corporate office had launched a new Just-in-Time program which would require eliminating most of the work-in-process inventory; instead of an elaborate system for monitoring work in process, what the factory really needed was a system for eliminating that inventory.

In both these examples, the process design engineers had little insight into the connection between their decisions and corporate strategy and hence little basis for decision making beyond their own judgment about the output of the particular project. In the WORAL project, the engineers viewed the electromechanical pump purely as a potential process improvement for a furnace, not as a step toward building a capability for handling large amounts of recycled aluminum cans. In the circuit board plant, the designers of the software system had no connection with the business decision to eliminate as much inventory as possible in all corporate plants. In both cases, resources were wasted. In the WORAL example, only one step in the whole production process was addressed, and the project participants decided the innovation was not worth making all the systemwide adjustments that would have been necessary to fully evaluate its feasibility. In the circuit board factory, resources were wasted because the new work-in-process handling system addressed the wrong problem. In both cases, a project vision linking project members' decisions to the line-of-business strategy would have provided a view into the future and could have resulted in very different decisions (see Fig. 6.1).

In the context of new process development, a guiding vision is a clear picture of an operational future, a project destination that serves as a referent, and focal point for local decision making. It places the project within the context of the line-of-business strategy. Such visions are truly useful only if they are shared by all the people working on a team and in support of the team. Guiding visions are flexible in that they emphasize what must be accomplished and why, but they often leave room for individuals to pragmatically determine the details of how the vision is to be realized. Effective guiding visions promote learning and organizational "reach." A project-guiding vision directs team members in understanding not only what they need to learn during the project for successful completion but also what they need to learn from their execution of the project so as to run follow-on projects more smoothly and what the company must learn from the project to enhance internal capabilities. Such visions place bound-

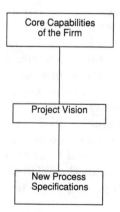

Figure 6.1 Project visions link process specifications to capabilities.

aries around the project so that project members know how much must be accomplished in this particular step toward a new capability.

For example, the management of the highly successful Chaparral Steel minimill in Texas wanted to produce very large structural steel beams. Rather than casting in the traditional rectangular shape, which had to be "worked" into the desired shape by many iterations through rolling mills, they wanted to cast steel in a dog-bone shape that approximated the final I-shape of the beams—right out of the mold. The first step in building that very advanced process capability was a project with the relatively modest objective of casting medium-sized beams in a "profile" shape, indented on the sides. The experience gained in this project propelled the company toward the revolutionary "near-net-shape" casting process that lowered costs drastically by eliminating the need for more than a few energy-intensive passes through rolling mills.[3]

A project vision guiding the development of a new product can also have substantial implications for process development. For example, when Digital Equipment Corporation decided to challenge Sun Microsystems and enter the workstation market, two critical components of the project's guiding vision were that the new line of DEC workstations be developed based on Unix language software rather than DEC's proprietary VMS and that the product reach the market within 9 months. The decision to use Unix forced a recognition that the company would need to develop as substantial a competence in Unix software as the company had in the VMS software upon which Digital's main stream of minicomputers was based. The factory where the workstation was built needed to change all its testing equipment over to the new standards—and of course to train personnel in running it. The decision to hit a particular market window forced product and production designers to develop a new level of coordination between their groups and new procedures for speedy design transfer. During the project, 35 software engineers and several manufacturing engineers were temporarily transferred to Palo Alto to accelerate product development. Moreover, the dialogue between design and manufacturing intensified and the quality of the exchange improved over the life of the project, with resulting benefits for follow-on projects. These investments in new software, new equipment, new skills, new procedures, and new interfunctional relationships were worthwhile because the DEC3100 was only the first product in a whole new line of workstations. The project vision positioned the DEC3100 in this stream of future projects for the development team so that its members could make informed design choices affecting future products and processes. The net effect of having a guiding vision was not only the effective development of a single product but the development of systemwide product development capabilities which could then be applied to future product and process generations.

Project leadership

While leadership in development projects is clearly critical, the proper way to structure project management to provide this leadership is unclear. A decade-long comprehensive study of the design projects in the automobile industry involving 29 design projects in 20 automobile manufacturers (all the major automobile manufacturers in the world) identified four primary modes of management organization in design projects:[4]

> *Functional structure.*   A traditional hierarchical structure where communications between functional areas occurs at the functional manager level and is usually codified in strict rules and specifications.

> *Lightweight product manager structure.*   A traditional hierarchical structure to which a product (project) manager has been added to play a coordinative role between the different functional areas within the design organization.

> *Heavyweight product manager structure.*   A matrixed structure led by a product (project) manager. The manager has extensive influence not only on the working-level engineers in the functional groups involved in the project but also has powerful influence extending into other groups critical to design (such as marketing and manufacturing) and is responsible for product planning and concept development.

> *Project execution teams.*   A structure where working-level engineers leave their specialized functional areas to join project teams controlled by a project manager. Such teams have been called "Tiger Teams," or at Sharp Corporation, "Gold Badge" teams in reference to the color of the identification badges worn only by team members—and by the chairman of the corporation.

The study identified the use of heavyweight project managers as being "an effective strategy for meeting diverse, uncertain, and inarticulate customer needs" (Clark and Fujimoto, 1991, p. 342). A heavyweight product manager not only champions a vision of the desired result within the technical subfunctions but also communicates and coordinates that vision between other less-involved (yet also important) stakeholder groups such as the final consumer and general management. The heavyweight project manager thus provides conceptual, technical, and managerial leadership. In a well-understood environment where there are minimal needs for handling contingencies, the heavyweight project manager's role is less critical (Clark and Fujimoto, 1991).

Another decade-long study looked at 289 separate new product innovation projects of all types in 53 firms,[5] probing the likelihood of the success of a new product innovation as a function of project management structure. Forty-six percent of the projects studied met or exceeded their commercial expectations, but the probability of success apparently differed depending on the management structure. Like Clark and Fujimoto, Souder (1987) found that new product innovation projects managed functionally through the normal chain of command of either the marketing or the R&D department were less successful than average. Projects managed by departments dedicated to new ventures and new projects were also less successful, as were projects both managed and executed by someone from either the marketing department or the R&D department.

The projects most likely to succeed had one of two forms of leadership: a commercial project manager or a new product committee structure. In the first of these, the project manager was a technically competent and technically respected functionary drawn out of the marketing department, one whose characteristics are similar to Clark and Fujimoto's "heavyweight" project manager. Such effective project managers were given a broad range of authority, had influence with multiple departments, and encouraged communication both within the multifunctional project team and between the team and the involved departments. However, while in many projects a technically capable marketing person functioned effectively as a bridge between marketing and R&D, technical personnel from R&D were much less successful in the same role (Souder, 1987).

The other form of successful project management identified by Souder was the new product committee structure, which was used primarily when the project was high risk. In this structure new product innovation projects were managed by a combination of cross-departmental standing committees and ad hoc task forces from the various functional departments. While this type of organization was effective, the study concluded that they were "cumbersome giants" to be "used sparingly" (Souder, 1987, p. 147). That is, although the communication afforded by these committees was quite *effective,* such a structure was not found to be *efficient.*

There is not complete consensus, then, on the most effective project management form. However, both of these large-scale studies found that important projects require much (early) top-level attention; both suggest that a "heavyweight" manager with strong organizational credibility improves the chances of success.

Cross-project learning

If development projects are to serve the dual purpose of both achieving the project objectives and also enhancing or creating new capabilities, cross-project learning is imperative. A tool more often used in new product development projects than process development is the project "audit" (see Wheelwright and Clark, 1992, Chap. 11). After a new product development project is completed, companies may devote as much as 2 days to an exhaustive review by all members of the team. These can be painful, of course. In explaining why his organization resisted such an audit, one engineering manager observed, "We lived through it once—barely—why would we want to live through it again?" However, well-conducted project audits offer the opportunity to reveal barriers to team coordination and cross-functional coordination, resulting in a number of action items and usually including a few different project management procedures to be used on the next project.

For example, the senior author met with the vice presidents for engineering and manufacturing at a firm producing high-performance printers, who decided to review the development of their latest model. While the product was highly successful in the marketplace, the project audit revealed some important opportunities for development process improvement. Marketing input was late into the project and inadequately communicated. The product had reached the market later than desired. The engineers felt they had been forced to accept an unrealistic schedule. Consequently, they had skipped an important prototyping step, and some quality problems could be traced directly to that omission. The audit provided them an opportunity to argue that had their realistic schedule been accepted to begin with, they would not have omitted the prototyping step, they would have avoided the quality problems, and thus they could actually have reached the market earlier. The audit also revealed a discrepancy between the resources available for development and the number of projects undertaken that draw on those resources—a frequently encountered problem in new product and process development. As a consequence of conducting the audit, the managers made numerous changes in their project management practices.

As this example suggests, such audits can be a valuable tool in developing new operational capabilities. They reinforce the need for project visions and strong, appropriate project management. Most important, they legitimate learning from experience in the organization—from mistakes and failures as well as from successes.

Managing the Interaction between Process Developers and Users

This section discusses interactions between the sources of new process technologies and their users in operations. Operations may obtain new processes from sources external to the corporation such as university researchers, research consortia, or small technology-based start-ups as well as from vendors. Therefore, this section begins by examining the relationship between the type of alliance set up by a company with external technology sources and the expected benefits from each type. The transferability of technology is also discussed. Then attention turns to internal sources of technology, describing a small exploratory study inquiring into the perceived advantages and disadvantages of internal versus external vendors of new software for manufacturing and engineering processes. Finally, there is an in-depth look at developer-user interactions in internal transfers of technology.

External sources of new technology[6]

Technical alliances of all kinds are proliferating. In the 1980s, U.S. corporations formed over 2000 alliances with European companies alone (Kraar, 1989). These corporate marriages are undertaken for many reasons. One is that corporate laboratories are no longer considered the only source for cutting-edge technology; instead companies are seeking innovations from sources as diverse as universities or competitors (e.g., see Hamel, Doz, and Prahalad, 1989). Carefully constructed alliances can produce such benefits as learning otherwise unavailable technical knowledge (such as discussed by Reich and Mankin, 1986), as well as access to technological products beyond the capabilities of one of the partners (Hamel, Doz, and Prahalad, 1989).

In general, the type of alliance into which companies enter depends upon the potential benefits to be realized. Companies have a wide menu of possible alliances to choose from, ranging from the relatively light commitment of research grants and research and development contracts for specific technology development to the relatively heavy commitment of joint ventures and mergers or acquisitions. This notion is depicted in Fig. 6.2. These alliances may be thought of as matching potential benefits ranging from a window on the technology (that is, some insight into the future possible benefits of a new tech-

Type of Alliance

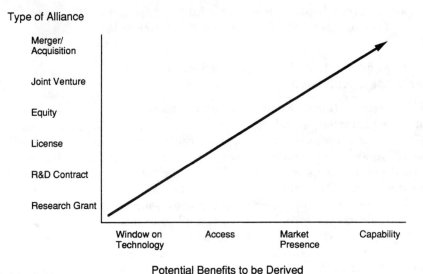

Figure 6.2   Investments in alliances and potential benefits.

nological development) to major positions in the market acquired through joint ventures or mergers.[7]

A company is unlikely to enter into a joint venture or merger if the only benefit is a window into the technology. Similarly, a major position in the market requires a strong technological capability, which is likely to be developed through alliances in the form of large equity positions, long-term collaboration, joint ventures, or mergers. There is therefore a rough logic to the match between the type of alliance entered and the benefits to be derived. However, problems arise when a firm's expectations do not match its level of investment in the alliance.

Many alliances fail—some of them spectacularly. AT&T's alliance with Italy's Olivetti, consummated in the hope of becoming an international force in computers and office automation, lasted only a few years, and AT&T's 50-50 joint venture with Philips was troubled from the beginning (see Kraar, 1989). An IBM spokesperson has estimated that only 10 percent of their vast number of alliances actually deliver as expected. One reason is the failure of both parties involved to reach agreement about what will constitute a successful outcome. Other reasons for failure (suggested by Hamel, Doz, and Prahalad, 1989) include the parties' sharing similar objectives and thus competing with each other during an attempt at a cooperative venture and asymmetry in size, independence, or ability to learn from the other partner. Most Americans are comfortable with outright (usually arm's-length) technology purchases, but other types of agreements make them uneasy because of both U.S. legal structure and business culture. However, Reich and Mankin (1986) observe that many Japanese firms favor intimate hands-on technical relationships and have mastered the art of learning from technical partners. It seems clear that the firm most skilled in learning comes out ahead in alliances.

Therefore another issue important in purchasing or acquiring new technologies from outside is the degree to which the receiving organization invests internally in the capabilities needed to customize or further develop the technology. Most buyers assume that the technology source is totally responsible for fitting the technology to the receiving organization. When the technology source is not a vendor in the traditional sense, the issues of what constitutes success and who is responsible for developing the technology to the point that it is a fully functional production process are often sources of friction.

Consider, for instance, the U.S. response to the Japanese ability to target certain industries for government-corporate investment. One such response was the Microelectronics Computer Technology Corporation (MCC) in Austin, Texas. In 1982, 16 major U.S. semiconductor and computer companies met to construct a strategy to respond to the Japanese challenge. The CEOs of these companies founded MCC as a privately owned research venture "marked by intellectual risk taking that would lead to innovations—a place where some of the world's best researchers would cooperatively extend technologies to place the United States in the forefront of a highly competitive world."[8] MCC was the first industry attempt to develop a consortium through which member companies would share the costs of mutually beneficial research and development tasks. Despite a number of technological inventions, MCC has generally not lived up to its promise. The major problem encountered was one anticipated from the beginning, yet not solved: the difficulty of transferring technology developed inside MCC to stakeholder companies. Who had the responsibility for developing the innovations to the point of large-volume production? Since all the companies owned access to the technology, no one company wanted to take on the task of development. Moreover, the CEOs of the stakeholder companies who set up MCC had seen their relatively minor investment as providing an early window into cutting-edge technologies. However, as responsibility for relationships with MCC moved down in the corporation, payment for MCC membership came out of research budgets—sometimes owned by internal corporate developers of competing technologies. Some stakeholder companies which had originally joined MCC with the idea of obtaining access to new technology through their equity position began instead to want an outright purchase of developed technology—technology that would be of advantage to them in the market. That is, a mismatch developed between the type of investments and the expected benefits shown in Fig. 6.2.

Similar problems arise when large corporations invest in small ones to acquire new process technology. For instance, when five corporations invested in a small artificial intelligence start-up firm spun off from Carnegie Mellon University, the five had very different expectations and desires. Ford Motor Company regarded Carnegie Group as a vendor who would deliver finished software applications ready to plug in and use in manufacturing diagnostics, whereas Digital Equipment Corporation desired access to cutting-edge technology, which their internal artificial intelligence group would further develop and convert into applications. Clearly it is difficult for one small company to serve both of those stakeholders.[9]

Nor is outright acquisition a guarantee of major marketplace advantage. Even if the parties agree upon a desired outcome, transferring technical capabilities in from an external source is still likely to be difficult. For example, in 1987, E-L Products Company considered acquiring a division of one of their competitors (Grimes) with whom they shared a market in the manufacture of lamps, components, and systems using electroluminescence technology. To assess the value of the proposed acquisition, a team consisting of the R&D manager, the manager of engineering, and the manager of sales and marketing visited Grimes for 2 days. They toured the silk-screening, lamination, lamp assembly, panel production, drafting, and sales departments, asking innumerable questions. Impressed by the large number of mechanical and other trained engineers in the company and by the equipment they saw, they agreed to pay $4 million for the company—a price that included the value of finished goods inventory. After the acquisition, E-L management learned that they had not received what they had expected. A sample of the finished goods inventory revealed that 80 percent of the lamps had lamination problems, rendering them valueless. Much more critically, E-L managers found that they had not acquired the scientific knowledge they thought they were purchasing—in large part because some of Grimes' employees who possessed the tacit knowledge about operations in their heads did not transfer with the sale. This expensive surprise could have been avoided had E-L management conducted a more thorough analysis.

Two questions that are often inadequately addressed by financial analysts assessing alliances and acquisitions are (1) how *desirable* is the knowledge that purports to make this technological capability competitively valuable and (2) how *transferrable* is that knowledge?

With regard to the first, the more state of the art and proprietary the technical systems, the more unique the managerial systems (such as incentive plans, project planning methodologies) and the more experience-based and scarce the skills, the more *desirable* is the technological capability. Such characteristics imply tacit knowledge, not easily purchased in the open market or imitated and hence competitively valuable. A capability is more easily *transferred* the more fully documented and operator-independent are the technical systems, the more formal and codified are the managerial systems, and the more readily available and concentrated in a few people are the skills.

Obviously, these two dimensions of desirability and transferability often work against each other. The more easily the capability is transferred, the more available it is to any comers and hence the less competitively valuable. On the other hand, capabilities that are deeply ingrained, tacit, and nonobvious are difficult to transfer. Moreover, as will be addressed in the final section of this chapter, these different capabilities are all interdependent. Unless technological capabilities are carefully assessed for both their competitive desirability and their transferability, and the trade-offs between those two considered, the transfer is likely to disappoint the recipients. Had the E-L managers explicitly addressed the issues of desirability and transferability of the knowledge they were buying, they might have either paid a good deal less or have anticipated the problems that they ran into when they discovered that most of the knowledge was not embodied in the equipment but in the people.

Internal sources of new technology

Recent research consisting of 37 in-depth case studies of the development and transfer into operations of software packages within four large electronics firms provided an opportunity to explore the advantages and disadvantages of using internal technology vendors compared to external vendors. Thirty of the cases were transfers within the United States[10]; the other seven were transfers (of some of the same software tools) to subsidiaries or partners in Japan. Data on relationships with internal, compared to external, vendors were gathered as part of the larger study in personal interviews of 1 to 3 hours with two informants for each case: (1) the manager of the development project and (2) the manager of the first work unit to receive the new software (i.e., the manager of the initial users, who had the most opportunity to interact with developers).

When queried in open-ended questions about advantages and disadvantages, the informants focused on four criteria for judging the success of the relationship: (1) cost and risk, (2) speed of the development activities (development lead time), (3) user support, and (4) efficiency. Typical comments are listed in Table 6.1.

Cost and risk. Internal vendors had a clear advantage on cost and risk, primarily because, as Table 6.1 suggests, many users did not pay for the development costs; the developers were paid out of "somebody else's budget." One of the implications of "free" development is reduced risk for the user department (see comments in Table 6.1). However, a potential downside pointed out by one of the developers was the recipients' lack of commitment to the project: "The single biggest problem was getting a single person from the receiving organization to carry out and support [the project]." Moreover, the focus on costs was very local. Neither users nor developers were concerned about the overall corporate costs for "free" development services or about developing internal services that might not be justified if purchased outside.

Development lead time. Although both users and developers saw internal development as posing fewer risks and less cost to internal "customers" than developments purchased outside, there was little consensus among informants regarding which type of vendor is superior on any of the other criteria. Some users found internal development to be faster because of less formality, the ability of the vendor to send prerelease versions of the software out earlier, or, more broadly, closer communications than with external vendors. However, some users had the opposite experience. "*Outside* developers are faster," one declared flatly. Others complained about internal vendors dragging out the schedule and about conflicting internal goals: "It is harder to negotiate schedule priorities," said one, while another complained about having to "fight the internal bureaucracy" for resources. As the comments in Table 6.1 suggest, users felt that lack of direct financial or contractual control over internal vendors affects lead times adversely.

User support. While developers and users alike identified user support as an important criterion in judging their relationship, neither internal nor external vendors had a consistent advantage. One user described the excellent support his group received from the developers, working "hand in glove," while another complained that he got "zero support" from his developer. Some of the internal vendors themselves believed that "an external vendor is expected to do a lot more hand holding."

Efficiency. Internal vendors received mixed reviews on efficiency, as well. Several developers discussed the benefits of working with users

TABLE 6.1 Comments by Developers and Users Regarding Internal versus External Developers

Internal developers are more costly: "The development costs can become high (when using an internal vendor) because we don't have the leverage of an outside customer."	**Internal developers are less costly:** "(The service of the internal vendor) was free, at least in the sense of not having to spend dollars to get it." "Outside development is costly when you consider financial costs, long-term health of the organizations, and the loss of proprietary information." "We couldn't have justified it (our new software tool) in a proposal." "We have to justify our purchases. If the software is from (a different division of the firm), however, it is ours for the asking."
Internal developers are more risky: No such comments.	**Internal developers are less risky:** "There was no real risk involved; it worked, great—if it didn't, no real problem."
Internal developers have longer lead times: "Internal vendors can drag out the schedule because you have no dollar clout or other forms of leverage." "You are certainly not sure when you are going to get (the product). The lag in getting fixes was quite long"	**Internal vendors have shorter lead times:** "Internal vendors have quicker response and have the capability to send prerelease versions out earlier." "The advantage (of an internal vendor) is faster turnaround because of easier communications."
Internal developers offer less user support: "Outside vendors give more contact." "We were a little closer to the users because we are in the same company, but we were not as nice to them as an (outside) vendor (would have been)." "Sometimes (external) vendors are very hungry for business, and it is easier to get (external) vendors to give you what you want."	**Internal developers offer more user support:** "If something breaks on a Saturday, I have their home phone number." "Representatives from the development group were very close if you needed them."
Internal developers are less efficient: "The expectations are squishy. You might get what you want, or you might not, and you are certainly not sure *when* you will get it."	**Internal developers are more efficient:** "There was no need to be concerned with contracts, etc., and the associated administrative chores. The physical proximity is an advantage (as well)." "There can be a closer working relationship (between internal vendors and users) since there is not a big confidentiality issue to worry about." "We went into functional details on algorithms with the users. We were almost in each other's back pocket. An external vendor would have been more formal."

as codevelopers (experiences reported as extremely uncommon with external clients). The users saw efficiency benefits from physical proximity and from not having to worry about contracts or proprietary information. However, several informants reported that the informality of the close relationships led to uncertainty about both the content and timing of the projects.

Intermediary factors. Rather than demonstrating any consistent advantage to operating within or outside the boundaries of the firm on all of the criteria, the interviews suggested that the perceived success of the vendor relationship was affected by two major factors: accountability/control and level of communications. These implications are diagrammed in Fig. 6.3.

Accountability. Many users complained that a lack of financial leverage over internal vendors removes accountability, as implied by several of the quotes above. However, some internal vendors felt that "our accountability is higher [than that of external vendors] because we use [the tools] internally," and others referred to us as "customers."

Financial leverage is thus not the only way to secure vendor accountability. When a process tool is on the critical path to deliver a product to market, the internal vendor clearly has high accountability. As one internal-vendor user stated, "if the project fails, we are not

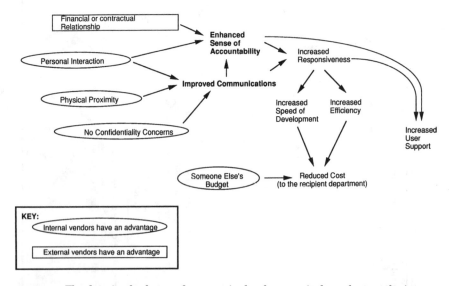

Figure 6.3 The data imply that performance in development is dependent on the intermediate variables of accountability and communications.

going to hide it [from people higher up in the firm]!" In at least one firm, review systems within the organization heighten accountability: "The user has more control over the outcome because the developer has a harder time getting a product released if he is internal rather than external."

Communications. As reported in Table 6.1, a number of informants referred to communications between vendor and user as important to development lead times, costs, user support, and efficiency. Developers felt the exchange of proprietary technical details sped up problem solving and innovation. Physical proximity ("representatives of the development group were very close if you needed them"), trust stemming from a shared work environment ("we had a rapport with the internal vendors as if we had been dealing with them for many years"), and "relationships" ("it is possible to have closer relationships with an internal development group") benefited the project. Physical proximity and a tradition of continuous communication also increased flexibility. For instance, one user reported, "I could change my mind at the last minute. [The developer's] methodology was one of constant interaction and prototyping."

Still there were some organizations that reported better communication with external vendors, as implied by the user who stated, "Outside vendors give more contact." The comments from the corresponding internal vendor justified the user's frustrations: "We only check on the users every 6 months."

Implications of the study. This small study suggests that a distinction between internal and external vendors is less important to the users of the process innovations in terms of lead time, efficiency, cost, and quality of a development project than are (1) the level of communications between vendor and user and (2) vendor accountability (see Fig. 6.3). Internal vendors have a potential advantage because of the shared work environment (or culture), long-term personal interactions, physical proximity, and an absence of confidentiality concerns, while external vendors can leverage financial and contractual relationships.

Surprisingly, only two of the informants in the study discussed another factor potentially affecting the choice between internal and external vendors, namely retaining internally the capabilities gained from development. "This is a business [our company] wants to be in," said one user. "We could have contracted outside with another vendor, but we didn't even consider looking outside because (1) we had to understand the technology, and (2) we wanted proprietary use [of the tool developed]."

The next section of this chapter draws on this same study of 37 cases to take a deeper look at the interactions between these internal vendors and the users of their new process technologies. Special attention is given to the types of knowledge that were transferred between the two groups.

Modes of interaction between internal vendors and users[11]

As suggested above, internal vendors are not all alike. Corporate research laboratories have very different reward systems and corporate mandates from those of dedicated groups set up specifically to support operations. The larger study from which the data on external versus internal vendors was taken revealed that just as a superficial distinction between vendors based solely on internal versus external status proved to be unenlightening, so too the relationships between internal developers and users could not be meaningfully categorized according to the organizational source for the technology. At times, laboratory researchers fulfilled the stereotypical role of ivory-tower thinkers, unwilling to descend to deliver a working process tool; at other times they worked hand in hand with users. Nor did user control over the development budget always predict the amount of influence they had over the design of the tool they received. Development groups set up specifically to support a production unit sometimes sought representation from production on the design team and at other times worked at arm's length. Rather than categorizing internal technology transfers according to the technology source and that source's formal mandate, it is more fruitful to examine the types of relationships that arise in individual projects as results of the relationship deliberately sought by the particular managers involved. Four categories of technology transfer emerged in this study, although the boundaries between the four sometimes blur: delivery, consultation, codevelopment, and apprenticeship. As described below, the four differ as to the *amount* and *closeness* of developer-user interaction as well as the type of knowledge transferred and the conditions necessary for success.

Developer-user interaction.   The interactions between developers and users of the new technical systems appeared to vary along at least two prominent dimensions: (1) the *amount* of user involvement at different periods of the technical system development and (2) the *closeness* of that interaction (i.e., whether users and developers work face to face or at a distance, directly or through intermediaries). As Table 6.2 indicates, the mean amount of user involvement differs most among the

TABLE 6.2 Characterizations of the Four Types of Internal Technology Transfer

Types of internal technology transfer	Amount of user involvement during development*			Closeness of the interaction	Representative comments by developers and users of technologies—Ghost, Twig, and others
Delivery mode: Developers serve as vendors, delivering completely designed system to users.	Almost None	U.	D.	Arm's length.	"Ghost" developers: "The users were not directly involved in its development at all." "Twig" developers: "It ran. It was not a big deal. There haven't been any bugs in it." "Twig" users: The developers "tossed us a [software] tape and off we went. We didn't know what was going on during 90 percent [development]." "The software behaved as advertised."
	Prototype	1.7	2.1		
	Pilot	2.0	2.8		
	General release	2.4	3.2		
Consultancy mode: Developers consult with users about functions and features.	Little; periodic	U.	D.	Often through user representatives; occasionally direct.	"Layout" developers: "We wrote the specification [for the system] but the users reviewed and approved it." "Simulate" developers: "Users validated the algorithm. They requested more speed, at a sacrifice of accuracy." "Diagnose" users: [The developers] "talked to us now and then. We saw a prototype."
	Prototype	3.7	2.4		
	Pilot	4.5	5.1		
	General release	5.4	5.3		
Codevelopment: Developers and users function as a design and implementation team.	Much	U.	D.	Continuous; usually through direct contact between users and developers. Sometimes users write actual code.	"Hats" users: "Everybody had a role and contributed to the whole. It was just a dedicated, focused, small team trying to get something done within a time frame." "Adam" developers: "We generated 300 thousand lines of code; one group of users wrote 100 thousand lines; other user groups together wrote another 100 thousand." "It was very much a joint development." "Adam" users: "We worked hand in hand with the developers. They didn't need to educate us about what to expect. We knew what the capabilities were." "The key to our success was our early involvement with the developers. I'm not a big fan of waiting for something to be officially released. When you get involved early, you can tailor the system to your particular needs."
	Prototype	5.6	6.2		
	Pilot	6.6	6.5		
	General release	6.0	6.7		
Apprenticeship: Users travel to developer site to learn how to create the system.	Much	U.	D.	Often users give functional specifications screened for feasibility. Continuous; always through direct contact between users and developers.	"Alien" developers: The users "resolved what were the real user requirements, balanced the actual needs of user organization with their wants and unreasonable desires." "Alien" users: It was "a good working team." "They talked to the guys in the trenches—not just senior management." "Roller" developer: "I spent a phenomenal amount of time in training [the user/developer]." "Adept" developers: "Since [the users] built the system, there wasn't really a transfer."
	Prototype	5.7	7.0		
	Pilot	6.0	7.0		
	General release	7.0	7.0		

*The numeric amount cited here is the average of the user managers' (U) and the development managers' (D) responses about user involvement at three points during development, measured on a scale from 1 to 7, where 1 = no user involvement at all and 7 = very heavy involvement. Each response was obtained after a series of open-ended questions on the nature of user involvement in development at that point.

first three types of transfer: delivery, where there was uniformly little user involvement; consultancy, where the level was low at first and quite a bit higher in later stages; and codevelopment modes, which are characterized by heavy user involvement throughout.

The comments (some of which are listed in Table 6.2) made by the developers and users about the *closeness* of their interactions are even more enlightening than the arithmetic measures shown in the table. In the delivery mode, the users were not involved at all in the development of the system; in the consultancy mode, the users were only involved indirectly and/or sporadically. However, codevelopment and apprenticeship modes were characterized by intense user involvement. In the codevelopment mode, the users contributed directly to the designs of the systems, interacting continuously with the developers. In the apprenticeship mode, users were always physically colocated with developers, so as to learn about the technology.

Kinds of knowledge.   Three kinds of knowledge were noted in the technical systems studied here: (1) the principles and scientific knowledge that underlie the system (*technical knowledge*), (2) knowledge about the task and work processes that the system is to aid (*domain knowledge*), and (3) knowledge regarding the ability to operate the system (*operational knowledge*).[12] These kinds of knowledge are usually resident in different people. Developers are expected to be the experts in software design, users to be the experts in their own work processes. The third kind of knowledge, operational knowledge, is a blend of the first two: knowledge about software interface design and an understanding of the technical skills of the particular users targeted. This third kind of knowledge therefore usually derives from both developers and users. As will be described below, the four modes of developer-user interaction differ in terms of the type of knowledge transferred and in the direction of the transfer.

Criteria for technology transfer success.   In the research reported here, none of the four modes of technology transfer was found to be uniformly more successful than the others on all dimensions of interest (such as user satisfaction, perceived benefits from implementation, and so forth). Since no absolute failures were included in the sample, all of the projects succeeded on some dimensions and fell short on others. Within each transfer category, some of the software packages highly pleased initial users; these were not necessarily the same ones that satisfied the next, different group of users nor were they the ones developed most efficiently in terms of time and money.

However, there were discernible patterns of better and worse management practices *within* each type of transfer. That is, certain conditions, assumptions, and management behaviors appeared to be associated with relative success and failure within each technology transfer type, as noted in Table 6.3. To understand the differences in conditions, it is useful to understand the kinds of knowledge transferred in each of the development modes and who is the principal source of that knowledge.

Assumptions, hazards, and conditions for success. In the delivery mode, the developers are the source for all three of these kinds of knowledge, as noted in Fig. 6.4. The assumption is that they need no user input to deliver a tool built on adequate technical knowledge and incorporating domain and operating knowledge that exactly fits users' needs. Moreover, no knowledge is transferred to users beyond whatever is embodied in documentation and whatever they can glean from use of the new software. The hazard for these projects is the possibility that the tool does not in fact reflect user needs. Delivery projects were successful when one of two conditions existed: (1) the users were self-selected, that is, free to accept or reject the tool depending on whether the tool offered actually matched their needs closely or not or (2) the users had essentially the same skills as developers and therefore could remedy any shortcomings of the tool as delivered. Projects in which the users had no input, no choice about using the new software, and no ability to change it if it failed to suit them were high risk.

An example is "Ghost," a computer-aided design system developed for engineers to use in laying out circuit boards. The engineers, who had no say in design or implementation of the new system, fought its initiation for years, forcing very costly redesign and enhancements. Even 4 years after its initial implementation, the development group that had introduced it was still referenced with derision and dislike by the engineering group. This estrangement was very expensive for the corporation, for although this group was their designated source of software, the engineers avoided dealing with them by constructing their own tools and negotiating with other software development groups instead.

In contrast, an expert system to help engineers estimate the costs of plastic molding parts that they designed was offered at large through the corporate network to engineers in all locations. Those engineers who decided to use "Estimator" did so because they were highly computer literate (otherwise they would not have even known about its availability) and because it fit their needs exactly. They

TABLE 6.3 Assumptions, Hazards, and Conditions for Success in Technology Transfer

Types of technology transfer	Assumptions	Hazards	Conditions for success
Delivery mode	No need for user input.	System does not fit user needs.	Users either self-select or users have all three kinds of knowledge.
Consultancy mode	Users not necessarily the source of domain knowledge but are the source of operating knowledge.	Users not typical.	Domain knowledge already codified or the experts who provide it represent users well.
Codevelopment	Users are source of domain knowledge and of operating knowledge.	Users risk-averse or have no forward vision.	Users who are willing codevelopers; users at several levels in organization who are forward looking.
Apprenticeship	Developers transfer technical knowledge to users, who have other two kinds of knowledge already.	User organization fails to support.	User organization views project as investment, not merely end in itself.

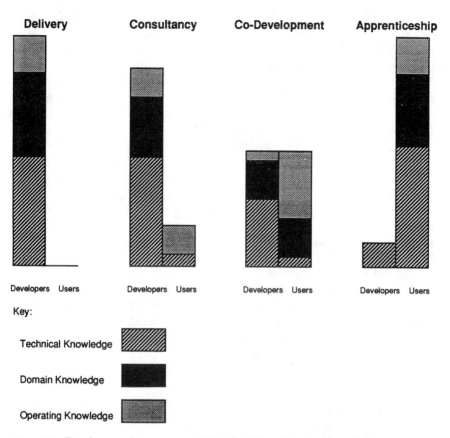

Figure 6.4 Developer and user responsibility for different types of knowledge.

required no knowledge from the developers. These users were very satisfied with the system, and it saved the corporation many times the cost of its development.

In the consultancy mode, the developers are the source of the technical knowledge and responsible for most of the domain knowledge (albeit sometimes it is embedded in prior software or obtained from other experts). However, they share with the users the responsibility for the operational knowledge. In a couple of cases, developers who tried to hand their software off in a delivery mode were forced into consultancy by users who refused to accept the software as originally designed. For instance, "Monkeys," a computer-aided design tool for semiconductor chip designers, required that engineers enter data in the form of numerical description. The design engineers refused to use the tool, telling the developers to "come back when you have a graphical interface." The developers were thus forced to interact with the users periodically as they redesigned the interface.

The consultancy mode seemed to work best for the development of tools that were to be diffused across the corporation to serve reasonably stable work processes, often for the objective of standardizing the work as well as computerizing it. The hazard of the consultancy mode, however, is that there is a tendency for developers to focus on the needs of the most vocal, most powerful, or most competent users; the resulting process tools may not match the needs of the more typical user.

In the codevelopment projects, the developers were still solely responsible for the technical knowledge in software construction but shared responsibility with the users for developing the other two types of knowledge. These projects were usually explorations into uncharted territory. Almost none of the work processes in these projects had been computerized before, and therefore developers had no prior systems to serve as models. In codevelopment, developers assumed that users were the real experts about their work procedures and were therefore the appropriate source for both domain and operational knowledge. Users had to be willing to take on the role of codeveloper, with all the concomitant risk and experimentation involved in creating or capturing the domain knowledge.

One hazard observed in codevelopment was that users are sometimes so wedded to present skills that they push developers to exactly duplicate current work processes. As one developer commented, "There's a tendency for users to be fixated on what they're using today instead of thinking about features they'll need in 3 years." In one case studied, the design was obsoleted by changes in the users' working environment even before it was completely operational. The developers had automated history. In contrast, in another case the interaction of developers and users was an act of creative extension, each group pushing the other to think beyond current capabilities. Far from requiring the developers to duplicate current functionality, the users kept trying to use the tool in applications not envisioned by its creators, thereby altering the design of both the software tool and their own work processes. Such opportunities for collaborative design can produce real leaps in productivity, service, and quality.

In the apprenticeship mode cases, users set out with the specific objective of taking responsibility for all three types of knowledge. They already possessed domain knowledge. They traveled to the developers' site to obtain enough technical knowledge to build new systems, including user interfaces. This mode was employed when users wished knowledge independence from the developers. For this mode to succeed, the developers had to be willing to serve as tutors, and the users had to invest enough time and resources to become expert in the underlying technology. The projects falling in this category succeeded to the degree that those two conditions held.

For example, "Adept," an expert system to identify and diagnose problems in circuit boards during manufacture, was built by users with no prior software programming experience. The software developers, who were expert in artificial intelligence, taught a manufacturing test engineer with a long history of troubleshooting in circuit board manufacture and a technician how to run a proprietary expert system shell. They then stood by as mentors and advisors as the two manufacturing personnel wrote "99 percent of the code themselves." The developers were eager to turn responsibility over to the users. "We took the lumberjack approach: it's your axe—you keep it sharp." Adept was a great success, allowing the firm to cut the scrap rate by 50 percent in within a mere 6 weeks, and then to cut it by an additional 47 percent over the next year. Moreover, the manufacturing test engineer went on to build other expert systems, using his new-found software programming skills to benefit the firm elsewhere.

Not all apprenticeship-mode projects were so successful. Those projects that were not adequately supported by managers when the newly trained user-developers returned to their own organizations were ultimately a waste of money. The whole purpose of the apprenticeship mode is to transfer desired process capabilities that are still not well enough codified to be transferred except as embodied in people. Unless the user-developers can continue to develop their tacit knowledge and diffuse it to others, these new process capabilities eventually atrophy and the investment is wasted.

Implementing New Processes: The Five R'S

The above section on developer-user relationships focused on initial implementation and on internally sourced technologies. This section enlarges the discussion in two ways: (1) looking beyond the initial implementation of a new process to consider how to manage its internal diffusion to multiple sites throughout a corporation as, for example, when a new manufacturing procedure is transferred from a corporate laboratory to multiple factories sited throughout the world, and (2) including purchased as well as internally developed systems. Five critical management tasks are discussed—the five R's of new process implementation.

1. *Communicating a business reason.* Just as development team members need to understand how their design activities link to their business unit strategy and to corporate capabilities, so do users of a new technical system. Surprisingly, new operations systems are often implemented without any real attempt to communicate the business reason driving the innovation. The plant manager surely knows. The process engineer may know. However, the operators are often simply handed a new piece of equipment or new procedures without any context. Innovations that benefit the corporation as a whole are often suboptimal for local operations. They may be so costly in terms of new skills or time away from production that their selection seems arbitrary and illogical. Of course, in fact it may be, and the designers or purchasers of the new technology may not be adequately aware of the local ramifications of implementation. However, even when there is a good rationale, it is often inadequately conveyed to workers. As noted earlier, the implementors of the new electromechanical pump were unaware of an overall corporate strategy to move toward more recycling of aluminum cans.

Because of the inherent difficulties in communicating vision throughout an organization, some Japanese companies have evolved a formal hierarchical set of objectives to ensure that an entire organization is in step (see Akao, 1991). *Hoshin Kanri* (alternatively translated into English as "Hoshin Management," "Policy Management," or "Policy Deployment") starts with the multiyear *hoshin* (vision, objectives, and business philosophy) of the company president (usually after approval by both the external board of directors and the internal executive board), which is published throughout the company. The various division managers establish division *hoshins* based on the company president's *hoshin* and publish these to the people in their own divisions. Similarly the department managers, section managers, subsection managers, and work leaders all establish their own *hoshins* based on the *hoshins* the next level up (and receiving reviews and approval from the next level up). They publish their *hoshins* to all subordinate levels. The aim is to help workers on the bottom levels see how their efforts directly tie into the realization of their group, section, department, division, and corporate objectives and to give all members throughout the firm the tools to make local trade-offs which are in harmony with global business principles.

2. *Reengineering the process.* All implementation situations differ somewhat—even for the same technology. Moreover, technology advances almost inevitably outpace organizational redesign. Therefore some lack of alignment between the new technology and the work environment into which it is introduced is always likely. If such misalignments are recognized during initial implementation, an

opportunity exists for mutual adaptation of technology to work environment and vice-versa (Leonard-Barton and Sinha, 1992). That is, the technology can be somewhat customized to enhance business purposes and complement work structures at the same time that the work environment is redesigned to fully exploit the potential of the technology. An important constraint on the extent to which the adaptation is mutual (i.e., is of both technology and of workplace) is the degree to which the needs of the user environment are known at the time the technology is designed. Technologies developed in a delivery mode are already completely designed by the time they are introduced into the user environment, and therefore the users must adapt their work practices to the technology design (after shopping around for the technology which fits best). In the consultancy mode, the users have a bit more influence over design, but only codevelopment offers a full opportunity to cocreate both technology and work environment simultaneously. In the apprenticeship mode the user-developers usually customize the technology to fit the organization rather than alter the work environment (see Fig. 6.5). Therefore, the extent of *mutual* adaptation undertaken depends upon the mode of technology transfer.

Recently, information systems practitioners and consultants have introduced a very similar concept into practice but have extended the scope of adaptation to include transformation of an entire operations process rather than just the one step directly affected by a new technical system. Many of our operational processes were originally designed 50 years ago—before computers. Therefore our organizations, our procedures, our structures are all designed to facilitate the

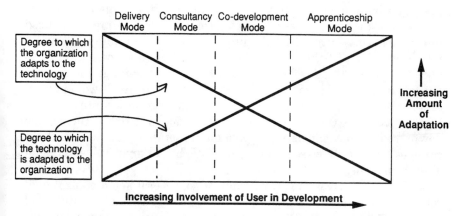

Figure 6.5 The relationship between the types of adaptation and the development mode employed.

flow of physical embodiments of information—papers and objects. Given the immense information creation, transfer, and monitoring capabilities available to us today, it is reasonable to assume that we should design operations very differently to take advantage of those capabilities. Whether the new technical system is taken off the shelf (delivery mode) or (at the other extreme) is totally customized to the users' work processes, the business processes need to be "reengineered" to enable entirely new ways of operating.

Jim Hall (1992) asserts that one problem experienced by many firms is that they "retrofit" (i.e., overlay old systems with new technology) instead of "reengineer" (i.e., create new systems based on new technology). "For most manufacturers to achieve optimum benefit from technology, genuine new functions, practices, and procedures must be designed and built" rather than merely systemizing functions through creating a "cohabitation" of the new technology with the old system without completely integrating the technology into the system (p. 16). Ronald Arenson, Wang's senior manager of image marketing, says, "Don't lay new technology on old cow paths" (Verity and McWilliams, 1991, p. 69). For example, Ford Motors used computer equipment to reengineer their accounts payable system. Rather than giving their accounts payable clerks PCs to assist them in their work of matching vendor invoices with orders placed ("retrofitting" computers into existing processes), Ford designed a computer-based "invoiceless system" using technology to eliminate the need for invoices altogether. This system resulted in a 75 percent reduction of accounts payable labor (see Hammer, 1990).

The concept of reengineering potentially applies to the introduction of any new technology—not just computer systems. Increasingly, managers understand the interdependency of the systems that they manage—both across functions and within them. For instance, in the case of the previously described electromechanical pump introduced into a furnace at a WORAL aluminum foundry, the pump itself was a very minor part of the production process. Yet to successfully introduce the pump, the developers would have had to change the furnace design, the materials mix in the furnace (which in itself would entail different procurement and raw materials storage procedures), operators' procedures (and the incentive systems that drove them), the downstream truck capacity for the molten metal, etc. In fact, all these changes eventually were made as WORAL began massive recycling of aluminum cans.

3. *Key roles.*   Over the past decade or so, several key organizational roles have been identified as important to successful implementation of new technologies, especially project sponsors, project champions, pro-

ject stakeholders, and expert first users. Explicit recognition of these roles and careful selection of people to fill them affect implementation success at least as much as does the quality of the innovation itself.

In the literature on managing technological change, a *project sponsor* (sometimes called an "executive champion") has been identified as someone with "direct or indirect influence over the resource allocation process and who uses this power to channel resources to a new technological innovation..." (Maidique, 1980, p. 64). Such an individual screens investments for their potential to contribute to the corporate strategy—either current or emerging. Even if the sponsor is someone who is just a high-level "bootlegger of funds" (Roberts, 1977), the support of such a backer is critical to the project's survival.

Sponsors do more than finance fledgling ventures, however. Roberts (1977) notes that a sponsor is "the in-house senior individual" who has the role of "protector and advocate...so that innovative technical ideas survive...to gain the confidence of the technical organization" (p. 27). Sponsors know the organization well enough to anticipate opposition; they know where possible pitfalls lie. An effective sponsor believes in the project vision, that is, understands how this project is to contribute to the corporate weal and can articulate that vision to his or her peers and above. However, sponsors do not usually originate—and definitely do not run—the project.

The daily operations of a project require a *project champion,* which Maidique (1980, p. 64) defines as "a member of an organization who creates, defines, or adopts an idea for a new technological innovation and who is willing to risk his or her position and prestige to make possible the innovation's successful implementation." Project champions develop and communicate a clear and compelling vision for the project but may or may not actually manage the project on a day-to-day basis. As Schon (1963, p. 84) notes, "the new idea either finds a champion or *dies*"; only a champion with vision and a "heroic" commitment to the project is willing to fight to overcome the resistance encountered by new ideas. Kanter (1983) points out that champions must "communicate strategic decisions forcefully enough and often enough to make...intentions clear." This clarity of vision unites the organization and coordinates actions toward a specific goal.

Schon argues that a project champion cannot be designated or assigned by management; rather a champion with self-induced commitment must "emerge" spontaneously. The job of the firm is to create an entrepreneurial culture which nurtures such champions and to ensure that when a champion appears (as manifested by zealous support of a project), he or she receives adequate time and resources (and perhaps authority) to be a driving force. If no champion emerges for a new technology, it may be because the risks of championing it out-weigh the benefits. Usually three conditions are necessary to mitigate these risks: (1) the technology must show sufficient promise; (2) the rewards (which may be intangible) provided by the organization for successful projects must be perceived as adequate; and (3) the potential punishments for failure must not be perceived as inhibiting. Evelyn Berezin, a division manager of Burroughs, recounts her experience as a champion:

> We had developed a new peripheral device which was very sophisticated yet inexpensive. It was a new technology. It was, in a word, "gorgeous." Yet corporate engineering said no, they were working on something else and this project could not be supported... What did I do? I took the device to top management. What happened? I had to fight for four months, but I won the fight. In a good large company, if you fight—long enough—you win. (Quoted in Maidique, 1980, p. 70)

Project stakeholders are persons who will be affected by project outcome, including operators who will need to learn new processes, product design engineers who will have to accept the capabilities of the new process as design constraints, and managers who will have to oversee (and eventually fund) the new processes. Many authors have identified the tendency for stakeholders to sharply resist the new ideas inherent to process design projects (for example, Lawrence, 1954; Schon, 1963; and Maidique, 1980). Kanter (1983) observes that "[one] kind of early information need is 'political': information about the existing stakes in the issue and needs of other areas that could be tied to the project to help sell it and support it" (p. 219). One type of stakeholder analysis employed by the authors identifies three categories of individuals: those who must *let* the new process occur (i.e., not oppose it), those who must *help* it occur (i.e., who have needed resources), and those who must *make* the innovation occur (i.e., must put in the extra hours and effort to achieve the targeted benefits). After identifying the stakeholders in the early phases of project definition, champions can gain their support through preselling, "horse trading" (Kanter, 1983), and coalition formation.

The role of *expert users* in implementing new process capabilities, especially "off-the-shelf" technology, is often overlooked. As suggested above in the discussion of mutual adaptation and reengineering, many adjustments—both to the workplace and to the new technology—are usually necessary when a new process technology is first introduced into an organization. Some misalignments between tool and work practices are identified only through actual use. Therefore, managers always face the problem of anticipating impacts of a new process—how to minimize detrimental effects and maximize beneficial ones. As suggested in a prior section, one way to ensure that a beneficial fit is achieved is to involve the users, either directly or through representatives. However, expert user involvement at an early stage of design and the *full* mutual adaptation of both technology and user organization are not always possible. Technical systems purchased off the shelf from vendors or developed in "delivery" mode by internal vendors for multiple sites are designed to fit a generic situation and cannot be easily customized. Therefore the receiving organization must adapt to the technology more than vice-versa, and the user expert plays an extremely important role in this situation. An in-depth comparative study of two large new software programs being introduced into multiple sites in the manufacturing operations of two companies in the same industry illustrates this (see Leonard-Barton, 1990). Both software packages studied were designed to aid purchasing departments and both were at an advanced stage of development when implemented. In both companies, the more successful implementations were characterized by the assignment of supervisors to become early expert users of the new technical systems and their release from other duties to experiment and learn. These significant managerial investments in knowledge creation paid off since these early users served as advance scouts into unknown territory for the rest of the organization. They performed a type of organizational prototyping, that is, experiments with various organizational responses to the new process (see Chew, Leonard-Barton, and Bohn, 1991). In the less successful sites studied, the user managers abjured such costly hands-on learning for the sake of less costly, more abstract training through lecture classes for all prospective users and extensive planning based on expected (but not experimentally prototyped) organizational impacts. Their financial conservatism was counterproductive, as discussed further in the next section.

Expert users not only discover what alterations may be necessary in current work procedures, but they also identify those adjustments that may still be made to the technology, including customization of user interfaces. They probe the adequacy of documentation and identify essential links to existing equipment. They also serve as role models, trainers, and process champions for later users, demonstrating to their colleagues the value of the new technologies, establishing organizational expectations that the new technologies be used, and tutoring others. Such readily available, credible sources of help smooth and speed technology introduction.

4. *Marshalling needed resources.* The need for adequate resources to support the implementation of a new technology seems obvious. However, many implementations fail because managers are penny-

wise and pound-foolish. Skimping on resources at a critical point almost certainly results in overruns later.

In the study of purchasing systems software described above, two plants in one corporation contrasted startlingly on this point. In one, managers not only did not invest in hands-on learning but naively assumed that no additional resources would be needed at the time of cutover. The plant was thrown into chaos when the systems brought on-line interacted in totally unanticipated ways. The lack of knowledgeable staff to cope with the problems brought production to a near standstill for 3 very expensive weeks. In contrast, the other plant not only allocated resources to intense hands-on training up front but instituted a 24-hour-a-day "roving patrol" at cutover to address problems. Here the cutover was smooth, and operations were stable virtually from the start.

Similarly, in the second corporation studied, those managers who allotted adequate resources to develop a cadre of "expert users" to "over-staff" the introduction period and to train others close to the time of use actually saved money. In one of the factories, where resources were extremely scarce, the manager wisely staged the system introduction, gradually phasing it in over a period of 12 months so that the resources on hand were always adequate to handle the inevitable production disruption and to address training needs.

5. *Rewards.* Since people first began to regard management as a science, issues of compensation, rewards, and motivation have been researched. Managers tend to focus on wages as motivation because financial rewards are such readily accessible levers. However, individual knowledge-based capabilities are difficult to measure; hence "extrinsic rewards" (pay, promotions, recognition by the firm) are difficult to apply in a knowledge-based firm. "Intrinsic rewards" (from within oneself) are often much more important.

For example, while studying productivity in an extremely successful semiconductor manufacturing plant in Japan, the second author asked a 20-year-old equipment operator what productivity improvements he had made in his own work area. The operator pointed somewhat hesitantly to a system of signs and markers he put in place to prevent misprocessing batches of wafers and responded, "This may not be very important, but it was my idea." When asked what other ideas he had suggested and had seen implemented, the operator pointed out another 10 or 12 improvements. With the description of each of "his" improvements, his enthusiasm and pride in the impact he has had on his area grew until he was nearly running around the factory pointing out his innovations. While the simple improvements aggregated to significant increases in quality, productivity, and cost reductions, of even more importance were the pride and motivation that the operator derived from having had a significant impact on his work and the known factory objectives.

Such intrinsic rewards are critically important in creating a competitive advantage through process knowledge. Management's job is to figure out how best to stimulate them. Management can make work more "motivationally engaging" (i.e., more intrinsically rewarding) by (1) setting up an organizational context with challenging, specific performance objectives with positive (extrinsic) consequences for excellent performance, (2) designing regular and trustworthy feedback systems, and (3) designing tasks such that workers can use a variety of high-level skills; their efforts result in a whole meaningful, visible piece of work; the work produces significant consequences for other people (and thereby becomes more meaningful in the social context); and group members have substantial autonomy (i.e., task ownership).

Learning as a Process Capability

As the preceding sections suggest, the development and implementation of new operations technologies is inherently a learning process. What would an organization look like that is designed to learn, to absorb new technology, to create new products rapidly? While no perfect archetype exists, Leonard-Barton (1992b) suggests that we can learn much from looking at a highly successful minimill in Midlothian, Texas: Chaparral Steel.

CEO Gordon Forward claims as one of Chaparral's core capabilities "the rapid realization of new technology into products." He adds, "We are a learning organization" (Leonard-Barton, 1992b). This claim is substantiated by a close look at this steel mill, where problem-solving activities and experimentation dominate. Employees from the CEO to the operators on the floor (separated by only two levels of hierarchy) pride themselves on both continuous improvement to current processes and continuous exploration of new processes.

Problem solving at Chaparral is not the exclusive domain of process engineers but is everyone's business. For example, during the introduction of the new near-net-shape caster mentioned earlier in this chapter, cooling hoses began bursting. A group of operators, a welder, a foreman, and a buyer immediately convened to discuss the problem and then spontaneously dispersed to seek a solution. Each person telephoned someone he or she thought might have some insight and within a few hours suggestions were pouring in. "If it had been just one guy, probably a foreman, and everyone [else] had walked out," observed an employee, "it would have taken him 10 times longer to find a solution" (Leonard-Barton, 1992*b*). Chaparral employees are expert experimenters, both with formal experimental design and with cut-and-try methods, although they tend to use the latter more, especially if relatively swift or inexpensive trials will provide an answer. When they were designing the novel mold to produce the near-net-shape beams, they used easily formed, almost pure copper rather than more expensive copper alloys. The experimental molds withstood the intense heat of the molten steel just long enough to prove the concept viable. Similarly, they used wetted plywood to explore the potential of proposed metal splash boards. The boards held together just long enough to confirm the feasibility of the idea.

When one studies this unusual factory, which functions simultaneously as a research laboratory and a highly productive manufacturing facility, one is struck by the culture that has been built to support the learning activities. The employees are highly skilled. Much of their knowledge is developed internally through an extensive apprenticeship program and large investments in all sorts of formal outside education as well. The workers are selected not so much for the experience they bring as for their interest in learning and enthusiasm for the work. The formal rewards, which include profit sharing and some stock holdings, align well with the desired behaviors.

However, perhaps most impressive are the values that underlie the managerial systems, including intrinsic rewards. Egalitarianism is reflected in superficial (but symbolically important) practices such as no reserved parking spaces in the lot, no corporate dining room, and deliberately understated offices. Egalitarianism is also reflected in daily behaviors. Anyone can make suggestions, and even the CEO is fair game for questions as he walks through the factory. Risk taking is regarded positively (unless of course it compromises safety). For example, a mill superintendent championed a $1.5 million electric arc saw for cutting finished steel beams. After over 18 months of very expensive and frustrating experimentation, the saw was pronounced a failure. What happened to the superintendent who risked his career on this innovation? He was recently promoted to vice president of operations.

These values and managerial practices interact to produce superior technical systems. Rolling mill equipment assumed by its vendor to be limited to 8-inch slabs is currently turning out 14-inch slabs—and the vendor has tried to buy the redesign. The two electric arc furnaces designed to melt 250,000 and 500,000 tons annually are producing at least twice that amount. Only one other steel mill in the world has a "hot-link" between the cast steel and the rolling mills. Chaparral's horizontal caster is unlike any other, and they hold patents on a number of processes, including the new near-net-shape casting.

One of the key lessons to be drawn from a close examination of Chaparral Steel as an example of a learning organization is that their core capability is both multidimensional and intensely interconnected. That is, the skills of the people, the managerial systems designed to select and train them, the technical systems that embody their knowledge, and the values that underlie the practices encouraged and rewarded in this factory all comprise one interdependent process system (see Leonard-Barton, 1992a). Chaparral employees learn readily because they were selected for their attitude, because they are constantly given the opportunity, and because they are given both extrinsic and intrinsic rewards for doing so. The employees innovate constantly because innovation is rewarded and constantly reinforced. The technical systems are among the best in the world because everyone in the company is focused on making them so—and constant change is in the employees' own best interests.

Conclusion

Two key themes in this chapter are learning and knowledge transfer. The chapter has discussed project guiding visions as links to developing technological capabilities, cross-project learning through audits, knowledge exchange between technology developers and users when developers are external or internal vendors, and learning during implementation. Mutual adaptation, reengineering, and the role of expert users in organizational prototyping were covered as examples of learning during implementation.

All these practices can be brought to focus in one organization (as, for example, in Chaparral Steel) so that continuous learning and continuous process innovation become core capabilities. A primary message of this chapter is that managing operations requires attention to investments in both technical and organizational design. Moreover, managing operational innovation is more like growing a garden than building a brick wall. That is, a manager must constantly revisit the processes in place because yesterday's solution can be today's problem. Use of project visions and project audits, attention to structuring relationships between the developers of new process capabilities and their users, careful selection of sponsors, champions, and expert users, thoughtful design of extrinsic and intrinsic rewards—all are procedures that will help. However, the most important aid to continuous process innovation is the realization by the manager that he or she is building long-term capabilities, not just getting the product out the door.

References

Akao, Y. (ed.), G. Mazur (trans.). *Hoshin Kanri: Policy Deployment for Successful TQM,* Cambridge, MA: Productivity Press, 1991.

Chew, B. W., D. Leonard-Barton, and R. E. Bohn. "Beating Murphy's Law," *Sloan Management Review,* 32: 3, 1991, 5–16.

Clark, K. B., and T. Fujimoto. *Product Development Performance,* Boston: Harvard Business School, 1991.

Hall, J. "Integrating Technological Upgrades with Reengineered Processes," *Industrial Engineering,* March 1992.

Hamel, G., Y. L. Doz, and C. K. Prahalad, "Collaborate with Your Competitors—and Win," *Harvard Business Review,* Jan.–Feb. 1989, 133–139.

Hamilton, William F. "The Dynamics of Technology and Strategy," *European Journal of Operational Research,* 47, 1990, 141–152.

———. "Corporate Strategies for Managing Emerging Technologies," *Technology in Society,* 7, 1985, 197–212.

Hammer, M. "Reengineering Work: Don't Automate, Obliterate," *Harvard Business Review,* July–Aug. 1990, 104–112.

Hayes, R. H., S. C. Wheelwright, and K. B. Clark. *Dynamic Manufacturing: Creating the Learning Organization,* New York: Free Press, 1988.

Hitt, M., and R. D. Ireland. "Corporate Distinctive Competence, Strategy, Industry and Performance," *Strategic Management Journal,* 6, 1985, 273–293.

Itami, H., and T. Roehl. *Mobilizing Invisible Assets,* Cambridge, MA: Harvard University Press, 1987.

Kanter, R. *The Change Masters: Innovation for Productivity in the American Corporation,* New York: Simon and Schuster, 1983.

Kraar, L. "Your Rivals Can Be Your Allies," *Fortune,* March 27, 1989, 66–76 passim.

Lawrence, P. "How to Deal with Resistance to Change," *Harvard Business Review,* Jan.–Feb. 1954.

Leonard-Barton, D. "Implementing New Production Technologies: Exercises in Corporate Learning." In M. A. Von Glinow and S. Mohrman (eds.), *Managing Complexity in High Technology Organizations,* New York: Oxford University Press, 1990.

——— "Core Capabilities and Core Rigidities in New Product Development," *Strategic Management Journal,* 13, 1992a, 111–126.

———. (1992b) "The Factory as a Learning Laboratory," *Sloan Management Review* 34(1), 1992b.

———. "Modes of Internal Technology Transfer and the Growth of Capabilities," Harvard Business School Working Paper #92-088. Presented at the IBEAR Research Conference, Strategies and Change in the U.S. and Japan, May 10–12, 1992c Graduate School of Business, University of Southern California.

——— and D. Sinha. "Exploring Developer-User Interaction in Internal Technology Transfer," Harvard Business School Working Paper #91-029, 1992.

Maidique, M. "Entrepreneurs, Champions, and Technological Innovation," *Sloan Management Review,* Winter 1980, 59–76.

Pavitt, K. "Key Characteristics of the Large Innovating Firm," *British Journal of Management,* 2, 1991, 41–50.

Prahalad, C. K., and Gary Hamel. "The Core Competence of the Corporation," *Harvard Business Review,* 68: 3, May–June 1990, 79–91.

Reich, R. B., and E. D. Mankin. "Joint Ventures with Japan Give Away Our Future," *Harvard Business Review,* Mar.–Apr. 1986, 78–86.

Roberts, E. B. "Generating Effective Corporate Innovation," *Technology Review,* 80: 1, 1977, 27–33.

Rumelt, R. P. *Strategy, Structure and Economic Performance,* Harvard Business School Classics, Boston: Harvard Business School Press, 1974, 1984.

Schon, D. A. "Champions for Radical New Inventions," *Harvard Business Review,* Mar.–Apr.1963, 77–86.

Snow, C. C., and L. G. Hrebiniak. "Strategy, Distinctive Competence, and Organizational Performance," *Administrative Science Quarterly,* 25, 1980, 317–335.

Souder, W. *Managing New Product Innovations,* Lexington, MA: DC Heath, 1987.

Verity, J. W., and G. McWilliams. "Is It Time to Junk the Way You Use Computers?" *Business Week,* July 22, 1991, 66–67.

Weick, K. E. "Technology as Equivoque: Sensemaking in New Technologies." In Paul Goodman, Lee Sproull and Associates, *Technology and Organization,* San Francisco: Jossey-Bass, 1990.

Wheelwright, S., and K. Clark. *Revolutionizing Product Development,* New York: The Free Press, 1992.

Endnotes

1. This concept is not new. Various authors have called them distinctive competencies (Snow and Hrebiniak, 1980; Hitt and Ireland, 1985), core or organizational competencies (Prahalad and Hamel, 1990; Hayes, Wheelwright, and Clark, 1988), firm-specific competence (Pavitt, 1991), and invisible assets (Itami, 1987). These authors base their work on earlier research by Rumelt (1974).

2. See *New Technology at World Aluminum Corp.: The Jumping Ring Circulator.* Harvard Business School Case #9-687-050.

3. See *Chaparral Steel: Rapid Product and Process Development.* Harvard Business School Case #9-692-018.

4. Clark and Fujimoto, 1991.

5. For a full description, see Chap. 9 of Souder, 1987.

6. This section draws on material developed by Dorothy Leonard-Barton in *Developing Strategic Technological Competencies, Module Three: Acquiring Technology.* Harvard Business School Teaching Note #5-692-064.

7. For exploration of these issues, see William Hamilton 1985, 1990; Leonard-Barton, *Developing Strategic Technological Competencies, Module Three.* Harvard Business School Teaching Note #5-692-064.

8. Grant Dove *Vital Speeches,* February 15, 1989, quoted in *MCC: The Packaging and Interconnect Program.* Harvard Business School Case #9-692-020.

9. See *The Carnegie Group,* Harvard Business School Case #6-690-033.

10. Four more cases were studied but excluded from consideration in this paper since they represented a special situation of less interest, namely when users discovered a specific use for an already partially developed software system. For an account of all 34 within-U.S. cases, see Leonard-Barton and Sinha, 1992.

11. The following section of the chapter draws heavily upon Dorothy Leonard-Barton, 1992c.

12. Weick (1990) has cited Berniker in distinguishing between technology and the technical system built upon such knowledge.

7

Product Launch and Follow-On

Roger Calantone

Mitzi Montoya
Michigan State University

With this chapter, we arrive at the end of the new technology development cycle (refer to Fig. 1.1). Getting the product into the user's application, following up to ensure its continued routine use, and providing for future generations of products are the final steps in the process that began in Chap. 2. Launch and follow-on are not trivial matters. The launch portion of the overall process is often the most expensive, most risky, and least well managed. What should the launch plan contain? How is it developed? What strategies can be used to manage the customer? How will the product be conveyed to the user? What is the appropriate pricing strategy? These are some of the important questions answered by this chapter.

WM. E. SOUDER AND J. DANIEL SHERMAN

Introduction

Product launch and commercialization, of all the steps in the new product process, require the largest commitment in time, money, and managerial resources (Urban and Hauser, 1980). The launch of a new product presents major risks to the achievement of profit goals because launch activities are among the most expensive things a firm can do. A full-scale commitment of resources across distribution, sales, and advertising makes the risk of failure considerably more dollar sensitive than the risk of development. Coordination of efforts and the timing of the launch are the crucial elements that must be monitored and controlled so that the project and profits will not be threatened.

Let us assume that multiple cycles through the development stage, based on results from various Alpha and Beta tests of concept, use, function, and other elements of the design, have produced a physical product. Thus, all the decisions to date have been somewhat devoid of final reality in that these tests have been conducted under less than typical market conditions. Furthermore, pretesting is usually carried out without the final commercial version of the product as it will appear on the shelves or in the catalog. Thus, one major issue in commercialization is that very often, up to this point, the product has been produced by a production process that is not representative of the one necessary to satisfy the full-scale commercial market. In other words, the products are either produced as prototypes or laboratory specials or on a small-scale pilot production process.

Commercialization and launch make the new product operational and test it in concert with the actual marketing objectives (Crawford, 1991). Marketing and production are simultaneously scaled up. The launch process aims to achieve a satisfactory distribution of the product or service, provide adequate provision of all necessary auxiliary services, and determine an acceptable price that meets market demand conditions, cost recovery requirements, and profit goals. Furthermore, launch and commercialization will test the impact of the advertising and promotion strategy and determine the specific levels of each that are necessary to satisfy return on investment objectives, as well as intermediate and final market penetration goals. The commercialization process requires continued update and revision as the new product or service is launched (Crawford and Tellis, 1981). Expected and unexpected consumer reactions, competitive responses, changes in technology and/or the economy, changes in company strategies, and the unfolding of the pattern of growth for the new product can all cause perturbations that may require modification of the launch plan. Thus, a certain amount of tailoring of both the production and the marketing plan is necessary as it is rolled out across the country. Figure 7.1 illustrates the conceptualization of the concurrent commercialization process that will be discussed in this chapter.

Marketing Plan Finalization: The Final Spin into the Market

The basic marketing strategy for the new product evolves in the development stages of the project. The next step is to advance to the level of tactical planning in an effort to reach as final a stage as possible for each component of the marketing plan. Marketing plans should specify three things: (1) marketing objectives, (2) marketing strategies, and (3) specific programs to achieve these objectives and strategies. The marketing plan is the plan of action for the new product introduction.

Before beginning the discussion of the actual steps of the marketing plan, the issue of timing must be considered. Marketing planning is

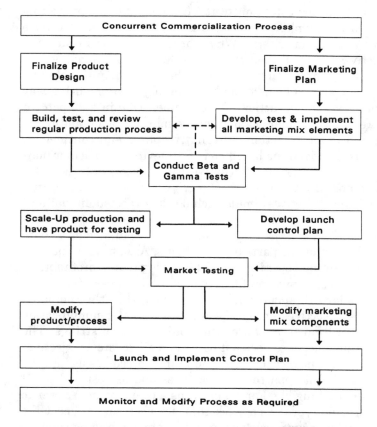

Figure 7.1 Conceptualizing the commercialization process. (Adapted from Crawford, 1991, p. 34.)

an ongoing activity that should occur both formally and informally throughout much of the new product process. Informally, it begins at the very first stages of the process, right after ideation (Chap. 2), and continues to evolve during the remaining stages (Chaps. 3 through 6). It more formally begins in the market definition stage (Chap. 4). Timing is critical. It is critical in the sense that starting work on the marketing plan early is more important than launching on a particular date. Starting too late or waiting for the perfect opportunity or waiting for a go-ahead from the results from prior stages has inherent penalties. It results in an uncoordinated launch, delays in the introduction, wasted marketing resources, and a loss of momentum for dealers and sales force alike. Marketing activity should begin as early as feasible. In general, it is appropriate to begin marketing planning activities when the product enters the concept development phase (see Fig. 1.1 and Chap. 3).

Now, let us turn our attention to the components of a marketing plan. First, marketing plans must have a set of objectives. Objectives play the role of decision criteria. When faced with several courses of action, objectives set forth the rules available for use in making that decision. Secondly, marketing objectives should be quantifiable and measurable. One should be able to phrase an objective in such a way that it becomes a guide to action. For example, "Obtain a 30 percent market share in a particular segment of the market" is a quantifiable objective and a guide to action. Alternative plans and contingency plans can be assets given the likelihood of achieving a particular market share relative to another alternative way of achieving the same goal. It is important to measure objectives in equivalent units. Typical marketing objectives should include the expected annual unit and dollar sales of the product, the projected annual gains and positions, market shares, and the target markets to be pursued as a share of the whole market or the particular segments. Also, from a financial perspective, marketing objectives should include product profitability, margin percentages, annual profits, payback periods, cash flows, and any other standard financial guidelines used within the company. These financial objectives encourage interfunctional communication and elicit cooperation from those who hold the purse strings. The process of setting objectives does not occur in a single step. Objectives are recycled during the development and launch processes as integration between different departments increases and as cost and profit goals become more tangible. For example, as a company gets closer to launch, the specific types of payback that are necessary to meet the requirements of the equity markets, or other project funding sources, become clearer. Commercialization infuses reality into the marketing and financial objectives which, in turn, induces various modifications and adaptations.

Situation analysis

The marketing planning process begins with an awareness of the situation. A situation analysis generally comprises two major elements: (1) market analysis and (2) product analysis. We will begin with a discussion of market analysis, which consists of five components. First, the market analysis consists of an analysis of the consumer, user, or buyer of the product. The purpose of this investigation is to discover what is important to the consumer. This is usually accomplished via benefit segmentation analyses which identify the main buying criteria, or core benefits (Urban and Hauser, 1980), for each major group of buyers. For example, we may want to know which buyers value appearance more than cost, and vice versa. This may involve various complex models relying on conjoint or trade-off analysis. Part of this analysis also seeks to uncover the actual buying process consumers go through, as well as how they physically consummate the purchase.

The second component of the market analysis is a competitor analysis. This is an analysis of each competitor's role in the marketplace and should include their positioning, a summary of their product offerings, a review of their behavior, comparative pricing, a summary of their distribution systems, the effectiveness of their sales force, and identification of the market segments they target. Furthermore, a thorough competitive intelligence analysis should be performed to elicit, as specifically as possible, the strategies the competition appears to follow. This is generally accomplished through a competitive size-up. A competitive size-up evaluates the comparative advantages of one's competitors in specific target markets. Generally, two major dimensions are evaluated: structural advantages and response advantages. Structural advantages are generally input or efficiency advantages such as better technology, lower cost, better trained labor, or other input advantages relative to your own processes. Response advantages are those that accrue to specific decisions. These are generally based on the leveraging of strategic phenomena at work in the business. One must compare how key competitors are set up to compete. This includes their head-to-head competitive stance as well as their reactive stances and capabilities.

The third component of the market analysis is an analysis of exogenous factors affecting the situation. Such factors include legal requirements as well as general economic and sociocultural factors. Legal requirements generally pertain to product performance and how the product is sold, made, or used. Legal restriction on new products typically come from federal agencies such as FDA, FTC, or OSHA. An analysis of these exogenous factors is important since they set the tone of the general environment. Furthermore, if legal restrictions require product modification, it would be much better to know this prior to production ramp-up and full-scale launch rather than after.

The fourth component of market analysis is a market share forecast. This should include projections of sales, profits, and eventually a pro forma budget. The product design should be described in detail as it evolves. It is important to describe the objective of the product or service in the marketplace in specific, quantifiable terms. Following this, the internal and external constraints and restraints of the new product project should be delineated. These should be stated in terms of the expected- (normal), best-, and worst-case scenarios. Finally, an initial statement detailing the support required from other functional areas (i.e., other than sales and marketing) should be included.

The fifth and final component of the market analysis section is an analysis of key market opportunities and threats. This should include the opportunity within particular segments as envisioned by the developers and proponents of the product. Threats must specifically include competitive launches that are expected or actually under way as well as other potential problems the new product may encounter.

Market opportunities can be found in various ways. Sales volume size and trends are important potential sources of opportunity. This includes industry sales figures and the sales figures of leading competitors, both of which can usually be obtained from secondary sources. It is important to isolate the trends as well as the extremes and to consider both with a critical eye. Why is a certain competitor doing very well in an area? Do they have a particular combination of assets or actions that causes them to be so successful? Are there regional areas of the market that stand out for some reason? These questions and more need to be asked and answered, in turn, to identify potential threats and opportunities. If competitive performance differences are identified, the next step is to determine the underlying cause of these differences. Is there a basic difference in the demographics or in the rate of growth, or is there some specific trade reason why sales are high in one area for a certain competitor? A company's findings may indicate a need for more in-depth research to specifically identify the particulars of that opportunity (or threat).

A hard and close look at the competition is a very important part of market threats and opportunity analysis. For example, it is important to determine whether or not the competition's general marketing practices are comparable. As a result, this analysis can lead to the identification of a need for up- or downgraded performance which, in turn, translates into market opportunity. Another factor that can serve as a signal of opportunities is a consideration of foreign vulnerabilities. For instance, there may be a particular foreign market in which the competition is rather weak or lazy, or there may be general consumption changes or political or ecological influences that create a special opportunity in some foreign marketplace. New demand, replacement demand, and diversification could all lead to trends and new directions. Each of these things needs to be substantiated by further research in order to reveal which opportunities to target and pursue.

Each of these sources of opportunity can be potential threats as well. It is critical that the situation analysis present an informed and realistic view of the market into which the new product or service is being launched. Launching and commercializing a product in an unfavorable market would be a very expensive way to learn this lesson. The type of resistance or reaction the product launch will meet is of great importance; it *may* dictate strong constraints at launch strategy.

These five separate analyses constitute the market analysis portion of the situation analysis. Upon completion, the new product team will have a thorough understanding of the targeted customer, the competitors, the external environment, and the performance goals and expectations. To complete the situation analysis, the product itself must be considered. A product analysis includes a detailed description of the product in terms of dimensions, attributes, and relevant test data verifying performance characteristics and functionality. Competitive product comparisons should be included in both tabular and prose format. Upon completion of the situation analysis, attention can be turned to the marketing strategy itself.

Marketing strategy

Marketing strategy begins with the definition of target markets and segments. In addition, the position of each product in each segment must be specified. This means identifying how the products that are currently in the marketplace address the needs of these different segments. This will help clarify which segments could be potential targets. Market targeting and segmentation require prioritizing the market by segment, end-use benefits sought, geography, demography, behavioral characteristics, psychographics, custom, or micro-needs within the segment. After defining the target markets and segments, an evaluation must be made as to whether or not a positive perception of the new product or service can be created relative to the competition. Thus, market segmentation with market targeting is essentially selecting which segments to go after, given the match between the new product and the needs of the customers, in light of the competitive offerings.

The second part of developing a marketing strategy is concerned with the general roles of the different aspects of the marketing mix: product, promotion, pricing, and distribution. The products—new, old, proposed, and planned changes—must coincide with market needs. The role of the product in the market must be carefully considered and planned prior to launch *and* development. In most companies' marketing plans, the product plays the central role in the marketing mix. This role must be decided first in order to effectively coordinate the other aspects of the marketing mix. The extent to which the product element is well defined and well matched to a segment of the market determines the ability to effectively plan and coordinate the remaining elements of the marketing mix—promotion, pricing, and distribution.

The next mix element to consider is promotion. Promotion includes advertising and personal selling. Advertising decisions must ensure consistency between the copy platform and the proposed positioning for the product in the target market segment. The advertising strategy must evolve so that each ad becomes a link in a long-term plan for the new product or brand or service. Other promotional tools that may be used in the launch, such as samples, free giveaways, or discounts and allowances, need to be considered ahead of time so that they are a part of the long-range promotion plan. Personal selling decisions must answer questions such as: To what extent is personal selling necessary to translate the needs of the customer into the specific use of the product; to what extent must the benefits and functionality of the product on the time value sequence be explained to the customer; should personal selling support the product by sales to the channel or by sales to the final customer; and is the product complex enough to require technical selling by sales engineers?

The role of personal selling is usually assessed during the development of the marketing plan. Very often, sales forces are depended upon to play a critical role in the launch of a new product, yet they are not involved early on in the development process. The general role of the sales force, the training requirements, and the amount of coordination expected must be outlined early in the planning process in order to gain commitment. If sales management is not made aware of these issues, failure to create a coordinated launch effort should be no great surprise. To an objective viewer outside of the firm, such failure seems to be a logical consequence of poor planning for sales force involvement. If different *types* of salespeople or different kinds of sales force *training* are needed, this can create earlier demands on the process of promotion planning. A reallocation of sales time and budgets may be needed in order to accomplish the launch as intended. As the product becomes disseminated in the marketplace, further sales support may be required. Too often, the personal selling aspect of the promotion decision is ignored or put off until the launch is underway. Personal selling is an integral part of the overall new product marketing strategy that must be taken into account in the planning stages in order to achieve a successful launch.

The third element of the marketing mix is pricing. Pricing policy includes discounts, allowances, the price schedule, and any changes planned during or after roll-out. Decisions must be made as to whether one wants to enter the market at a low introductory price (penetration pricing) and then perhaps try to raise the price later or, instead, start with a high price, skimming the market, and then bring it down or some variant of the two. Regardless of the resultant strategy, some well-thought-out pricing strategy is needed early in the launch. Price directly affects both unit demand and revenue. Too often, planners lose sight of this dual effect of pricing, and they focus only on the revenue that higher prices will generate with no consideration for the fact that higher prices generally imply reduced demand.

Many models are available to help determine the appropriate pricing policy, particularly in the diffusion literature (see Mahajan and Muller, 1979).

The final element of the marketing mix which must be specified in the marketing strategy is the distribution policy of the new product. Distribution policy decisions must address such issues as the need for structural channel changes, how the product should be distributed, who should do the distribution, what functions should the distributor handle, and what types of guarantee protections and margins need to be provided to the distributor to elicit cooperation.

Structural distribution decisions pertain to the number of members and levels in the channel and the logistical design necessary to distribute the product to the market. Channel policies need to coincide with the type of physical distribution required for a particular product. Physical distribution policy and structure contribute to future customer service relations and should therefore be flexible. Issues such as the handling of returns and servicing repairs need to be considered in advance. For example, will repairs be done at the factory or at the site of the distributor? All of these issues should be settled as early as possible in the planning stages because they require important budgeting considerations. For that matter, they should actually influence the product design to some degree. For instance, the extent to which whole units have to be replaced rather than just repaired can be determined, a need for modularized products may be identified, or a preference for easy-access low-performance parts may be uncovered. All of this information is critical input to the design and development of the product and must, therefore, be included in the early marketing plan, as well as back in design and development.

Distribution policy is inherently tied to customer service policy. The types of warranty, installation, refund, and repair policies that will be available to the customer are decisions which must be made in concert with the channels and logistics decisions in order to achieve consistency. If these decisions are not made simultaneously and with respect to one another, the entire distribution system will be wholly and negatively affected. While such issues as repairs and refunds seem like nonpressing considerations when a new product is just being launched, they are critical to the continued success of the new product throughout its life cycle in the marketplace. They are "after-the-sale" issues concerning channel and logistics design and structure that must be made prior to launch as a part of the commercialization plan to ensure long-term customer satisfaction.

Finally, to create an environment conducive to cooperation, a policy has to be derived which facilitates interfunctional relationships. All of the marketing mix decisions discussed above require different departments that may not be currently working together to communicate and cooperate. While interfunctional relationships may be handled through committees, this is *not* the only way. The goal is to achieve a successfully coordinated product launch, and the interfunctional relationships within the company, however facilitated, play a critical role (see Souder, 1987).

Financial summary

Turning now from marketing strategy, the next major component of market planning is the financial summary. Typical questions that must be answered include: what type of financial information has to be provided on a regular basis throughout the life of the project; how will these sales forecasts, in dollars and units, be achieved; and to what extent should sales forecasting be used as a replacement for planning? In answer to the last question, it is generally recognized that most sales forecasts for new products are woefully poor. Therefore, extensive planning and research are needed since forecasts are not highly reliable tools. Forecasting and planning affect the accessibility of funds for the company. The company must consider the incentives that forecasts and plans create for a draconian (forced) procedure of cash flow models (i.e., are people essentially being encouraged to be deceptive in order to get product development funds?). Financial projections and goals need to be subjected to routine reality checks in order to prevent costly new product investment mistakes.

The second major section of the financial summary is an expense budget, categorized by activity. Every effort should be made to complete this activity expense budget in as much detail as possible. Contribution to profit, pro forma income statements, and future capital expenditures with cash flows should be included in this summary as well. These cash flow projections should be developed to the extent that the costs are provided for each proposed action of the marketing and product development plans. Finally, a risk statement which indicates the major hurdles and uncertainties of the project should accompany the entire summary. The end result will be extensive plans accompanied by an average forecast situation which, in turn, is augmented by a risk and opportunity forecast.

Tactical plans

To complete the marketing plan, tactical plans, or explicit delineations for each marketing strategy component, must be developed. The development of tactical plans involves six steps. First, determine exactly what needs to be done. Second, state the objectives of the task. Third, identify the people responsible for each objective. Fourth, specify the time schedule for each action and indicate how all of the actions relate to each other. Fifth, specify the special teams that will be required, the interfunctional cooperation expected, and the escalation needed to get these special teams formed. And finally, make clear the operational budget within which each group is to accomplish its tasks. All six of these tactical components must be addressed and completed in order to make the marketing plan operational.

Control plans

In relation to launch and commercialization, control plans essentially delineate how management intends to control the launch process activities. Control plans should include four major parts. First, key control objectives must be determined for reporting purposes. Critical launch activities need to be identified for control. Second, the key internal and external indicators must be determined so that the control plans are tuned to the appropriate information signals. This would include information on measurement of the key indicators so that it is clear which characteristics are the triggers for the control process. Third, tolerance limits, benchmarks, and standards must be delineated so that measurements of the key variables may be easily interpreted relative to launch goals. These specifications would indicate how well the project should do on a variety of different items measured in terms of time, as well as performance. Fourth, an information generation schedule should be developed. This is a summary of the timing and inventory of the characteristics that must be measured (i.e., a compilation of what information needs to be collected at what point in time to keep the launch process in control). This will allow immediate attention to be called to those elements which are responsible for causing the process to deviate outside of the acceptable tolerance limits or benchmarks.

Once the four parts of the control plan are completed, a chronological schedule of the activities of the launch should be developed in the form of PERT charts à la Pessemier (1982). This schedule, or time line, will serve as the master plan for tracking the progress of the launch. It is important that all components of the marketing plan are included so that a fully coordinated and mutually supportive launch results. One or another of the mix components may lead, while the others support. Regardless of the phasing strategy, all components of the launch must be in agreement and managed consistently. Each marketing element should be used as a selective tool at the point in the launch where it is most effective.

Major support necessary from different areas

It is essential to specify what support for the marketing plan is needed from various other functional areas in order to make it successful. Keep in mind that the marketing plan being developed will eventually be linked with the product plan. Therefore, it is important to identify which major support groups will contribute to the success of the marketing and product plans, individually and interactively.

The major groups that are likely to be considered in this analysis of required support include processing, packaging, warehousing, technical services, finance, accounting, personnel, and public relations. It must be decided how these groups will interface and what the basis of the interface will be. The expected contributions of each different area should be explicitly specified. For instance, the support needs should specify expertise, material, personnel, and timing goals for the different groups. The support required from different functional areas needs to be made clear as early in the development process as possible in order to obtain solid commitment from them. This commitment is crucial for a successful launch.

Design Finalization

A combination of basic technical and marketing research will determine product development feasibility and the final product design. During commercialization, engineering will use the assessed capabilities to solve specific problems in ramp-up in order to produce the product that delivers the specific benefits desired by the market. These "core benefits" are the major result of the market analysis performed earlier. It is important to note how interrelated these seemingly separate processes really are.

It is important that marketing, R&D, engineering, and manufacturing decisions be integrated. In fact, it is absolutely crucial for initial product runs and production scale-up. After various iterations of Alpha and Beta tests, and prototype modifications, the development process culminates in finalized product specifications and a pilot plant product. From this, the manufacturing process can be finalized and the cost estimates and forecasts can be updated to reflect the final product design.

After finalizing the manufacturing process, regular production capacity must be committed and built as product requirements are augmented. The initial runs and review of the product and process should be conducted in concert with actual market testing to provide as realistic a setting as possible. It is necessary to take the production process to as final a stage as possible so that it can be proven in the testing process.

Launch Timing

At the beginning of this chapter, it was noted that the timing of planning activities is very important. Equally important is the timing of the launch itself. For products specifically, there is a major prelaunch effort required to reach full-scale production. All of the process equipment needs to be purchased, set up, and tested. People need to be trained on the equipment and the process if it is new to them. Material must be procured, or perhaps make-buy decisions have to be made. Tolerances and quality control standards and procedures have to be put in place for the use of the equipment and for the acceptance of materials, modules, and subsystems. All of these things take considerable time and energy, so the timing of the launch must be carefully planned with sufficient time allowed to complete all ramp-up activities. Otherwise, deadlines will be missed, the launch will be uncoordinated, and the project could be a failure.

A major concern in the launch is the timing of the start-up of manufacturing. While there must be sufficient lead time to fill the channel, start-up should not be so early that it creates excess inventory. Being late means unsatisfied demand and/or back orders, but being too early means large inventory carrying costs. It is a trade-off decision that is all a matter of timing. The level of start-up production and the rate of increase needed to satisfy demand is set based on earlier decisions and analyses—namely, the results of the market testing and forecasts. The problem with this chain of events is that the market test results are not necessarily reliable. For example, simulation models very often give rather glowing results which tend to be enhanced even more by the positive attitudes of the people involved with the project. Unfortunately, biased results such as these often lead to an unusually high forecast relative to actual start-up demand. Forecasting errors can have very serious effects in terms of process control.

In a plant that produces multiple products, people are usually trained on one particular line. In a unionized factory, people may bid for the new jobs to move from the product lines they have been working on to the new product line. If the new product forecast comes in too high, excessive inventory will be produced and the line will eventually be forced to shut down. As people are laid off or bid back to other lines, incredible costs and feelings of ill will are incurred that have nothing to do with actually producing a product. All of this could be avoided with careful planning and improved forecasting methods. The timing, level, and scope of manufacturing start-up are absolutely critical. They can very easily result in the failure of a product if not done properly.

There are some key things that companies can do to target the optimal launch window and avoid launch timing disasters. Joint plans with timing objectives and necessary lead times must be developed to synchronize product introduction and the marketing mix. Internal barriers to joint planning and interfunctional cooperation have to be overcome. A number of researchers have commented on this very issue (see, for example, Larson and Gobeli, 1988; Tushman and Nadler, 1986; and Souder, 1987). To overcome these internal barriers, firms must first create a positive culture conducive to cooperation. This should be in place at the beginning of any new product development process and should be continued, or transformed if necessary, into the launch. Secondly, shifts in responsibility that occur during the launch, such as from the new product team to the brand manager, must be smooth transitions. There must be a sense of continuity maintained with respect to the new product launch process, regardless of who has management responsibility. Those gaining responsibility need to be aware of their roles in the hand-off, and they must be committed to the plan. A third factor in overcoming internal barriers is the reward system. New reward systems must be put in place to encourage interfunctional cooperation. For example, if the production department is currently evaluated based on the amount of downtime, they will not be willing to incorporate a new process that requires multiple changes in the way they do things. Or, they may not be willing to make any changes to the process once it is operational. Reward systems must be created such that the different departments involved in the development and launch of the new product see value in cooperating.

Competitive pressures and threats can cause a firm to accelerate their original launch schedule. Very often, this pushes the launch back into the testing cycle. The market test roll-out becomes a step-by-step testing actualization foray into the marketplace. There are many studies that indicate that getting to market first is competitively advantageous (see Urban et al., 1986). On the other hand, other studies suggest that products launched a little later, in a slightly different form, succeed even better (see Schnaars, 1986). Thus, two competitive strategies, first to market or second but better, play a role in determining the appropriate launch timing for a given new product or service.

The speed of launch involves trade-offs. There is normal development, fast development, or crash development. A quantification of the expected gains of acceleration is necessary to decide which rate of development is best. The potential losses, costs, and risks of accelerating (or not) must also be considered. The decision is a trade-off between information and the amount at stake. There is always the chance of the product failing in some small way. The real risk is that a minor problem or delay could then be amplified by competitive pressure such that the product fails miserably. There are as many costs and inefficiencies of acceleration as there are benefits. Very often, firms see the need for speed but do not see the costs of speed. Products fail more often because of the cost of acceleration than because of the lack of speed. Launch timing is a complex and critical issue that weighs heavily on the success or failure of a new product.

Management and Control of the Launch and the Commercialization Process

The previous discussion on the timing of a launch gives some hint that there is significantly more to launching a new product than just putting it out in the market. In the past, firms selling industrial products would simply send out an announcement and list the product in their catalog. No special activities were undertaken to launch the new product. Conversely, consumer product firms would launch their products with a lot of fanfare, including special media events, public relations hype, special roll-out programs for resellers and channel members, and so on. Studies by Cooper (1979) showed that many companies did not really spend any time planning the launch. They just put the product out in the market and hoped for the best. A diagnostic follow-up of the many failed products showed that executives rated themselves poorly on the launch of the product. While things have improved considerably, many firms still do not completely manage and control the launch to their advantage. In general, getting the product out to market, or launching it, is not so much an event as it is just another step in the entire commercialization process.

An exception to this generalization of the launch process as just another step in the overall commercialization of a product exists when the product is new to the market. In such a case, the launch is much more of a special event because of the entrenched status quo attitude that has to be overcome. The new product needs a splash in order to get the attention it deserves. Regardless of whether the launch is to be a major event or just the next step in the commercialization process, launch plans are necessary. Successful new product launches do not happen by accident. They must be managed and controlled. Coordinated timing and carefully planned execution in communication are really the cornerstones of a successful launch. Once the decision has been made to go forward, all of the production and marketing mix elements have to be brought on-line in order to roll the product out to the consumers in such a way that it is easy for them to make a purchase decision.

One part of managing the launch is managing the sales force responsible for promoting the new product. Generally speaking, the sales force has to be convinced that this product is worth its time and effort. Most salespeople already have plenty to do selling the current product offerings. As a result, the sales force often has to be persuaded, not ordered, to promote the new product. Thus, a plan must be well laid out before its support is needed in order to secure sales force commitment. In the event the sales force is representing several divisions of products, sales management commitment is necessary in order to secure the time and resource requirements for training and motivating the sales force. An interesting situation arose with a leading electrical components manufacturer. A new machine control safety switch was developed and put into pilot production before anyone spoke with the head of industrial sales or inquired whether the distributors would carry the new product or even requested information on how to list the new component in the company industrial products catalog. These "oversights" were discovered by an accountant in the pricing department (which was jointly administered by sales management and accounting). While this is an extreme example, the resulting poor response by the sales force was understandable and, in fact, justified. The sales force is just one of the many disparate groups that must be brought together and coordinated in the actual launch process.

The hand-off aspect of the launch is the point at which the new product team turns the project over to the brand or product manager. The people who are generally responsible for the product line to which the new product will belong begin to take more and more responsibility, since the launch is specifically relevant to their day-to-day activities. The procedures that are going to be necessary in the launch process should be made known to as wide an interdisciplinary group as possible in order to spot potential problems or weak spots. Specific critical variables that need to be controlled should be identified. Furthermore, contingency plans for failed or delayed intermediate plans should be set up so that immediate action can be taken to keep the launch moving forward. While contingency plans take over, attention and resources can be focused on the weak spots or the actions that are lagging behind.

The final aspect of managing and controlling the launch is the tracking system. The tracking system should monitor consumer reactions to the product, competitor actions and reactions, promotion and price effectiveness, distribution channel performance and compliance, and general economic variables that affect the affordability and profitability of the product. Monitoring and tracking the critical variables of the launch create a basis from which the process can be managed and controlled. To be effective, it is imperative that the tracking system be in place prior to the actual launch.

Distribution, Sales Forces, Promotion, and Customer Service Strategies

It should now be clear that a successful launch requires coordination with support organizations. This includes decisions in distribution, sales force design, advertising, platform development, and customer service. The decisions required in each of these areas will be considered more specifically in the next three sections.

Channel of distribution and customer service

Figure 7.2 illustrates the sequential flow of the new product channel decisions that must be made. The design and development of the channel begins with channel objectives. The desired levels of performance and service by the channel for the new product must be explicitly set. This may involve using channels already in use or perhaps new channels of distribution. The channels can be wholly owned by the firm or the arrangement can be contractually based. Ownership and novelty affect the amount of control the firm retains over the channel, the financial support required by the channel, and other operational factors. All of these concerns should be identified and specified to reflect the distribution requirements of the market segments defined for the new product.

It is not just the institutions of distribution but the actual movement of the goods that is of interest. Since customer service is an integral part of distribution, service goals have to be established for the

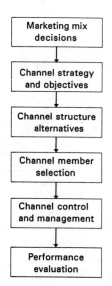

Figure 7.2 New product channel of distribution decisions. (Adapted from Hisrich and Peters, 1991, p. 373.)

logistics operations. There are three measures of customer service: availability, capability, and quality (Bowersox and Cooper, 1992). Availability is concerned with inventory levels and whether or not demand is being met. Capability is concerned with the speed, consistency, and flexibility of order fulfillment operations. Service quality refers to the ability to perform all order-related activities error-free or within the accepted industry service window. Objectives for each of these measures of customer service should be identified and tested during the test market phase of the launch. These objectives cannot be afterthoughts if they are to be achieved. If they are to be realized, they must be an integral part of the development and finalization of the marketing plan.

Properly designed distribution arrangements are a potential source of competitive advantage and customer satisfaction (see Stern and Sturdivant, 1987). For example, consider Frito-Lay and their guarantee of a 99.5 percent service level for handling such things as replacement stock and keeping stock fresh. This type of service promise, if fulfilled, creates a barrier to competitors who might try to confront Frito-Lay in their market segments. The important point here is that Frito-Lay could not achieve such high levels of customer service without a properly designed distribution system. The entire link must be fully supportive of the customer service goals or they cannot be consistently achieved.

Channel structure and strategy analysis consists of five categories (Hisrich and Peters, 1991): the degree of directness, the degree of selectivity, the criteria for selecting intermediaries, the number of channels, and factors influencing the selection of the best channel.

Degree of directness and selectivity

Degree of directness refers to the extent to which the manufacturer distributes the product directly to the consumer without the aid of intermediaries. This depends on such factors as market conditions, the geographic density of the target market, and the number of potential end users. Product attributes can also affect this decision if the product is perishable or bulky, thus making storage difficult. Another factor affecting the degree of directness is the promotion strategy design. For example, whether the strategy is push or pull will affect the degree of directness, as does the advertising media that will be employed and the personal selling efforts required. The distributing firm must consider what services are desired from the channel and what value is added at each level. The size of the distributing firm, its financial strength, previous channel experience, current product mix, and marketing policy all affect the length of the channel.

Furthermore, the amount of service a customer expects can increase the need for dealers. Lastly, environmental factors such as economic conditions, competitive distribution techniques, and legal regulations play a role in this decision. All of the factors listed above must be considered because they affect the length of the channel, and they determine how much direct control the company has over the channel.

Degree of selectivity refers to the number of distributors there are in the channel. Distribution can be intensive, selective, or exclusive. Intensive distribution means using every possible distribution outlet which can carry the new product. Selective distribution imposes requirements that have to be met in order for a distributor to carry the product. Exclusive distribution arrangements tend to be contractual and can be based on geography or type of store or some other characteristic. The degree of selectivity decision is affected by, and in turn affects, the positioning of the new product. Furthermore, the degree of selectivity will necessarily affect the availability of the product in the market. The ability to fill the channel must be factored into this decision to ensure that the firm's manufacturing capabilities are not overdrawn. Finally, the firm's current level of selectivity may factor into this decision. It would serve no purpose to cause negative feelings in the channel by unnecessarily disrupting the current flow of goods. Firms must consider all of these factors when deciding the degree of distribution selectivity for the new product.

Criteria for selecting intermediaries

There are numerous criteria by which to select intermediaries in the channel. Some potential criteria include: the mix of services provided and the relevance to a new product, the complementary nature and consistency of functions and selling efforts between the intermediaries and the manufacturer, the reputation of the distributor, the distributor's knowledge or expertise of the particular product category, the level and attitude of cooperativeness, and the ability to exclude competitive products from the channel. For example, it may be of interest to know whether or not the potential channel member has contacts in the specific markets the firm is targeting. If so, their reputation in that market could provide insight as to their potential value for the firm's product. Other points of interest include the intermediary's past performance with other new products, their overall quality of services provided to the manufacturer, the compatibility of their marketing mix strategy with that of the manufacturer, the ability of the manufacturer to maintain control over the product, the probability of an effective long-term relationship, the history of their turnover with other manufacturers, their flexibility to changing customer needs, the impact of this potential intermediary on future new products, the amount of investment required to meet objectives, and the extent to which the company must financially support the intermediary (i.e., do they have sufficient cash available to get started and to basically make their money from higher margins?).

Potential intermediaries must pass a litmus test of key criteria before being selected. Haphazard choices are extremely difficult to correct after the launch is in progress. Firms need to delineate the requirements for participation in the distribution of the new product and then they must screen all potential intermediaries according to these requirements. The extra time and effort spent up front will pay off later in a smoothly functioning channel of distribution for the new product. For example, in the industrial electrical components industry, it is extremely important to cover not only all of the potential contractors geographically with well-placed dealers and suppliers but also to have suppliers who aggressively seek to supply bids from local and regional contractors. This requires constant monitoring of the dealer corporations.

Number of channels

A firm can use one or multiple channels of distribution, depending on the target market(s) and the company objective for the new product. A major factor in this decision is the firm's ability to support distribution in more than one channel. This may mean distributing to different areas of the country at the same time at many different service levels, which would necessarily increase the complexity of the launch. One option is to roll out, not geographically, but by channel in order to maximize access to the market. Some firms begin with very exclusive channels using a skimming price strategy and then gradually begin using less selective channels as the product begins to disseminate in the market and production volumes can be increased without sacrifices in quality. These new channels command lower prices, but they increase the penetration of the product. This example of going from an exclusive single channel to a less selective channel form of distribution illustrates the fact that the number of channels is a strategic decision which must be considered prior to as well as during the launch phase.

Selection of best channel

Finally, the selection of the best channel is limited by the financial investment required, the timing and distinctiveness desired, strengths and weaknesses of the manufacturer, and the bargaining power or control of the manufacturer and of the channel members. These factors represent trade-offs which must be balanced in order to achieve the desired level of distribution for the new product. All of the previous channel decisions must be tempered by these factors. For example, if the firm has limited power or control over channel members, it may choose to distribute the product directly to the consumer, assuming this is financially feasible. The best channel available for the distribution of the new product must be determined based on each individual firm's resources and constraints. After selecting the channel structure most appropriate for the firm and the new product, the actual channel members must be selected. This decision is based on the criteria previously determined, as discussed above. Essentially, the choice is based on the firm's experience, the potential channel members' experiences, judgment, and selecting firm's marketing objectives.

Management and control of channel

The conflicting interests of intermediaries and manufacturers create a need for motivation on the part of the manufacturers. It is essential to actively create a cooperative environment. Rarely can dealers and distributors be overpowered. Quite often, the manufacturer has to offer some kind of benefits or incentives to the distributor or dealer in order to get them to cooperate and comply with the strategic goals of the manufacturer with respect to the new product. Common channel strategies for motivating channel members include: slotting allowances, trade discounts, higher margins for members performing more or key functions, cooperative advertising and display assistance, technical assistance, training and technical seminars for salespeople, special licensing or franchising agreements, financial assistance, market research assistance, contests, and push money. Recently, there has been increasing interest in forming distributor advisory councils which allow dealers and distributors the opportunity to provide input to the manufacturers on product design and distribution decisions and concerns.

The relationship arrangements that result can be very effective in helping to manage the channel. There are three types of channel arrangements (Rosenbloom, 1991). First is voluntary cooperative arrangements which often provide enough incentives to get extra effort from channel members in the promotion and distribution of the products. A second type of arrangement is the partnership or alliance approach. This relationship involves more of a vertical management system because there is a contractual tie between the manufacturer and the dealer. In such cases, cooperation is less of a voluntary issue and more of an expectation of the relationship (Bowersox, 1990). The third type of relationship arrangement is distribution programming (McCammon, 1970). Here, a comprehensive set of policies for the promotion of a product through the channel is set forth. These policies include the manufacturer's marketing goals and channel requirements as well as the various distributors' requirements. Essentially, the goal is to formalize the relationship as clearly as possible so that all expectations are understood by each party to the agreement. The program also includes punitive policies to deal with channel members who violate the agreement. These three types of channel arrangements vary in their respective abilities to manage and control the channel. Each has its advantages and disadvantages which must be considered before entering into any arrangement. For example, increased formalization of channel relationships does not necessarily imply increased cooperation. It is important to determine how much control of the channel is required to achieve the distribution objectives of the new product.

There are five bases of power from which to control the channel: reward, coercion, legitimate, referent, and expert power. The availability of any particular source of power depends on the channel structure, the size of the firm, the scarcity or uniqueness of the product, and other such factors. When a firm is dealing with a new product in a new type of channel, power-based methods of control are usually not as effective as bases of compliance which are derived from cooperative arrangements. In such cases, there are too many unknowns to use power as a motivator. Instead, taking the initiative and doing things like training the sales force, offering financial and economic rewards, providing channel members with electronic linkages that allow rapid order entry, helping to minimize inventory, and other such positive actions that provide real value to the dealers and distributors tend to increase the level of cooperation. These types of arrangements ultimately become sources of power which increase a firm's ability to manage and control the channel.

Finally, communication linkages play a crucial role in channel management. Information technologies are enabling improved management and control of the channel by making available real-time information on critical variables. Information technology is creating new partnerships in the channel and is increasing the need for cooperation. Communication and feedback on the launch of a new product is essential for tracking its progress and correcting deviations. The channel is a source of critical information related to problems or roadblocks the new product may face in the market. It is critical that linkages are established to facilitate the flow of this information and to monitor channel member activities in the distribution of the new product.

Evaluating channel performance

Once the channel is set up and being managed, the firm must determine how it will evaluate channel performance. Based on specific objectives and customer performance expectations in the four areas of distribution channel service, the quality of performance can be monitored and measured during and throughout the launch so that corrective action can be taken to ensure a satisfactory distribution arrangement (Bowersox and Cooper, 1992). The first major area of channel performance, product availability, is generally measured by stock-out percentage, fill rate, and orders shipped complete. Secondly, distribution system capability must be monitored. This is measured in terms of order cycle time, cycle time consistency, system flexibility, shipment presentation, product damage, shipping errors, and error correction. Thirdly, the information support system and quality should be evaluated. This would include such things as inventory status, order status, and shipment tracing order documents. Finally, the performance of the life-cycle support of the distribution system can be measured in terms of repair parts and service, technical service installation, and reverse distribution capabilities.

Monitoring these four aspects of the distribution system leads to a gap analysis (Crawford, 1991), that is, the identification of points where the system is meeting or falling short of its expectations with respect to the new product. This gap analysis should be done during testing phases and throughout the launch to allow early detection and correction of distribution problems. An evaluation system should be in place well before the product is launched so that attention is focused on the critical variables. All of the variables mentioned above are directly related to the level of customer service achieved by the distribution system. As such, customer service programs for the new product must be designed on the basis of cost-benefit trade-off analysis so that optimal channel and service structure can be determined and evaluated.

New Product Channel Issues

In addition to the considerations and decisions discussed above, there are other factors that affect new product channel decisions (Rao and McLaughlin, 1989). First, channel member input to new product planning creates positive feelings in the channel and can potentially be the source of new product ideas, modifications, or improvements. The firm should listen to the channel members who are close to the market. They may have important information about the market that should be incorporated into the new product design. Secondly, chan-

nel member acceptance of new products must be fostered by addressing the final channel member concerns. Common concerns include
such things as how the new product will sell, their enthusiasm in selling the new product, whether or not the new product will be easy to
stock and display, packaging concerns and input, and whether or not
the new product will be profitable for them. The third factor, which
affects new product channel decisions, is the current channel member
assortments. This means ascertaining the need for an entirely new
channel or for intensive persuasion efforts. Fourth, channel members
must be educated about the new product. Naturally, the extent of the
education required varies by industry and technical complexity of the
new product. However, it is important for the firm introducing the
new product to make sure that the channel has sufficient knowledge
of the new product to support it. Channel members will be reluctant
to carry or promote a product that they do not understand. And finally, the firm introducing the product must make every effort to make
sure the new products are as trouble-free to handle as possible with
respect to defects, warranties, guarantees, and so on. If the distributors perceive the product to be more trouble than it is worth for them
to handle it, certain avenues of distribution may unintentionally be
eliminated from further consideration.

In summary, channel strategy essentially determines the likelihood
that distribution and customer service will be a boost or a hindrance
to the penetration of the market by the new product. The more efficient the channel is, the more available the product will be. If the
product is not available, it obviously cannot be purchased. Channel
strategy not only determines the level of market coverage but also the
level of customer service and, thus, the level of customer satisfaction.
The distribution channel plays a critical role in the launch of a new
product. Distribution involves many complex decisions that have a
significant impact on the success of the new product.

Sales Force Strategy for New Products

As has already been mentioned, the sales force is one of the key factors affecting the promotion of a new product. This is especially critical for high-tech, industrial products and short-channel situations.
There are several alternatives in structure: the current company
sales force, industrial distributors, independent representatives, a
special multifunctional setting, telemarketing, or trade shows. Very
often, more than one of these alternatives is used in combination to
launch the new product. For instance, special teams may be set up
initially but then phased out as the more usual sales channel takes
over. In the pharmaceutical industry, when a new drug that is not

familiar to the current sales representatives is introduced, special sales people are added to the force to call on those clients to whom the drug might be an important prescription item. Furthermore, so-called spot team members from central areas are sent out to the field when a very large sale is on the line (e.g., to a large hospital group). Using multiple sales channels is a common way of bolstering the sales force during the introduction of a new product. Generally, some of the sales channels will be phased out as the product begins to take hold and establish itself in the market.

The criteria used to determine the feasibility of using independent representatives as an alternative to an internal sales force are based on several concerns (see Anderson, 1985; Powers, 1988; and Hisrich and Peters, 1991). First, there are market characteristics affecting the decision. The market potential must be large enough to justify the expense of training and maintaining an internal sales force. If the product is new to the target market, the prospects will probably be new to the company and, therefore, not familiar to the current sales force. In other words, independent representatives may already have the skills and expertise necessary to sell the new product. They will probably be familiar with the market and already have contact with key prospects. All of this must be factored into the decision. Another market characteristic is the expected frequency of purchase of the new product relative to the number of sales calls required. If this ratio is rather low initially, such that numerous sales calls are required prior to purchase, this would place an enormous burden on the current sales people. Such a situation would require them to forego selling the products they know they can sell and earn a commission on to spend time pushing the new, unknown product. There would have to be great incentives for a salesperson to do this. In addition, the company must realize that asking its sales force to concentrate on the new product necessarily implies reduced concentration on existing products. Both of these issues must be considered when evaluating the feasibility of using independent representatives to sell the new product.

The second group of factors that affect the decision of using independent representatives are product characteristics. Sometimes the firm needs the prestige of an independent representative in order to gain acceptance in the market. This is particularly true if the product is new to the manufacturer or if it does not match well with the manufacturer's reputation. Sometimes products require a technical specialist who is not available internally and who would be too costly to train. If the product requires long periods of technical negotiation before the transaction is completed, it would divert too much time from the current sales forces' selling activities. Often, this drawn-out negotiation process is due to the complexity of the product and the need for very detailed product specifications.

Finally, there are some company characteristics that can create a preference for independent representatives. If there are insufficient financial resources to build an adequate internal sales force, or if the new product is insufficient to compensate a direct internal sales force, the best alternative may be to utilize independent representatives. When a new product requires different market and distribution channels than the current markets and channels, the current sales force may not have sufficient knowledge to sell the product. Sometimes firms are having problems with the current sales force and simply want to move some things away from them or keep certain opportunities out of their control. In such situations, independent representatives provide a reliable channel through which the product can be promoted with considerably fewer strings attached. Finally, a company may desire a quick and immediate distribution and servicing of the new product without an extended training period. Using competent independent reps is one way to achieve this objective. For example, several firms that sold a variety of electrical products for both homes and commercial buildings used independent reps to approach architects and high-profile developers when they entered the "smart home" systems market. In this industry, this was the necessary approach in order to get an initial foothold in the market.

Market, product, and company characteristics all affect the internal versus external sales force decision. Budget and compensation are perhaps the two key overriding factors. The budget determines the training funds available and, therefore, affects the decision to use an internal or external sales force. If an external, already competent representative or agent is not going to be used, money must be allocated to the training and conversion of the internal sales force. Salespeople must be made aware of any technical information relevant to the sale of new products, as well as the firm's expectations and the personal rewards and benefits of selling the new product. Thus, the usual cycle of assessing the training needs, setting the training objectives, evaluating training alternatives, designing a training program, performing the training, and conducting follow-up evaluations are long and costly, yet imperative steps of choosing an internal sales force.

Compensation decisions are an integral part of motivating the sales force. The basic goals are to provide an acceptable ratio of costs and sales force output in volume, profit, and other such measures. Secondly, compensation encourages specific activities that are consistent with the firm's overall marketing and sales force objectives and strategies. For example, a reward and incentive system may be used to encourage selling a new product or to conduct important follow-ups after the sale. A third goal of compensation is to attract and retain competent salespeople, thereby enhancing long-term customer relations. The impact of launching new products on the firm's ability to retain competent salespeople and the high producers must be considered. If new products encourage turnover, other sales channel alternatives might be considered. Finally, the compensation system must facilitate adjustments in the administration of the reward system. A clearly stated, reasonably flexible plan greatly assists its administration. Creative compensation plans, including financial and nonfinancial compensation, can be developed to ensure sales force motivation to sell the new product.

Advertising Decisions

Advertising decisions are usually critical to the new product process. They are *vital* to consumer goods and most commercial goods launches. Long lead times for print media as well as the necessity of testing copy and photographs for effectiveness require an early start, sometimes well before the final product specifications are in place (e.g., during the market definition step; see Chap. 4). In some cases legal issues strongly influence advertising development (e.g., pharmaceutical products). The distribution and sales strategies are supported by advertising which can presell the prospects, refer them to the proper channel entity, or act as a "door-opener." At the same time the level of sales force and distribution support needed to close the sale and deliver the product may critically limit the ultimate success of an advertising program. Certainly the "reach" of *each* marketing mix element should be commensurate among them.

At a high strategic level the specific role of advertising must be delineated. Often a primary initial objective of advertising in a new product launch is *awareness* followed by *comprehension.* Quite simply, awareness is enabling the potential customer to correctly identify the specific new product as a relevant purchase alternative for a particular need or set of needs. Comprehension, a later objective hierarchically, is enabling customers to understand the attributes and functions of the specific new product and how or why it can satisfy their needs. To achieve these two objectives, advertising creators must be directly linked to the development process to understand what the product is, what it does, how it does it, its distinctive properties relative to competitive products, test data, etc. The quality of the advertising effort depends directly on the quality of the information given to the advertising craftspeople (copywriters, account executives, etc.).

There are three major advertising decisions: budget, copy, and media. The budget decision generally is paramount since it limits the other two efforts and is usually set by an objective-and-task approach in most successful settings. The copy decision is what the firm desires to communicate to the target customer and how it wants to communicate it (humor, reason-why, benefits, etc.). The media decision involves selecting a schedule of media and vehicles that will effectively carry the copy (message) with sufficient reach and frequency to accomplish the communication goals for the target segments. The following steps outline the operationalization of the advertising decisions to support a new product introduction (Burnett, 1984; Longman, 1971):

1. Conduct appropriate research—consumer, product, market analysis, etc.
2. Set advertising objectives for the new product—communication device, basis for decision making, criterion for evaluating performance
3. Define target market(s)
4. Determine advertising budget
5. Design the advertisement—creative strategy (basic appeals), creative tactics (techniques), and copy platform
6. Pretest the advertising campaign in conjunction with a market test of the product in as realistic a setting as possible so that the results are meaningful
7. Develop the media plan completely
8. Select sales promotions to support the advertising effort
9. Evaluate the advertising effort against the objectives and any changes in the product or other marketing mix elements that may affect advertising

For a new consumer product, all of the steps of developing advertising strategy indicated above are necessary. For industrial goods, less emphasis is placed on broad media placement and more is directed toward vertical books, trade magazines, and sales force support.

These steps must be planned and implemented in such a fashion that sales force and distribution efforts work efficiently with the advertising. Coverage, timing, and the content of persuasive efforts must be coordinated across all three sectors. Distribution and advertising are often carried out by organizations not wholly owned by the innovating firm, and thus, compliance cannot be assumed. Also, sales force problems, mentioned earlier, can spawn resistance from this sector. Therefore, sufficient lead time, training, and cross-functional communication are necessary. One successful manner of dealing with these interfunctional conflicts is to bring in brand or product management from successful current products. Due to their longer relationship with these groups, they are often able to coordinate the different groups effectively. For example, sales force training materials, brochures, mailing pieces, leave-behind displays, and other support materials should be highly coordinated with the other advertising decisions. In fact, they should be integrated with the advertising platforms and themes. Obviously, this requires strategic coordination at a high level. These types of details are often uninteresting to the technically oriented marketing, R&D, and manufacturing people, so a hand-off procedure or blending in of regular marketing management personnel into the team must occur prior to this point. The team-building aspects and interorganizational coordination efforts are critical challenges to a successful launch.

Conclusion

This chapter has discussed the launch and commercialization process of new product or service development. This is the stage requiring the greatest commitment in time, money, and managerial resources. It therefore presents the greatest risk to the organization. A successful launch requires the simultaneous development, finalization, and ramp-up of the marketing plan and the product design. This concurrent activity is a process of revision, updating, and recycling as new information is learned from and incorporated into the launch process. Commercialization and launch are the first true market test of the production and marketing aspects of the new product. The physical product and its production are tested, as are the promotion and distribution strategies.

This chapter has presented the elements of the marketing plan that must be finalized prior to launch and revised in process. These elements include a situation analysis, marketing strategies and objectives, the financial summary, tactical plans, control plans, and support requirements. Design finalization involves the product design itself, as well as the market testing of this design. Key decisions that must be made regarding the launch and commercialization process relate to launch timing and the management and control of the launch. Timing is a critical issue in today's highly competitive markets. Important trade-off decisions exist that must be made when deciding on the appropriate speed of launch. Successful new product launches do not happen by chance; they must be planned and then carefully managed and controlled. Systems must be in place to monitor the launch and to keep the process in control.

Finally, there are specific distribution channel, sales force, promotion, and customer service decisions that must be made to ensure launch success. The channel of distribution decision is a pivotal factor in the launch of a new product. It affects the level of customer service and customer satisfaction that the firm can achieve. There are five categories of decisions that must be made regarding channel structure design: the degree of directness, the degree of selectivity, the criteria for selecting intermediaries, the number of channels, and the selection of the best channel. Sales force decisions must coincide with the launch and commercialization process. The firm must determine which type of sales force they will use to promote the new product based on various market, product, and company characteristics. Three major advertising decisions must be made in such a way as to create a consistency of marketing effort in the launch: the budget, copy, and media.

The production, distribution, sales force, and advertising decisions must be planned and implemented in a complementary way that causes them to function concurrently in an effective and efficient manner. It is critical that launch timing is coordinated across each of these areas. This coordination requires high levels of interfunctional communication and cooperation. It is essential that linkages are established to facilitate the flow of critical information between functions. In order to successfully launch a product, firms must realize that commercialization is a concurrent, *not* a sequential, process of marketing and product decisions. Although launch and commercialization may be the final step in the new product development process (Fig. 1.1), it is clear that the elements of this step have their roots in many of the earlier steps. In most new products, commercialization decisions arise from the moment the idea for the new product arises and continue throughout its life cycle.

References

Anderson, E. "The Salesperson as Outside Agent or Employee: A Transaction Cost Analysis," *Marketing Science,* Summer 1985, 234–254.

Bowersox, D. J. "The Strategic Benefits of Logistics Alliances," *Harvard Business Review,* July–August 1990, 36–45.

―――― and Cooper, M. B. *Strategic Marketing Channel Management,* New York: McGraw-Hill, 1992.

Burnett, J. J. *Promotion Management: A Strategic Approach,* New York: West, 1984.

Cooper, R. G. "The Dimensions of New Product Success and Failure," *Journal of Marketing,* Summer 1979, 93–103.

Crawford, C. M. *New Products Management,* Homewood, Illinois: Irwin, 1991.

―――― and Tellis, G. J. "An Evolutionary Approach to Product Growth Theory," *Journal of Marketing,* Fall 1981, 125–132.

Hisrich, R. D., and Peters, M. P. *Marketing Decisions for New and Mature Products,* 2d ed., New York: Macmillan, 1991.

Larson, E. W., and Gobeli, D. H. "Organizing for Product Development Projects," *Journal of Product Innovation Management,* Fall 1988, 180–190.

Longman, K. A. *Advertising,* New York: Harcourt Brace Jovanovich, 1971.

Mahajan, V., and Muller, E. "Innovation Diffusion and New-Product Growth Models in Marketing," *Journal of Marketing,* October 1979, 55–68.

McCammon, G. C. "Perspectives for Distribution Programming." In L. P. Bucklin (ed.), *Vertical Marketing Systems,* Glenview, Illinois: Scott, Foresman, 1970, 32.

Pessemier, E. A. *Product Management: Strategy & Organization,* 2d ed., New York: John Wiley, 1982.

Powers, T. L. "Switching from Reps to Direct Salespeople," *Industrial Marketing Management,* 16, 1988, 169–172.

Rao, V. R., and McLaughlin, E. W. "Modelling the Decision to Add New Products," *Journal of Marketing,* January 1989, 80–88.

Rosenbloom, B. *Marketing Channels: A Management View,* 4th ed., Chicago, Illinois: Dryden Press, 1991.

Schnaars, S. P. "When Entering Growth Markets, Are Pioneers Better Than Poachers?" *Business Horizons,* March–April 1986, 27–36.

Souder, W. E. *Managing New Product Innovations,* Lexington, Mass.: Lexington Books, 1987.

Stern, L. W., and Sturdivant, F. D. "Customer-Driven Distribution Systems," *Harvard Business Review,* July–August 1987, 34–41.

Tushman, M., and Nadler, D. "Organization for Product Innovation," *California Management Review,* Spring 1986, 74–92.

Urban, G. L., Carter, T., Gaskin, S., and Mucha, Z. "Market Share Rewards to Pioneering Brands: An Empirical Analysis and Strategic Implications," *Management Science,* June 1986, 645–659.

——— Urban, G. L., and Hauser, J. R. *Design and Marketing of New Products,* Englewood Cliffs, New Jersey: Prentice-Hall, 1980.

8

The Emerging Paradigm of New Technology Development

David Wilemon

Murray Millson
Syracuse University

This chapter summarizes all the steps in the new technology development process discussed in Chaps. 2 through 7. The reader will also note that this chapter includes some discussion of the product R&D step (see Fig. 1.1). The focus of this chapter is on contrasting the traditional paradigm with the newly emerging paradigm for new product development. The authors make a case for the necessity of an entirely new paradigm. What is wrong with the old paradigm? How will the new paradigm be an improvement? What are the implications for managers? How can traditional organizations and traditional managers participate effectively in the new paradigm? These are some of the important questions answered in this chapter.

WM. E. SOUDER AND J. DANIEL SHERMAN

Introduction

This chapter addresses and integrates four important new product development (NPD) topics. First the traditional NPD process is described. This discussion will provide a platform from which to describe a shift that is taking place in the manner by which new products are developed. The "old" NPD process is a prescriptive model for the development of new products. While many firms develop new products, fewer companies are following the rigid, linear NPD path so frequently depicted in the literature.

Second, in addition to a description of the traditional NPD process, this chapter examines several problems and limitations associated with its application. It is noted that these problems can hinder new product success even when the traditional NPD process is explicitly followed. Many of these problems have been identified previously (Wind and Mahajan, 1988). This chapter summarizes them to demonstrate how they are fueling what is described as a "paradigmatic shift."

Third, a detailed assessment of an emerging NPD paradigm is presented. This chapter discusses the current state of NPD paradigm change and characterizes the attributes of a new paradigm, along with developing a perspective on its future evolution and growth. Attributes of the traditional NPD paradigm, an emerging NPD paradigm, and paradigmatic shift indicators are displayed in Table 8.1.

Finally, what firms can do to employ this new process is described and implications relative to the actions management must take to participate in this emerging process are suggested. This chapter describes how to achieve "new product successes" in and through the use of the emerging NPD processes. This discussion is summarized by presenting an integration of the traditional NPD process, NPD process shift indicators, the emerging NPD paradigm, and dimensions and elements of firm participation in the NPD arena of the future.

Traditional New Product Development

The traditional NPD process depicted in Fig. 8.1 (Crawford, 1991; Cooper, 1992; Cooper and Kleinschmidt, 1991; Booz, Allen and Hamilton, 1982) is now seen as the standard, normative process employed to develop new products. It is a process typically performed by a single business organization for new product development. Figure 8.1 represents an aggregate version of the traditional NPD process. Each step in the figure can be envisioned as a synthesis of many microelements which better depict all of the NPD activities. The NPD process has been described by only a very few steps such as the two steps (stage 1: creation and development and stage 2: adoption and implementation) of Knight (1967) or the four steps (idea conception, idea proposal, adoption decision, and implementation) of Daft (1978). Such depictions have evolved into many more steps as researchers and practitioners have more precisely defined the activities that take place within the overall process and in each step.

The process portrayed in Fig. 8.1 represents a paradigm of NPD. It implies certain "rules" that are accepted by many who study and participate in the NPD process. As such, this figure prescribes what is to be done to develop new products. These steps are described in the following sections.

New product strategy development

A first broad step in the traditional NPD paradigm is the generation of "a strategy" to guide the creation of new products. "Strategy" also

TABLE 8.1   Transition from the Traditional to the Emerging NPD Paradigm

Traditional	Paradigmatic shift indicators	Emerging
Linear	Changes in organizational structures	Iterative; overlapping
Functional separateness	Globalization of new product development	Functional integration
Intermittent projects	Accelerated new product development	Continuous innovation
Efficiency/cost focus	Heightened concern for product quality	Effectiveness & efficiency
R&D driven	Enhanced communication for information technologies	Customer solution driven
Single-loop learning		Double-loop learning
Single firm development		Network of firms
Domestic perspective		Global perspective
Quality is expensive		Quality reduces cost
Short-term sales & profits used to measure success		Long-term customer satisfaction used to measure success

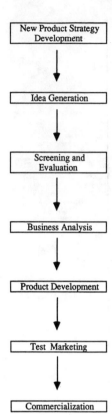

Figure 8.1 The traditional new product development process.

can describe the parameters within which the NPD process is to operate. Crawford (1991) depicts NPD strategy as a Product Innovation Charter (PIC). PICs perform two important functions. They integrate corporate strategy with plans for proposed new products and they delineate guidelines for new product planning. PIC-based guidelines include such details as the product class within which a new product(s) will fit, customers for whom new products will be developed, and process or product technologies through which new products are created. Strategy formation and PIC development provide the foundation for the remainder of the NPD process by providing a structure and context within which additional NPD decisions can be made.

Idea generation

Idea generation is the second step of the NPD paradigm. During this step new product ideas and concepts are discovered and created.

Typically, large numbers of ideas and concepts are generated. Firms attempt to obtain ideas from many sources such as customers, sales personnel, marketing, and R&D. Moreover, firms employ many methods or approaches to obtain new product ideas from these sources. Idea-gathering approaches encompass activities such as eliciting new product suggestions from customers and others and brainstorming sessions and other creativity practices that can assist new product producers.

Screening and evaluation

The third step of the traditional NPD paradigm involves screening and evaluating product ideas and concepts. This step acts as a filter or choice-making process for the many ideas and concepts gathered. Firms can employ several procedures to screen and evaluate new product ideas and concepts: rudimentary design and producibility studies can be made; preliminary customer reaction tests can be performed; and standardized checklists that scrutinize such factors as product idea fit with (1) the sales force, (2) desired new product lines, and (3) the type of customers the firm wants to serve.

Some firms perform extensive customer concept testing during this step. This entails developing a new product idea to the point that its attributes are clear and its product technology (how the product delivers customer benefits) is defined. Concept testing usually is only performed on those new product ideas that have passed more rudimentary filters like a demonstrated fit with the firm's new product strategy. It is important to screen out poor new product ideas through evaluation because as soon as customer evaluations, business potential assessments, and prototype construction begin, costs rapidly increase.

Business analysis

The next step in the traditional NPD paradigm is business analysis. During this step much effort is applied to ascertain a new product's financial viability. A product's sales potential is assessed in both the current period and some time into the future. Product concepts that demonstrate the potential to become financial successes are allowed to progress beyond this point. Marketing plans are developed. Costs and prices are also estimated. The opportunity to perform an in-depth assessment of the process technology (how the product is to be fabricated and assembled) is pursued, along with an analysis of the resources required to design various versions of the new product under scrutiny.

Product research and development (R&D)

Product R&D follows the successful completion of business analysis. During this step, the investment made relative to each product concept significantly increases. The product concept evolves into a tangible product during this step. The manufacturing equipment needed to fabricate and assemble a new product must be constructed or purchased. Equipment obtained early in the product R&D step is used to make the first prototypes of a product concept. Later in this step, additional equipment is obtained to establish processes for generating larger quantities of the product (pilot runs). Pilot runs provide insight into the eventual full-scale production operation necessary to commercialize new products.

As in other earlier steps, all functional groups need to work together. It is highly important that design, development, manufacturing, and marketing closely coordinate their activities during product development (Hise et al., 1990) because, as the prototype takes form and pilot runs are planned, important customer-desired attributes can be overlooked.

Test marketing

Test marketing is next in the traditional NPD process. For consumer goods, this traditionally implies a bona fide test market in which products are sold through typical channels of distribution, promotions are employed that have been created for the new product, and "everyday or expected" prices are sought. Alternative forms of market testing that firms employ include computer simulation of customer preferences and simulated test markets. Computer product test simulations offer advantages such as reduced test time and lower costs than traditional test marketing efforts. Another alternative is simulated test marketing. Unlike computer simulations, simulated test marketing employs actual customers and, therefore, mitigates one problem associated with computer simulations. For technology-based products "test marketing" may be implemented by placing a new product with a few important and trusted customers.

Commercialization

The final step of the traditional NPD process involves new product commercialization. The ability to commercialize arises from an integration of product and market capabilities and activities. During this step the product can be launched or offered to all parts of a national or world market simultaneously or it can be rolled out regionally. A roll-out strategy allows firms to be more conservative in their marketing approach, but it also allows the competition time to imitate.

TABLE 8.2   Traditional NPD Paradigm Limitations

Linear and fragmented

R&D as driver/limited marketing and customer involvement

Time crunch during final steps

High probability for total redesign/compounding of errors

Tendency to develop me-too products

Low transfer of learning between new product projects

Traditional NPD Paradigm Limitations

Although the traditional NPD model noted in Fig. 8.1 represents the current NPD paradigm for most firms, it has received significant criticism. As noted in Table 8.2, a number of limitations exist that preclude this paradigm from fulfilling the need for greater success in the development of new products. Wind and Mahajan (1988) note:

> ...most new product entries are "me too," slightly modified products, or line extensions. There are relatively few truly innovative new products, and, most disturbing, the success rate of new-product entry is still abysmal (p. 304).

The first significant limitation to be found with the traditional NPD process is embedded in its linearity and the isolation of its steps. Linearity projects a simple picture onto a very complex process. It implies, for instance, that later steps cannot begin until prior steps are completed. Additionally, the sequential character of this paradigm suggests noniteration—the notion that earlier steps cannot be repeated once later steps have begun.

The traditional NPD paradigm gives the appearance that some functions or organizational groups could successfully perform the tasks of each process step without integration with other functions. One might think, for example, that senior management would be responsible for new product strategy, engineering for product development, finance for business analysis, and marketing for commercialization. Such paradigmatic implications are far from the truth. Not only does interfunctional integration need to occur during each step of NPD but multifunctional approaches are necessary to bridge gaps between steps, which the traditional NPD paradigm fails to address. In addition, with judicious planning, NPD steps can be overlapped to complete the entire process faster, which represents another process aspect that is not adequately addressed by the traditional NPD paradigm.

The traditional NPD paradigm appears to relate most closely to organizations that are engineering driven or technology-push oriented. These firms have a strong R&D team and a management philosophy that supports the development of new products through internal engineering "breakthroughs." Marketing is likely to have limited influence on NPD. Consequently, customer desires may not be incorporated into new products. Such companies rely on engineering to make the best product choices for customers. This results in an NPD process that is heavily biased toward new technology-driven products. The results of such a process or paradigm may not be acceptable to a firm's customers and, hence, denote meaningful limits to the usefulness of such an approach.

In addition to applying or translating limited marketing input, early NPD steps of the traditional NPD paradigm tend not to be well performed. Especially in engineering-driven firms, product idea screening with customers and customer concept testing tends to be minimized. The portion of the NPD process that is most familiar to engineering and with which engineering has the most interest typically is product R&D. Therefore, it is not surprising that engineering or technology-push firms employing the traditional NPD paradigm tend to excel at product development but falter during other critical NPD steps. Even in technology-driven firms, product development steps based on the traditional NPD paradigm can be problematic. The most technically proficient products may be created, but they might not sell if customer needs are not met.

In addition to early steps of the NPD process being rather improperly performed, the steps at the end of the NPD process also do not receive their rightful share of attention in traditional NPD paradigm analysis and enactment. Later NPD steps are often shortened to meet scheduled launch dates, and product test marketing of any kind can be seriously compromised. Also, since the traditional NPD paradigm can be viewed as a series of steps requiring functional specialization, steps in which the firm has little or no expertise may be inadvertently omitted or be ineffectively performed. A final problem with the traditional NPD paradigm involves product or design faults found late in the process. In the traditional paradigm there is little provision for looking ahead. Therefore, when a design or marketing problem is encountered, a total product or marketing plan redesign can be required. Such major schedule slippages can cause market windows to close before products are launched.

An emerging NPD paradigm can offer solutions to these limitations in the form of paradigmatic changes such as overlapping steps, intelligent up-front planning, and continuous new product innovation. The next section presents an examination of such an emerging NPD paradigm.

An Emerging NPD Paradigm

Due to the limitations associated with the traditional NPD paradigm, the complexity of the NPD process itself, the structural changes taking place in many firms, and the complex environments in which new products are developed, the paradigm for the development of new products is evolving. As a paradigm, the traditional NPD process represents the way firms have produced new products. Because this paradigm is viewed as the standard approach to develop new products, rules have been created to support this approach. These rules have become the pillars upon which the traditional NPD paradigm rests. Given the view that the traditional NPD paradigm is "the right way," new-product-developing firms typically attempt to follow this methodology.

Many firms realize their NPD opportunities within the boundaries of these existing rules and their associated organizational structures. Even when difficulties like performing the linear NPD process faster or fusing better connections among the serial steps of the NPD process are discovered, firms often attempt to solve those problems within their known paradigmatic rules. Attempts to realize new opportunities through traditional NPD processes have not produced solutions that fulfill current NPD challenges. Consequently, the major objective—to increase the rate of new product success—also is unmet.

Since many problems are not solved within the traditional NPD paradigm, more and more firms are "breaking the rules" to find new solutions. There are some very compelling reasons to break the rules. These include the abysmally low new product marketplace success rate, the high cost of new product failure combined with rapidly changing consumer tastes, and the knowledge many managers have that NPD skills comprise an important strategic advantage in a fast-developing global economy.

Firms that break the rules of the traditional NPD paradigm are creating a new paradigm, a revolution, or a "paradigmatic shift." As Kuhn (1970) notes:

> The decision to reject one paradigm is always simultaneously the decision to accept another, and the judgment leading to that decision involves the comparison of both paradigms with nature *and* with each other (p. 77).

In addition, he points out that:

> Political revolutions are inaugurated by a growing sense...that existing institutions have ceased adequately to meet the problems posed by an environment that they have in part created...scientific revolutions are

inaugurated by a growing sense...that an existing paradigm has ceased to function adequately in the exploration of an aspect of nature to which the paradigm itself had previously led the way (p. 92).

Paradigmatic shifts can be likened to earthquakes. Sometimes they are slow and rumble along and other times they are swift and cataclysmic. In the case of the NPD process, the paradigmatic shift appears to be more of a slow, yet steady rumble. Both the slow and the fast processes have important implications for new product developers. It may appear that a slow change process is better for firms so that they can remain abreast of the situation and adjust accordingly. However, a slow paradigmatic shift may be more disastrous than a sudden one. An analogy that depicts this situation is the timeless anecdote about boiling the frog. A frog is more likely to jump out of water that is rapidly heated, whereas a frog may feel confident of adapting to water that is heated slowly. Either way, the frog is in trouble. The only difference is that in the former case the end comes more quickly and perhaps more mercifully. In the case of NPD paradigm changes, the situation is similar. The firm that notices a rapid change in the way new products are developed may "jump" onto new methods. When the process changes slowly, firms may be reluctant to change or not notice the change until the competition has "boiled them." A description of an emerging NPD paradigm and an exploration of the indicators that a paradigmatic shift is actually under way is presented next.

A new model for NPD

Figure 8.2 portrays the model of a new NPD paradigm. It not only depicts the core processes, as Fig. 8.1 does for the traditional NPD paradigm, but it includes the support concepts without which the NPD process cannot successfully function. This model implies that successful NPD must be driven from a sound customer and technical knowledge base. This means that for a product-developing firm to be successful it must be both market and technology driven. Figure 8.2 points out that a firm's new product strategy is a synergy of its customer orientation, technical know-how, and vision of the future.

A conscientiously crafted new product strategy is implemented through simultaneous employee empowerment and concomitant involvement and innovation of the workforce. This synergistic process typically has been described as involving members within a particular function and only for the duration of that function's NPD task. This new model suggests that the "players" are management, an NPD team, the workforce, customers, and perhaps other organiza-

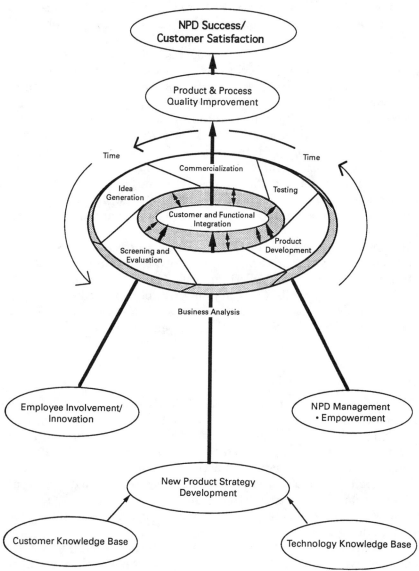

Figure 8.2 An emerging NPD paradigm.

tions such as suppliers or partners. Moreover, this model implies that synergy or integration must occur for the duration of the NPD process.

Another important aspect of this model is that it portrays the traditionally linear NPD process as a circle with its center supported by the integration of the internal and external organizations required to implement an NPD strategy. The circle of NPD tasks indicates continuous innovation linking the iterations of the development of a new product to one another and one NPD project to another. Moreover, the model suggests all of the integrated NPD personnel may have an input into each task of the NPD process. A final difference between this emerging model and the traditional model is that the tasks of this process can be overlapped. This is indicated, for example, by business analysis beginning before screening and evaluation has been completed.

The output of the model for the emerging NPD paradigm is improved product and process quality, which translate into increased customer satisfaction and NPD success. Thus, this emerging NPD paradigm can help fulfill a primary objective of new product development—market success. The following describes organizational and environmental activity that foretells the emergence of a new NPD paradigm.

Paradigmatic shift indicators

If NPD process change is depicted as a slowly evolving paradigmatic shift, how can management recognize that a shift is occurring and that this condition is not simply a trend within the traditional NPD paradigm? First, managers need to be able to distinguish between trends which are continuations of activities that are controlled by the rules of the traditional paradigm and activities that appear to be controlled by forces that break the traditional paradigm's operating tenets. Minor discontinuities in trends also must not be mistaken for paradigmatic shifts.

Trends are outcomes of decisions that are derived from the interaction of behaviors of many actors in the NPD process, such as the firm itself, the competition, and the state of the economy in general. Paradigms embrace the rules guiding the behaviors of those actors. Managers must be able to trace their way backward to the causes of the behaviors and separate those behaviors that are generated by actors who change their governing rules intentionally or unintentionally. They must also do this with those behaviors that are created within the scope of the traditional paradigm's rules but have interacted in such a way as to cause a change in an outcome trend. Five orga-

TABLE 8.3   Paradigmatic Shift Indicators

Changes in organizational structures

Globalization of new product development

Accelerated new product development

Heightened concern for product quality

Enhanced communication and information technologies

nizational and environmental factors that are indicative of an NPD paradigmatic shift are proposed. They are itemized in Table 8.3. The five factors include organizational restructuring, new product development globalization, NPD acceleration, concern for product quality, and the advent of new communication and information transfer technologies.

Organization Structure Changes and Organizational Restructuring

This section presents two general types of business restructurings that are indicative of a paradigmatic shift. The first is the development of new products by firms with flatter organizational structures caused by fewer people than previously, especially managerial personnel. The second is the development of new products by multiple firms.

Flatter organizations

Growth in profits has often been associated with growth in firm size, especially in terms of assets and personnel. It can be said that a correlation between profit and personnel growth has been a "rule" of the traditional NPD paradigm. Managers typically would not consider the development of a new product or product line without a considerable increase in personnel. Increasingly, organizations are creating and developing superior new products without adding to their workforces.

Since a paradigm involves rules, it is somewhat like a very high-level strategy or plan. If the strategy or traditional paradigmatic rule of "large organizations are needed to generate large profits by developing significant new products" is changed to "smaller organizations can develop significant new products and generate equally large profits," firms that have flatter, smaller organizations will operate under new paradigmatic rules. Currently many companies develop new products in organizations restructured to "do more with less." From this perspective firms have become flatter and the span of control of many mangers has far exceeded that recommended by classical organization theorists. The flattening of new-product-developing organizations has made way for a greater delegation of responsibility to people at lower levels in organizations. Moreover, managers now responsible for many more personnel do not have time for detailed instructional sessions with each employee—therefore, people within organizations increasingly are empowered to make and implement important NPD decisions.

Multiple-firm NPD

A second indication that the NPD paradigm is shifting vis-à-vis organizational structure is the use of networks to develop new products. Under the rules of the traditional NPD paradigm, firms typically choose to "go it alone." Under the rules of the traditional paradigm, managers are unwilling to share the profits of new product success with other, external new-product-developing organizations. It has been common for firms to build up large R&D departments to search for ideas and develop new products. Today it is common for firms to share the profits and risks of new product ventures with partners who employ their expertise to compensate for weaknesses within other NPD organizations.

Network-type organizations can be fashioned in many ways. Joint ventures in which a third firm is created to develop and market new products have been very common, especially with firms doing business crossing national boundaries (Harrigan, 1985). Another approach is strategic partnering. In this environment, the participating firms retain their own identity, typically creating contracts that stipulate the degree and type of interaction desired between or among partnering firms.

Whereas NPD under the traditional paradigm typically was based on a high degree of secrecy, NPD under the rules of this emerging paradigm is based on information sharing by partnering firms. New product information sharing causes organizational boundaries to blur. Such blurring is due to the sharing of people and resources needed to develop new products that are successful in the marketplace. Although sharing NPD information is critical to this emerging NPD paradigm, it is important to note that *not everything* can be shared. Each firm's critical core expertise needs to be protected to secure each firm's competitiveness and place in the network structure.

Global new product development

In the traditional NPD paradigm, it is believed that large markets, especially those of economically developed nations, should be targets for newly developed products. Therefore, firms should design, manufacture, and market their products close to such markets. In this emerging NPD paradigm, firms are actively seeking personnel, funds, other resources, and customers from all parts of the globe. This shift is occurring because foreign firms with high-quality products and the ability to obtain low-cost labor and materials have stretched beyond their borders to gain market access. Many firms that previously operated under the rules of the traditional NPD paradigm are now taking a "global" NPD perspective.

In this emerging NPD paradigm, new product developers are increasingly becoming firms without countries; they place their operations in whatever part of the world they believe is the best from cost, customer, and competitive perspectives. Such firms develop new products for any customer that needs their products, manufacture in parts of the world that offer economical labor and materials, and perform research and development activities where there are high concentrations of scientific and technical skills.

New product development acceleration

Another sign that a new NPD paradigm is emerging involves the interest of firms in the speed with which they develop and market new products. Many firms are finding that the traditional, serial NPD process does not allow them to become product pioneers in their chosen marketplaces. The rules of the traditional NPD paradigm indicate that NPD steps should progress in sequence (e.g., one step should not begin until a prior step is completed).

Firms previously took several years to conceptualize, design, manufacture, and market new products. Today many firms have less than 1 year to develop a new product, and some companies are even striving for new product launches in as little as 6 months. The shortening of the NPD process allows firms to more precisely time their new product introductions. For example, Honda, AT&T, Hewlett Packard, and Xerox have taken steps to reduce the duration of their NPD processes for new automobiles, telephones, printers, and copiers, respectively. They are thereby entering markets earlier and satisfying customers' needs sooner.

This emerging NPD paradigm recognizes that products need to be developed in a timely manner. In order to do so, firms are simplifying products and production processes, eliminating unneeded process activities and delays, paralleling and overlapping process operations, and speeding up NPD tasks to accelerate their NPD processes (Millson, Raj, and Wilemon, 1992). The traditional NPD process simply takes too long. When firms had less competition and product life cycles were much longer, the traditional NPD process was adequate.

Today, competition can arise from many points around the world. This makes it far more difficult to assess the length of new product life cycles.

New-product-developing firms have become very proactive in the replacement of their current products with new offerings—this is in response to concerns about accelerating NPD to meet launch windows and preempt competitive efforts. Firms have become particularly challenged by the need to replace their own product offerings when customers want new product solutions to their problems and for meeting business opportunities.

Product quality

In addition to accelerating the NPD process, firms must assure the design and manufacture of high-quality products. More than ever, customers expect new products to meet high quality standards. In the traditional NPD paradigm, product quality was sought by each organizational function that performed a step along the linear NPD process. "Quality" typically was measured by a product's conformance to engineering standards and specifications. These measurements usually did not occur until the new product had reached the product development step.

In this emerging NPD paradigm, process and product quality standards are developed from customer wants and needs. Product quality is measured by customer satisfaction throughout the NPD process. Again, important differences can be noted between the operating rules of the traditional and emerging NPD paradigms.

One program that is changing the shape of new-product-developing firms is Total Quality Management (TQM). TQM is associated with many important management functions and attributes such as education, communication, performance effectiveness and efficiency, and management style. TQM requires firms to "break down" communication barriers, develop measurable goals for individuals and teams, and focus on continuous quality improvement in all areas. The implementation of TQM typically necessitates dramatic social change within organizations. Many times the achievement of such major social change requires organizations to modify their organizational cultures (Scurr, 1991).

TQM also requires a firm to take a wholistic view of itself and its environment. The TQM process demands that all activities be performed in a high-quality manner. The emphasis of TQM is on the total organization; management communicates a basic premise to all employees which states that employees have "customers" for their product efforts "within" the organization, in addition to the "external" customers of the firm. Each customer's needs must be clearly understood. The supplier—an individual employee or the firm as a whole—must put forth every effort to satisfy customers' needs and desires. Quality Function Deployment (QFD), a more narrowly defined process, is a disciplined system for incorporating customer wants and needs into new products. A major objective of QFD is to incorporate customer requirements into new products at the most rudimentary level of the manufacturing process. QFD accomplishes the incorporation of customer needs in product designs by first understanding those needs and wants. The QFD process involves the creation of matrices that relate engineering solutions to customer needs. The correlation between engineering solutions and customer requirements is monitored to ascertain when and where trade-offs are required. When trade-offs are necessary, customer needs are reviewed to determine how to provide the highest customer satisfaction.

TQM addresses the social, educational, and communication aspects of the NPD process while QFD can coordinate the screening, evaluation, and product development steps of the NPD process. Companies who participate in this NPD paradigm need to study both processes and determine where each supports their NPD needs.

Communication and information technologies

A final paradigmatic shift indicator is the increased use of state-of-the-art communication and information technologies in NPD. Even relatively small firms are employing sophisticated communication technologies to develop new product offerings.

Some of the new devices that are speeding and easing the NPD process include the use of personal and laptop computers, facsimile machines, computer workstations, and networks and their associated software by new product designers, manufacturers, and commercializers. In some instances, these devices are eliminating the need for drafting personnel because the design process has been automated to permit the input and continual reprocessing of design information directly into NPD databases. These databases are also remotely accessible by manufacturing engineers and purchasing agents, thereby speeding the development of production equipment and the early procurement of production materials and capital equipment. Databases also are accessible to field personnel, such as sales and new-product-developing partners. This rapid access to new product information helps firms satisfy customer needs before their competitors do (Kalashian, 1990).

Travel costs and NPD partner remoteness have prompted increased reliance on sophisticated communication and information equipment. Devices such as video conferencing and Electronic Data Interchange

(EDI) have assisted in both timely new product decision making and receipt of raw materials. NPD partners can now interactively create and screen new product ideas where they previously shared pictures and concepts by mail or personal transport.

Communication and information storage and transmission devices are speeding the NPD process by allowing partners to form NPD network organizations. These developments violate the rules of traditional NPD because they allow steps to overlap and multiple firms to efficiently work together on NPD projects. We can anticipate that these developments in information and communication technology will increase the rate at which a revolutionary NPD paradigm emerges.

Summary of paradigmatic shift indicators

Employment of network organizational structures to develop new products and the implementation of product quality-enhancing processes are paradigmatic shift indicators. Such choices allow firms to change their new product development rules. The NPD success associated with flatter, smaller structures indicates that large organizations are no longer seen as necessary for achieving profitable new products. The globalization of the NPD process implies that firms no longer view the development of products solely within the confines of a single firm as necessary, nor does management believe that new products should always be designed, manufactured, and launched in a domestic market alone. Firms that accelerate their NPD processes are relying on a nonlinear view of NPD as a way to recognize and meet the demands of new product launch windows. TQM and QFD are complementary processes that can be employed to attain new and better products faster. The implementation of these processes requires a new perspective of the NPD process in that customers both external and internal to the organization become the focus of the NPD effort. Use of high-technology communication and information storage and transmission devices also is accelerating the transition to an emerging NPD paradigm. The development, adoption, and rapid diffusion of user-friendly software, electronic hardware, and more computer-literate personnel indicate that many firms are concomitantly creating a wholistic NPD perspective—a radically different approach to NPD than that prescribed by the traditional NPD paradigm.

New Paradigm Participation

This section describes how firms can perceive the emergence of a new paradigm for NPD, "learn" its attributes, and participate in it. Firms that are willing and able to participate must be prepared to make difficult, sometimes radical modifications in their organizations and

TABLE 8.4   NPD Paradigm Participation Recommendations

Encourage organizational learning

Emphasize continuous innovation/quality

Develop a prospector culture

Develop specific product technologies

Create an NPD acceleration capability

Develop an innovation infrastructure

Foster a wholistic NPD perspective

operating philosophies in order to move from their "old" NPD ways to new behaviors and ways of thinking. Table 8.4 presents recommendations for firms that wish to participate in this emerging NPD paradigm: (1) encourage organizational learning, (2) emphasize continuous innovation and quality improvement, (3) develop a prospector culture, (4) develop specific product technologies, (5) create an NPD acceleration capability, (6) create an innovation infrastructure, and (7) foster a wholistic NPD perspective.

Encourage organizational learning

There are three major methods by which learning occurs in organizations. The first level of learning can be described as the process of detecting "and correcting" errors (Argyris, 1977) or incremental innovation (McKee, 1992). The second level can be viewed as "double-loop" learning (Argyris, 1977) or discontinuous innovation (McKee, 1992). And the third and highest level of learning can be envisioned as "deutero" learning (Bateson, 1972) or organizational learning (McKee, 1992).

Organizational learning can help management understand the differences between the traditional and emerging NPD paradigms, provide information transfer between NPD projects and supporting organizational functions, assist in making sound NPD decisions, reinforce the importance of new organizational cultures, and enlighten people in their effort to create successful new products (Wheelwright and Clark, 1992).

The first product innovation learning method described by McKee (1992) may be the most common and easiest to undertake. This incremental learning model is a simple "closed-loop" system and can be likened to a temperature control system containing a thermostat as a control mechanism (Staudt, Taylor, and Bowersox, 1976). To operate this type of mechanism the thermostat is first set to a desired temperature. When the ambient temperature varies above or below acceptable limits, either a heater or an air conditioner turns on to "correct the error," which returns the area environment to the preset temperature. The error information comprising a temperature that is outside prescribed limits is the knowledge that actuates the system and inhibits further learning to take place in regard to the environment at temperatures outside the preset control limits. In this sense, the system has only learned to be in control by returning the temperature to the preset condition. Single-loop learning assumes the "correct" NPD process is known, and after an initial condition is set and a difference is detected, the learning process returns the system to the known condition, and the cycle begins again.

The second product innovation learning process described by McKee (1992) is much less common and far more difficult to employ. The description of this process, for example, begins much like the first one in that a temperature control system also can be used as the basis of the learning system. In this scenario the thermostat is set for a particular temperature, as in the previous example, and acceptable limits are set to allow the temperature control system to return the environment to the preset temperature when either the high or low limit is exceeded. Therefore, a single-loop learning system is the core of this process. In addition to the core single-loop learning system, this second learning methodology employs another control loop that embraces the entire single-loop system. This second loop system operates by occasionally asking the "operator" of the temperature control system whether the preset temperature is appropriate. Therefore, this system has the ability to learn about "errors" or "temperature settings" that are outside the single-loop control limits if the "operator" chooses to do so.

An example of such learning systems can be derived from the oil crisis of the early 1980s. Many people traditionally kept their homes near 80°F. Therefore, their thermostats were set near 80° and their temperature control systems would return their living areas to 80° when temperature deviations occurred. This is an example of a single-loop learning system in which no one questioned the validity of the preset temperature of 80° because it was a "normal" house temperature. With the oil crisis of the 1980s came much higher oil prices and higher energy bills. Those home owners who questioned the validity of operating their houses at 80°F were employing double-loop learning. Many of these home owners found that they did not need to have their homes at 80° to feel comfortable and that there were other ways to attain comfort at other ambient temperatures.

Another perspective of these learning systems can be found by analyzing the concepts of "doing things right" and "doing the right things." People are typically rewarded for doing things right and occasionally punished for doing things wrong. Moreover, errors are often associated with doing things wrong. At an early age, we are taught to do things right and not make mistakes. We are often reprimanded for behaviors that are deviations from the "normal" ways things should be done and occasionally rewarded for complying with others' wishes and desires. Therefore, "doing things right" is analogous to single-loop learning or incremental product innovation in which we are given limits regarding our new product methods and are rewarded for remaining within the limits of preset behavior or punished for deviating from that behavior. It is interesting to note that in such systems any deviation from the norm for any reason is regarded as an error. Obviously, this is a very confining system although simple and fairly easy to understand and operate.

On the other hand doing the right things requires a more complex control system. To do the right things, we must first find out what the right things are and then create and implement them efficiently. This form of new product learning is analogous to discontinuous innovation (McKee, 1992). A process that questions the activities that are being performed before attempting to perform them is double-loop learning. It is this system of learning that forms part of the basis for participation in this emerging NPD paradigm. The behaviors associated with double-loop learning lead to effective performance because the right things are being done. A note of caution must be made here. We also must perform the core, single-loop learning process because to be efficient we must do the right things right.

The third and highest level of learning in new-product-developing organizations is organizational (McKee, 1992) or deutero learning (Bateson, 1972). The organizational level of learning is concerned with institutionalizing NPD learning experiences to improve the effectiveness of future new product developments. This is the type of learning which firms that wish to fully participate in this emerging NPD paradigm must achieve.

This level of learning is not focused on the incremental enhancements of a single product or product line as in single-loop learning. Nor is this level of learning only associated with the development of a new-to-the-world product via discontinuous innovation. The organizational level of learning provides the learning that is necessary to carry a firm through the development of many new products whether they are incremental or discontinuous. This level of learning provides the "glue" that links NPD programs together in a continuous NPD effort.

To engage in organizational or deutero learning, McKee (1992) notes that "unlearning" may need to occur. To accomplish this firms can foster cultures that accept and learn from errors. In some instances firms may need to remove people to "unlearn." Firms need to provide individuals with error-learning skills to engage in single-loop learning to enhance their personal performance. Firms also need to reward the communication and the shared learning derived from this process. To move beyond single-loop learning, firms need to refrain from punishing failure and reward the "learning" that occurs when NPD tasks fail. These learnings need to be shared. The freedom to take risks and to learn from failure allows people to expand their limits; it can also lead to highly effective double-loop learning. The unconditional sharing of both successes and failures supported by an organizational culture committed to innovation can provide the basis for a learning organization (McKee, 1992).

Emphasize continuous innovation and quality improvement

The process of continuous innovation can be seen as a cultural attribute supported by a nurturing management philosophy. Continuous innovation connotes an ongoing renewal and quality improvement of products, processes, and organizations. The emerging NPD paradigm embraces the notion of increased customer satisfaction through product solutions that are more closely aligned with the changing needs of customers. Such new product-customer need alignment can only stem from the continuous development and timely introduction of new products, including the knowledge about products and their servicing.

Continuous innovation and development of high-quality, customer-satisfying products are important facets of this emerging NPD paradigm. Continuous NPD is important because it is easier to manage, fosters less hostility within organizations, is less disruptive, serves customers' needs better, provides continuous interaction with customers, and is a process that is synergistic with organizational learning and the strategy of prospecting.

In an environment of continuous change, it is vitally important for firms to create and communicate clear and compelling organizational directions. Such communication, based on serving customers well, can provide the context and structure for constant improvement and innovation. Within such a context the development of new products can become less stressful. The management of the development of new products is a challenging experience at best. To marshal the financial and personnel resources each time a new product is planned can be difficult. Continuous product innovation represents a way to start from a base or vision that provides the necessary explanations and justifications for NPD and a core of personnel from which to move from one product to another. Continuous product innovation can reduce the resentment that comes from selecting only a few personnel to be a part of a periodic NPD effort upon which a firm's survival rests. Moreover, continuous product innovation reduces the potential "we-they" attitudes that are often created when collateral organizations, elite groups, and "skunk works" are formed to develop new products.

Continuous product innovation can reduce the disruption that comes with intermittently removing people from their "regular" positions to be part of a segregated NPD effort. Continuous processes also mitigate situations in which "extra" or "undone" work from NPD personnel falls on others to either "clean up" or simply implement. If most personnel (some individuals will not want to participate in NPD) are selected from time to time to participate in the development of the next product iteration, the people left behind will not perceive themselves as the employees that get the "leftover" jobs.

To continuously innovate, firms must foster ongoing, highly productive interactions with their customers. This means organizational members must be in contact with customers before, during, and after product development and launch. Increased customer interaction provides firms with more in-depth understanding of customer wants and needs in addition to the increased trust that can be formed by customers who perceive employees to be genuinely interested in their problems. An important aspect of continuous product innovation and quality improvement—or operations—is that customers are served better. Customer service involves increased understanding of the problems customers want solved.

In these turbulent times, firms need to be doing all they can to stimulate customers and employees. Continuous innovation at all of the three previously mentioned learning levels is desirable. New, incremental changes and product enhancements can be made to current products and manufacturing processes to keep customers interested and costs decreasing. In addition, new product solutions to new and current customer problems need to be addressed. This entails employing discontinuous learning and major innovation. Continuous product renewal and development by different or overlapping groups moves the firm to the third, or organizational, level of new product learning. To facilitate this process, senior management needs to be committed to continuous innovation, give individuals and groups rewards that support continuous innovation, and develop a climate of intelligent risk taking where failures lead to learning and eventual success, not punishment.

Develop a prospector culture

Miles and Snow (1978) note four types of strategies that organizations employ to adapt to their environments. They point out that firms behave as either *prospectors, defenders, analyzers,* or *reactors.* Firms that select a prospecting strategy continuously search for market opportunities and create the change and uncertainty to which their competitors must respond. These firms "experiment" with new products, new markets, and new ways of thinking and behaving as their standard mode of operation. Due to such a constant state of flux, however, prospecting may not be the most efficient strategy.

A defender, on the other hand, typically has a limited product market (a specific product or product line targeted toward a particular market) and is an expert in this narrow market domain. The focus of such firms is on developing process adjustments to increase the efficiency of their processes, typically to reduce manufacturing costs. Miles and Snow (1978) describe analyzers as firms that develop some attributes of both the prospector and the defender. These firms are not as aggressive in the marketplace as prospectors and tend to create me-too products. Additionally, analyzers are not as introverted as defenders and, therefore, tend to be somewhat less focused on process development. The firms that comprise the fourth typology, reactors, are typified by inconsistent and unstable adjustments to changes in their environments. Therefore, these organizations are typically less successful than the other three types of firms.

The contrasting adaptation processes of the prospectors and the defenders can be seen as learning systems (McKee, 1992). The defender continually develops the process technology for its current product market. In this respect it develops limits relative to the quality of its product and creates and enhances process technologies to improve its cost position while attempting to hold its product quality constant. The defender's approach toward growth comes from its cost leadership position and a desire to gain greater market penetration and share. The defender is less likely to investigate developments outside its primary product market domain. The best defenders have well-developed single-loop learning systems to keep costs low and efficiency high. These firms have the capability to perform double-loop learning but are likely to find it more difficult. This difficulty comes from the defender's lack of market and environmental scanning. This lack of scanning lowers the potential for defenders to become compelled to question their governing policies and regulations. In other words, when a defender employs its NPD process, it is likely to be the traditional paradigm because defenders are less likely to detect shifts in the need to do things differently.

The prospector, on the other hand, is primarily interested in growth from the development of new products and the creation of new markets. This environmental adaptation strategy requires the continuous monitoring of the environment in domains remote from a prospector's current area of interest. Prospectors also are dedicated to finding new customers and new problems to solve for current customers. The implementation of this strategy is more likely to promote the questioning perspective required by the double-loop learning system and, therefore, detect and foster paradigmatic shifts. Such detection is important to prospectors and defenders alike, but prospectors appear to have an advantage. Therefore, in terms of the previous discussion, prospectors are more likely to do the right things and be more effective, whereas defenders are more likely to do things right and be more efficient.

It appears as though the business environment will be increasingly turbulent. Firms from all points on the globe will soon be competing in many markets as countries develop. The prospector that does the right thing and does the right thing right may be a new breed. This combination of a prospector with defender efficiencies is not an analyzer that brings forth me-too products at an acceptable cost. This new breed of firm brings forth innovative customer solutions and understands how to transport product innovation and learning to manufacturing processes and administrative systems to offer differentiated products in a highly efficient manner. Therefore, firms need to be prospectors first and defenders second.

A caveat is important here. It must be remembered that these two strategies are mechanisms employed to adapt to the environment. If the environment is stable and certain, a defender's perspective can be adequate. But, if the environment is uncertain and turbulent, a prospector's perspective can be more beneficial. In addition, a prospector can create and disturb an environment in ways that are incompatible with a defender's strategy. In recent years most organizational environments have become turbulent, uncertain arenas. This environmental phenomenon can be considered to be a force behind the NPD paradigmatic shift. Therefore, a prospector's NPD paradigm will more likely shift toward this emerging NPD paradigm before firms following other environmental adaptation strategies, such as defenders.

Develop specific product technologies

In order to innovate continuously, a firm needs to develop specific new product and process technologies. A product technology strategy can be defined as a particular way of providing customer benefits through the firm's product offerings. For example, some running shoe firms provide a more comfortable shoe by producing a wedge insert to support the runner's heel, whereas other firms have incorporated an air bag design to provide running comfort for their customers. The development of specific product technology strategies requires the choice of the best product solutions for customer problems and an organizational focus on those solutions. Therefore, the creation of specific product-technology strategies requires integrated inputs from all organizational functions. Organizational integration becomes a major factor in the development of specific product technologies and participation in this emerging NPD paradigm.

Process technologies can provide significant competitive advantages, deal primarily with the production of the product, and typically only affect activities internal to the organization. Product technologies also provide significant competitive advantages but extend beyond the organization to include how customer problems are solved and how information external to the organization can assist in the higher integration of organizational functions. Process technologies are primarily the domain of manufacturing, while product technologies fit and fulfill customer needs, mesh with marketing skill and expertise, and are supported by engineering's knowledge and aptitude. Moreover, product technology more than process technology requires that firms become prospectors. Therefore, for a firm to develop an effective product technology it needs a prospector's close customer contacts, an understanding of customer needs and problems, and alternative solutions for those problems. In addition, such firms need to possess a process capability that can be associated with the product solution which incorporates the chosen product technology.

Proprietary process technologies have given defenders competitive cost advantages. Such proprietary technologies were thought to have offered even more security than patents. Proprietary process technologies may no longer offer the competitive advantages they once did because of the unintended "technology transfer" of "process secrets" via layoffs and the increased mobility of personnel. Firms that develop specific product technologies and enter markets early with advanced solutions to customer problems can obtain competitive advantages in the 1990s.

Specific product technology strategies and their communication are important to a firm's participation in this emerging NPD paradigm because customers desire high-quality products *and* services. In addition to durability and reliability, product performance is a critical aspect of product quality (Garvin, 1984). Since product performance depends on the means by which the product solves customer problems and provides customer benefits, product and process technologies together become an important factor in the ability of a firm to participate in this new NPD paradigm.

Create an NPD acceleration capability

Another factor that must be considered by potential participants in this emerging NPD paradigm is the creation of a product acceleration capability. To participate in this emerging NPD paradigm, a capacity needs to be developed that provides firms with the ability to accelerate the activities of their NPD processes in order to take advantage of fast-closing customer-related new product launch windows.

There are few hard and fast rules to rely on to determine which new products need accelerating. Nor is it always economical to do so (Rosenau, 1988). It also is evident that not all new products require accelerating. It appears that each new product needs to be analyzed from a customer's need perspective, taking into account proposed and existing products and competitive positions. Firms that are continuously innovating, developing entirely new products, and iterating relatively new products are close to their customers and can ascertain which new product programs need acceleration.

The determination of strategic new product launch windows helps firms understand which new products need acceleration. A constant scrutinization of both the customers' needs and competitors' activities can be used to develop a product acceleration priority list. As a practical matter, NPD acceleration should become less important as firms develop cultures of continuous NPD and quality improvement. The reduced need for NPD acceleration should occur because firms will be continually focused on customers' needs.

In some cases, however, firms will need to accelerate the development of new products. To do this, five major acceleration techniques are recommended (Millson, Raj, and Wilemon, 1992). First, firms may be able to simplify products, the organizations associated with the development of new products, and the processes required for new product development. This can be achieved by reducing the number of parts and vendors used, creating effective linkages between NPD organizations, and assigning meaningful tasks to people. Second, firms need to eliminate all unnecessary activities associated with the development of new products, such as shortening the typically long list of approval signatures. Third, firms can minimize unnecessary delays between NPD activities and within major NPD tasks. This means that the personnel involved in succeeding steps of the NPD process should be informed when their efforts are required and materials and information need to be scheduled to eliminate delays in the NPD process. A fourth action that can be taken to accelerate the development of new products involves the paralleling and overlapping of NPD steps. This requires a review of the necessary NPD activities to determine which activities can be started before others are complete. This acceleration activity also requires risk taking, especially when the early steps of the NPD process are paralleled or overlapped. This technique requires that firms accept the risk that new product assumptions will not change significantly during the NPD process. The fifth action firms can take to accelerate the development of new products is to speed up NPD activities. This requires increased employee motivation and an investment in equipment to automate NPD processes.

Create an innovation infrastructure

This section addresses the often forgotten administrative actions, procedures, and systems required to support the development of new products, especially a process of continuous new product innovation. Examples of these administrative systems include firms' accounting processes, purchasing systems, training procedures, and human resource management (HRM) organizations.

First, managers need to understand how these traditional support structures operate and the information that they provide. Without this understanding, rational changes or additions to such administrative systems cannot be instituted. Second, managers need to understand what information they need in order to make effective decisions. Needless changes to existing systems only serve to aggravate individuals who operate them and reduce the trust of managers responsible for important system changes. Third, managers need to understand how their reward systems support or inhibit continuous product innovation. Well-intentioned reward systems can inhibit behaviors necessary to the functioning of continuous product innovation and product quality improvement. For example, rewarding personnel involved in the launch of a new product but not the people who made the critical changes to the accounting system (who provide the information needed to assess each separate iteration of a new product's development) is not enough to assure continued support during succeeding product iterations.

A final area to evaluate for its impact on continuous NPD is represented by the systems that support the hiring and training of employees. The interaction between new product managers and human resources managers can be highly beneficial to the development of new products. In situations where products are continuously developed and many personnel are employed during the development process, training is important. In organizations where new products are continuously developed, the transfer of information and learning experiences from one product iteration to the next is crucial for the acceleration of succeeding iterations and new product success.

Foster a wholistic NPD perspective

It is essential for managers to combine the foregoing concepts into an integrated package. The "winners" in the new NPD paradigm will eventually implement all of the previous suggestions. It is the "fit" among these elements that allows firms to become effective participants in this emerging NPD paradigm. Wholism implies that the process of developing new products is only as strong as its weakest link. Some of the links include the ability of personnel and organizations to learn and to transfer information among themselves and their markets through continuous innovation. Moreover, the environmental adaptation strategy of prospecting as a process fits well with a well-developed organizational learning environment. Developing specific product technologies which require close ties to customers also meshes well with the process of prospecting and learning and enables firms to better understand and serve their customers. Additionally, all of these concepts apply to the development of a culture of continuous product innovation and iterative product development which needs to be supported by a management philosophy that empowers people to do the right things. Morgan (1986) notes:

> Where corporate culture is strong and robust, a distinctive ethos pervades the whole organization: empowering employees to exude the characteristics that define the mission or ethos of the whole; e.g., outstanding commitment to service, perseverance against the odds, a commitment to innovation...(p. 139).

A wholistic perspective of continuous NPD needs to involve all of the NPD support functions and the NPD team, not just from a necessity point of view but from a total organizational team perspective. Including these support functions provides management with a better idea of how NPD information and learning flows through and is used by firms.

Implications for Innovation Managers

NPD managers need to become keenly aware of the factors associated with this emerging NPD paradigm. This new paradigm will undoubtedly influence NPD throughout the 1990s and beyond. Further, it has the potential to produce many of the answers that the traditional NPD paradigm has been unable to provide. A commitment to create a vision of the future that employs this NPD paradigm offers many important implications for managers. These implications stem from the performance of the activities associated with participation in this emerging NPD paradigm. They include "learning to learn," slowing down to speed up, doing more with less, focusing on customer-defined

TABLE 8.5   Implications of Emerging NPD Paradigm Participation

New paradigm behaviors	Implications and requirements
Encourage organizational learning;	Change culture; recognize and acknowledge errors
Develop a prospector culture	
Create an NPD acceleration capability	Plan, focus, and commit before speeding ahead
Develop specific product technologies	Focus to maximize resources
Emphasize continuous innovation/quality	Think like a customer, not a producer
Create an innovation infrastructure	Information is precious
Foster a wholistic NPD perspective	Think globally

product quality, developing a global NPD perspective, and effectively managing, storing, communicating, and using strategic information. The implications related to the performance of the behaviors required by this emerging NPD paradigm are summarized in Table 8.5.

Change culture: Recognize and acknowledge errors

Perhaps the most critical personal task managers have before them while attempting to participate in this emerging NPD paradigm is the struggle involving the process of relearning how to learn. Learning to learn implies that NPD managers and other personnel must be able to recognize and admit their mistakes. As children, people are given the inquisitiveness that is required to start them off in the world. Basically children are given "blank slates" with which to work. Children start to write on their slates as soon as they have experiences. Fortunately, they also are given another important tool: good erasers. When new experiences appear to conflict with the information children have detailed on their slates, they look for guidance and can readily make adjustments or delete the text altogether. At this time in their lives they don't know that they aren't supposed to make mistakes; they feel free to start over without external or internal impugnation. As children grow older, they unfortunately lose that precious quality of being able to start over, their erasers wear out, and they are often stuck with what is written on their slates.

The implication here is that someone else's eraser may be required to make adjustments and carry on. This change of culture and personal point of view can be difficult. But it is a basic premise of this emerging NPD paradigm. Change can occur. Old perspectives are not cast in concrete. Progress can be made.

Plan, focus, and commit before speeding ahead

The idea here is slow down to speed up. This apparent oxymoron is offered to point out that "speed" is not required all of the time. Stopping to smell the roses is an often-cited recommendation. This message should not be discarded. This emerging NPD paradigm implies that a new view of NPD should be taken. Tough decisions need to be made and risks need to be taken. And most of all, minds need to be opened to allow plans for the future to be made. Such plans will provide the capability to accelerate the development of new products. Anything short of a solid planning effort will typically result in progressing through the NPD process very fast, only to result in failure or an iteration of at least a portion of an errant process.

Focus to maximize resources

Focus to maximize resources implies doing more with less. The caption "do more with less" is very well used, but it does have importance in the context of this emerging NPD paradigm. In terms of the development of new products, doing more with less implies that management should strive for excellence in carefully selected new product categories. This means that firms need to focus their product markets and therefore their NPD efforts. This type of product-market concentration significantly helps perform the screening and evaluating activities described in the traditional NPD process, in addition to streamlining the firm's supporting infrastructure. Such a focus keeps the firm from wasting resources on developing concepts and products that detract energy from core businesses. A firm may continue to have multiple businesses and employ this emerging NPD paradigm. But each business must generate a sense of focus and commitment to its own product markets. The new product strategy step of the traditional NPD paradigm provides such a product-market focus.

A caveat regarding the focusing of effort in the NPD process and strategy formulation process needs explanation. Idea generation requires both focused and unfocused approaches. This means that during this particular task the participants need to be encouraged to wander as far and wide in their thinking as possible. This is another area where the child's clean slate metaphor applies. The idea genera-

tors for new products must have no restraints until they are required
to determine what to do with the new product ideas they have generated. An idea far afield may eventually be associated with a very profitable product at a later date. And product ideas that are not applicable now may become viable future alternatives.

Think like a customer—not a producer

Management needs to understand its customers' perspective of quality. This statement has implications for new product developers and
involves many aspects of this emerging NPD paradigm. Three important NPD aspects that are associated with the customers' view of
product quality include the concepts of a specific product technology,
continuous product innovation, and customer satisfaction or product
redesign after launch.

First, to become a leader in a product category, firms must be able
to develop distinctive product technologies that solve customer problems better than competitors. This means that new product developers need to intimately understand customer problems in order to be
able to develop new product solutions perceived as providing high-
quality answers to those problems. Second, the culture and process of
continuous product innovation instructs firms to "get close" to their
customers' problems, offer tailored solutions, and rapidly enhance
those solutions to move even closer to the "target" product.
Continuous product innovation also requires that firms remain close
to their customers "after" new products are launched. Third, customer
satisfaction needs to be monitored to ascertain which product benefits
need to be left intact, which ones need to be changed, and which ones
are so far off that customers actually make product changes themselves. Such customer redesigns need to be understood and assimilat-
ed into the new product technology.

Information is precious

Becoming a good prospector, creating tailored new product technologies, and developing a learning organizational culture implies that
management operating in this emerging NPD paradigm must manage, store, and use information effectively and efficiently. Information
is the glue that links the various parts of this emerging NPD paradigm together to form the wholistic picture required for NPD success.

Critical decisions associated with strategic information involve
important choices about what information to manage, store, and use.
These are not easy questions to answer and the answers can differ for
each product category and firm. In general, firms need to consider two

important factors related to new product information. These factors involve (1) understanding what information is important for developing successful new products and (2) training personnel to use that information.

There are two primary sources of NPD information. Much of the information that is important to new product developers comes from customers and represents the first source of NPD information. The second source of important NPD information comes from product and process technology specialists working on technologies that can support the products and production processes associated with both the ideas and concepts derived early in the NPD process. Given this information, management must instruct its new product development personnel in information use. It is most important that such information be readily available and easy to use. If a firm's important information is hidden and unwieldy, there will be much time wasted and many developers will cease to use it in favor of "personal databases" and "seat-of-the-pants" experience. This type of subjective information should always play a part in the making of good decisions, but the use of personal information makes it difficult to quickly obtain decisions with shared meanings. These problems can be solved through the implementation of shared databases and instruction for all new product developers relative to its use. Information access needs to be controlled but not to the extent that makes it difficult for developers. Data input and modification should be more stringently controlled. But even these processes need to be accessible to some so the NPD process is not unjustifiably slowed.

Think globally

The final implication for managers participating in this emerging NPD paradigm is to create a global new product perspective. NPD managers need to develop a world view to be aware of new markets, competitive forces in both foreign and domestic markets, and new, more economical resources such as materials, technology, and personnel. This can be achieved by developing first-rate environmental scanning capabilities.

Wholistic structures make internal what is all too often perceived and managed as external in centralized hierarchies, namely new markets that provide firms with additional outlets for new and existing products. These new markets also can provide additional customer and product information. Moreover, foreign countries can provide test markets for new products and the ability to experiment with various marketing plan configurations.

A second dimension of corporate activity typically perceived to be "external" becomes "internal" when corporations recognize that becoming aware of global competitive forces is more important than ever before. Even if a firm does not participate in global markets, it must ascertain the ability of competitors operating in those markets to attack its domestic markets. A firm cannot assume that its lack of interest in foreign markets will prevent foreign competitors from entering domestic markets, no matter how small. Circumstances in foreign "home" marketplaces may be such that they give foreign competitors incentives to attack even small markets in different parts of the world.

A final point concerning the creation of both a global and wholistic perspective involves an assessment of the most economical manner to develop new products over the long term. The economical development of new products can entail the employment of resources from many different parts of the globe. A critical issue related to the employment of global resources involves the interface between the personnel that perform the steps of the NPD process. The ease of NPD task performance may be hindered by the incorporation of very economical material resources from other geographic locations because the providers of the material may not be perceived as team players by the employees operating in the "home" office. Such hindrances can have devastating effects on the overall results of the NPD process. This implies that a wholistic NPD perspective is not only required but needs to be a deliberate undertaking.

Summary

This chapter has described an emerging NPD paradigm. An NPD paradigm is described and the implications of a paradigmatic shift are presented. An NPD paradigm represents the "right way" to develop new products by many new product developers. Therefore, an NPD paradigmatic shift is a change from the traditional NPD paradigm to a new paradigm. The traditional NPD paradigm was described and a variety of limitations associated with it were presented, including its linearity, lack of customer focus and involvement, the inconsistency with which it is implemented by new product developers, and its inability to offer consistently successful results. It was also noted how these problems might be solved by following the rules of this emerging NPD paradigm.

Changes in organizational phenomena were offered as evidence of the presence of an emergent NPD paradigm. Moreover, those indicators were closely linked to the solutions of problems that the traditional NPD paradigm has not been able to solve. Indicators such as flatter new-product-developing organizations, organizations with global perspectives, and organizations that incorporate NPD acceleration techniques into their NPD processes were pointed to as indicative of the emergence of a revolutionary NPD paradigm.

The requirements for firms to successfully participate in this emerging NPD paradigm were presented. Concepts such as double-loop and organizational learning were presented and the necessity of continuous innovation also was emphasized. Participating firms are expected to become learning organizations with an internal culture of continuous product development and product quality improvement and an operations flow involving the creation and communication of specific process and product technology strategies. To accomplish these actions, firms need to develop wholistic NPD perspectives which represent the final requirement for participating in this new NPD paradigm.

Finally, implications were offered for managers who decide to invest their resources and behave as prescribed by this emerging NPD paradigm. These behaviors include the need to change the firm's organizational culture, intelligently plan before accelerating NPD, obtain a comprehensive technology focus and operation, think like a customer to develop a sound product quality perspective, discover what information to employ in NPD, and develop a wholistic, global NPD perspective.

References

Argyris, Chris. "Double Loop Learning in Organizations," *Harvard Business Review,* 55, 1977, 115–125.

Bateson, Gregory. *Steps to an Ecology of Mind,* New York: Ballantine, 1972.

Booz, Allen and Hamilton. *New Products Management for the 1980s,* New York: Booz, Allen and Hamilton, 1982.

Cooper, Robert G. "The NewProd System: The Industry Experience," *Journal of Product Innovation Management,* 9, 1992, 113–127.

———— and Kleinschmidt, Elko J. "New Product Processes at Leading Industrial Firms," *Industrial Marketing Management,* 20, 1991, 137–147.

Crawford, C. Merle. *New Products Management,* Homewood, Ill.: Richard D. Irwin, Inc., 1991.

Daft, Richard L. "A Dual-Core Model of Organizational Innovation," *Academy of Management Journal,* 21, 1978, 193–210.

Garvin, David A. "What Does `Product Quality' Really Mean?" *Sloan Management Review,* Fall 1984, 25–43.

Harrigan, Katherine R. *Strategies for Joint Ventures,* Lexington, Mass.: Lexington Books, 1985.

Hise, Richard T., O'Neal, Larry, Parasuraman, A., and McNeal, James U. "Marketing/R&D Interaction in New Product Development: Implications for New Product Success Rates," *Journal of Product Innovation Management,* 7, 1990, 142–155.

Kalashian, Michael. "EDI: A Critical Link in Customer Responsiveness,"

Manufacturing Systems, 8, 1990, 20–26.

Knight, Kenneth E. "A Descriptive Model of the Intra-Firm Innovation Process," *Journal of Business,* 40, 1967, 478–496.

Kuhn, Thomas. *The Structure of Scientific Revolutions,* Chicago: The University of Chicago Press, 1970.

McKee, Daryl. "An Organizational Learning Approach to Product Innovation," *Journal of Product Innovation Management,* 9, 1992, 232–245.

Miles, Raymond E., and Snow, Charles C. *Organizational Strategy, Structure, and Process,* New York: McGraw-Hill, 1978.

Millson, Murray, Raj, S. P., and Wilemon, David. "A Survey of Major Approaches for Accelerating New Product Development," *Journal of Product Innovation Management,* 9, 1992, 53–69.

Morgan, Gareth. *Images of Organization,* Newbury Park: Sage Publications, 1986.

Rosenau, Milton D. "From Experience Faster New Product Development," *Journal of Product Innovation Management,* 5, 1988, 150–153.

Scurr, Colin. "Total Quality Management and Productivity," *Management Services,* 35, Oct. 1991, 28–30.

Staudt, T. A., Taylor, D. A., and Bowersox, D., *A Managerial Introduction to Marketing,* Englewood Cliffs, N.J.: Prentice-Hall, 1976, 49–52.

Wheelwright, Steven, and Clark, Kim. *Revolutionizing Product Development,* New York: The Free Press, 1992.

Wind, Yoram, and Mahajan, Vijay. "New Product Development Process: A Perspective for Reexamination," *Journal of Product Innovation Management,* 5, 1988, 304–310.

Emerging Issues: Cross-Cultural Issues, Strategic Issues, and Directions for Future Research

Part 3 focuses on several important emerging issues in new technology development. These issues include international differences between the United States and Europe in the management of new technology development, the emerging strategic management of technology paradigm, and new directions for future research.

Cultural differences in the management of new technology development between the European countries and the United States are explored by Professors Rudy Moenaert, Arnoud De Meyer, and Bart Clarysse in Chap. 9. Differences are observed based on differences in historical areas of strength, market heterogeneity, and culturally distinct managerial practices. Moenaert, De Meyer, and Clarysse demonstrate the need for greater understanding of the influence of socioeconomic and cultural contexts. These contexts include international differences in the level of radicalness of innovation, process improvement, formalization of innovation processes, and centralization of R&D operations.

Another important emerging area is the strategic management of technology (MOT). In Chap. 10, Professor

Michael Lawless discusses how this emerging area is differentiated from the traditional literature on strategic management. Lawless discusses how generic strategies like cost leadership and product differentiation link to different technology development strategies. He also explores how different technological events (e.g., radical innovations and the emergence of a dominant technology) and technological periods of development influence market responses. In this chapter, Lawless emphasizes the developmental status of strategic MOT and identifies a number of directions for the future evolution of this paradigm.

In the final chapter of this book, Chap. 11, directions for future research are explored. MOT is not a field in itself but rather a broad domain of research which crosses multiple disciplines. These disciplines include market research, new product development, strategic management, organizational behavior, organizational theory, entrepreneurship, organizational sociology, industrial engineering, and management science. Thus, Chap. 11 addresses a broad range of speculative directions in these fields which relate to the management of new technology and new product development.

9

Cultural Differences in New Technology Management*

Rudy K. Moenaert, Ph.D.
Free University of Brussels, Belgium

Arnoud De Meyer, Ph.D.
INSEAD, Fontainebleau, France

Bart J. Clarysse
Vlerick School of Management, Ghent, Belgium

In this chapter, the focus of this book changes from an examination of the detailed steps in the new technology development process to an examination of differences in processes. The focus here is on understanding the differences in management practices in Europe versus the United States. As the world moves to a more global economy, understanding the origins, rationales, and cultural differences in management practices throughout the world becomes vital to new product success. What distinguishes successful from unsuccessful new product developers in the United States and Europe? Do the same factors that inhibit success in the United States also inhibit success in Europe? What can be learned from other cultures? These are only a few of the interesting questions that are taken up in this chapter. The senior editor of this book and the senior author of this chapter are currently collaborating in joint research that will add to our understandings of cultural differences in management of technology (MOT). This chapter is a partial report on the results to date.

WM. E. SOUDER AND J. DANIEL SHERMAN

*The authors gratefully acknowledge financial support from the Kuijper Foundation of International Management of the Vlerick School of Management (University of Ghent).

Introduction

While the importance of technological innovation has been long recognized, the shortening of product life cycles and the success of some countries in speeding products to market has resulted in studies investigating the antecedents of product development performance. Recently, there has been a trend toward the investigation of international differences in the management of innovation and its impact on innovation success. The latest research efforts by Clark and Fujimoto (1991) have contributed to the notion that product development efforts are managed differently in different parts of the world.

This recent interest in international comparisons must be seen in light of the competitive decline of some nations and the relative competitive growth of others (e.g., the United States and Japan). Technological progress, historical studies have argued, is not to be taken for granted (Mokyr, 1990). Moreover, it takes more than scientific or technological genius to be a successful contender in today's global market economy. The United States is still associated with the invention of many of today's technological marvels. However, it seems that the benefits of commercialization are reaped by other countries, most notably the Far East (Florida and Kenney, 1990). While technological invention is undeniably a key asset for any country to compete, the management of technological product development assumes an ever more important role.

In this chapter, it is our endeavor to help the reader gain a better understanding of international differences on the product development front. Given the limitations in terms of extant theory and research findings, this is far too ambitious a goal to achieve within the scope of one chapter. Rather, it is our objective to kick off the start toward a more solid theoretical understanding of international differences with regard to innovation management and their consequences for a firm's competitiveness in a global market. In this chapter, we explore the effect of a country's socioeconomic and cultural context on innovation processes and outcomes. In the analysis, the emphasis is primarily on the factors that differentiate American and European innovation efforts.

In the first section, we review the framework we have adopted to structure the discussion on differences in terms of innovation management practices between the United States and Europe. The second section discusses the impact of socioeconomic and historical contexts on innovation. The third part of this chapter looks at the meaning of national culture and its consequences for innovation. Finally, the conclusion reviews remaining questions and identifies potential directions for future research.

Studying Innovations across Countries: A Framework

Studying differences between countries' innovation practices implies there are a number of dimensions along which these countries differ and which have an impact upon the product development process. In this chapter we will argue that there are at least two major groups of contextual variables: socioeconomic background and culture. Countries differ in terms of their *socioeconomic background*. While many scholars tend to agree that the socioeconomic context has a significant bearing on technological progress and development, ambiguities flourish. "The diversity of technological history is such that almost any point can be contradicted with a counter-example. Picking up empirical regularities in this massive amount of qualitative and often uncertain information is often hazardous" (Mokyr, 1990, p. 6). The variables we will study in our analysis are a country's technological heritage, its market structure, its administrative heritage, and its entrepreneurial climate.

Countries differ in terms of their *culture*. National culture has been defined as the collective programming of the mind (Hofstede, 1984). However, the relation between national culture and innovation is not a straightforward one. According to Smircich (1983), the concept of culture can be related to organizations in a number of different ways, serving diverse research purposes. The classical approach examines ethnic or national culture as an exogenous variable and corporate culture as an endogenous variable (Roger, 1990; Peters and Waterman, 1982). The modern approach, however, uses culture as an instrumental variable that may aid in the definition and exploration of ways of organizing (Rose, 1988). This approach adapts elements from cognitive, symbolic, or structural anthropological theories to define culture (Smircich, 1983). Our purpose is to explain when and how corporate cultural environments are perceived as stimulating innovative behavior. In order to have a better understanding of these surrounding influences, we will place this within a larger ethnic environment. Therefore, both elements from the classical approach as well as from the modern approach are judged to be appropriate in developing the present framework.

When thinking about the impact of these contextual variables, we can distinguish between *innovation processes* and *innovation success*. The organization can manage its innovation processes through basically three means: *organizational strategy* (e.g., project teams versus functional line management, top management involvement), *human resources* (e.g., choice of project leader, job rotation), and *methods and formal tools* (e.g., new product development plan, quality function deployment).

The outcomes of innovation efforts can be described in short- and long-term effects. The *short-term innovation success measures* relate to a project's outcome with regard to product quality ("the extent to which the product satisfies customer requirements"), lead time ("a measure of how quickly a company can move from concept to market"), and productivity ("the level of resources required to take the project from concept to commercial product") (Clark and Fujimoto, 1991, p. 68–69). The *long-term innovation success measures* relate to a project's contribution to the core competence of the organization (Prahalad and Hamel, 1990), that is, to the longer-term sustainable competitive advantage of the firm. Indeed, "...the impact of a single successful product on the firm's future is that of opening a window of opportunity, which means entering into new markets or new product categories. For a SBU, being prepared for the future means not only creating the opportunities but also preparing the technology needed and the right people for exploiting it" (Dvir and Shenhar, 1990, p. 293).

The guiding model for our analysis is summarized in Fig. 9.1. The present analysis must then be seen as a nonexhaustive treatment of potential relations between these contextual variables (socioeconomic context and national culture) and their effects on the innovation process and innovation success, and the relation between these two dependent variables. The framework must be seen as a potential breeding ground for hypothesis generation. The existing theory and

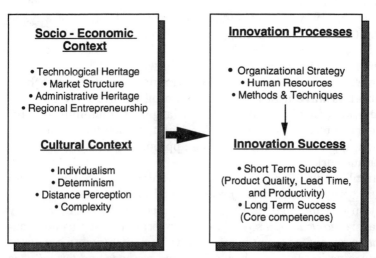

Figure 9.1 Framework for an international comparison of innovation management practices.

data do not enable us to launch a full-fledged testable theoretical model. While the international comparison on management of innovation is a rapidly growing field, many of the research efforts have, by necessity, remained quite phenomenological.

The Old versus the New Continent: The Impact of Economic and Historical Context

Innovation and technological development do not occur in a vacuum. The socioeconomic and the historical context in which a firm operates influences, to a large extent, the type of activities displayed by the innovating firm as well as the characteristics of the innovation process which is followed. Here, we will consider four types of influences which are typical of innovation processes and the ways they apply in Europe: the technological heritage, the market structure, the administrative heritage, and the entrepreneurial environment.

The technological heritage

Technological development and innovative activities are not to be identified with a process of information production where the generation is costly and the transfer and application is easy. On the contrary, industrial development is based largely on tacit knowledge. Most of this tacit knowledge has been created through applied activities such as product and process engineering, industrial development, and maintenance activities. Thus, technical knowledge tends to be firm specific or at least specific to a local supplier-customer network of firms that also collaborate with universities, research institutions, and government agencies.

Firms rarely innovate "out of the blue"; they build upon existing knowledge and explore fields and market areas which are contiguous to their existing activities. Atkinson and Stiglitz (1969) have already observed that when technology is localized in firms and cumulative in nature, decisions about investments in technology reflect both past patterns of innovative activity and expectations about the future. Both the historical record as well as future expectations are firm specific.

It is then merely a logical extension that what applies at the level of the single firm, in the case of innovation, equally applies at a more aggregate level of firms and institutions collaborating in the same local or regional network. Firms in a particular region will tend to choose a technology and an innovation strategy which builds upon a set of technological strengths that already exists in that region. We call this set of regional technological strengths *the technological heritage of the network*. This avoids confusion with terms such as core technologies and core competencies (Prahalad and Hamel, 1990), which are more firm specific. Technological heritage encompasses the technological strengths of a set of firms, the research tradition in a particular region, the existing networks between firms, universities, and government institutions, etc.

A few examples may illustrate this concept. When Japanese machine tool and automotive suppliers looked for locations in Europe for technology-intensive factories, the first iteration ended with sites in the triangle formed by Stuttgart, München, and Strasbourg. In this region, we find some of the most sophisticated European automobile producers. Furthermore, this region is characterized by a high-performing network of small and medium-sized suppliers, university laboratories in mechanical engineering, as well as a well-educated work force and an educational system that prepares for jobs in those industries. Another case is provided by the city of Basel (Switzerland). There, one finds three of the largest pharmaceutical companies in the world (Ciba Geigy, Roche, and Sandoz). Such a concentration cannot be explained by market reasons or government incentives. It builds upon a dense network of ideas that follows from the informal knowledge trading that goes on in these networks. Similar patterns have been observed in the U.S. steel and minimill industry (Schrader, 1991).

What is Europe's technological heritage that would influence European companies' innovation strategy both in the short and the long term? Patel and Pavitt (1986) have built their case of Europe's technological strength on several indicators: R&D investments, European patentings in the United States, analysis of specific innovations, judgments by technological peers, and scientific output. This diverse set of indicators suggested that western Europe was, in the 1980s, in a strong position in chemicals (including pharmaceuticals) and nuclear energy. In addition, this region was ahead of the United States in metals and automobiles and ahead of Japan in aerospace and in technologies for exploiting raw materials. In conventional industrial machinery and production engineering, it had a strong position, but that position was obviously increasingly challenged by electronics-based technology from Japan. Other evidence suggests that some of these technological strongholds may still exist today. Cool (1992) shows convincingly that the European pharmaceutical industry has consistently outperformed the United States and Japan in drug development.

One has to be very careful with a judgment about Europe's innovative capacities and technological heritage. Most of these evaluations are based on *past performance* and do not take into account sufficiently the current investments by firms (microlevel), the current national investments in R&D, and the human resources infrastructure (macrolevel). Furthermore, Europe is far from homogeneous. Also, its technological heritage is quite varied from one country to another. Some of these so-called European strengths may for instance be limited to specific countries. For instance, in the automotive industry, the strongholds are most likely to be Germany, France, and Italy.

The market structure

If we accept the above-mentioned analysis of Europe's technological strengths, we may be able to predict what kind of innovation and innovation processes the Old Continent is good at. The main European strengths are in industrial goods and products (e.g., pharmaceuticals) where market research is less important. European innovators are perhaps better than their colleagues in other regions (1) where there are *clear specifications* (such as curing a disease) or (2) when *close relationships exist with users and customers* of industrial goods and machinery. This can be understood if we take into account the *heterogeneous European market structure.*

In contrast to the expectations formed by the creation of a unified market—the process often called "Europe '92"—we cannot consider the European market to have the same homogeneity as the U.S. market. While Europe's submarkets have been steadily growing toward each other, and while the U.S. market is indeed less homogeneous than is often thought, Europe clearly outdistances the United States in terms of market diversity and market fragmentation. As a consequence, the economies of scale which exist in the United States are often impossible to achieve in Europe. Economies of scale are far more difficult to obtain, and *economies of scope* are necessary. One can hope that European integration will reduce the enormous variety in government-imposed standards and specifications. But the cultural, historical, and infrastructural context that shapes this market heterogeneity will not disappear overnight. The French still like washing machines with top loading in contrast to the British, who prefer front loading. Italian cities will keep their narrow roads and Italian drivers will remain inclined toward small, agile, and nervous cars. While these are merely a few of the anecdotal examples that illustrate European market heterogeneity, the broader consequences are compelling. Innovators who want to be successful in the European market will have to adapt to this need for differentiation, yet be competitive enough to survive in the global competition.

One of the responses to this challenge has been a trend toward product designs characterized by *delayed differentiation* (De Meyer, 1991). The idea of delayed differentiation is simple: redesign the product in such a way that the adaptation to different markets and market segments can be performed in the last step of the production process. Delaying the dyeing of clothes to the last step in the production process enables the production of standardized sweaters or jeans that can be adapted to the colors required by different markets. By designing electrical equipment such that the power cable can be disconnected from the product, manufacturers of consumer goods can standardize the production process for printers and TV sets. Customization for local plugs occurs at the very end of the production cycle by adding a cable set at the last step in the production and distribution process. The consequence of such a redesign is that components and subassemblies can be produced in large volumes and efficiently, while the cost of differentiation is limited to a single production step. While the concept is simple, it appears that the redesign of products and processes in order to adapt to this idea is far from obvious. Yet one sees many large European manufacturers of telecommunication equipment, cars, food products, etc., moving in this direction.

Administrative heritage

Bartlett and Ghoshal (1989) have introduced the concept of *administrative heritage,* or the heritage of administrative rules, procedures, and organization structures which are typical for a company. They observe that large European multinational companies have a tradition of confederated structures with strong national organizations and relatively weak centers or headquarters. One of the consequences is that organizations which operate as confederations have a tendency to excel at *local innovations for local applications.* Their capacity to innovate within a global market view or to innovate in project structures which encompass people or teams in different countries is fairly limited.

It would be an extreme caricature to suggest that European innovative efforts are limited to "local for local" innovations. But the tendency to create differentiated products in order to preserve independence from headquarters often goes beyond what is needed for pure market reasons. European engineers are often said to overengineer the product and to design specifications which hardly can be produced efficiently. In many cases, the R&D unit is not sufficiently challenged by its local management team. On the contrary, local management may very well see this overengineering as a way of preserving the distance with the headquarters.

Entrepreneurship's battleground

One of the major themes in the United States–based literature on innovation management is the need to develop *internal entrepreneurship*. Concepts such as "product champions" and "intrapreneurs" are conceived as cornerstones of the innovation process. Innovation in a European context requires the same entrepreneurial and risk-taking behavior. But the format of this entrepreneurial behavior may be somewhat different. Implicitly, the U.S. literature identifies this entrepreneurship with an individual. Even when Maidique (1980) refers to the need to embed the entrepreneur in an entrepreneurial network, the focus of the entrepreneurship act is with the individual.

It is our conviction that this implicit link is strongly influenced by the national environment. As is described elsewhere, perceptions of environmental uncertainty and organizational control influence strategic behavior (Schneider and De Meyer, 1991). Entrepreneurial behavior is a form of strategic behavior. The way entrepreneurship is expressed will differ not only from one culture to another but also from one legal context to another. The legislation on bankruptcy, the willingness to take risks by customers, and the way events are interpreted are all regionally different. The European environment, in comparison to that of the United States, tends to deemphasize individual risk taking and prefers the group to take the risk. Albert (1991) has suggested that in continental Europe a different type of capitalism is practiced than the traditional Anglo-Saxon capitalism in which individual entrepreneurship is rooted. He suggests that this "Rhine" capitalism is based on group decision making, consensus building, and objective functions which take other elements into account in addition to pure economic benefits. If one accepts the hypothesis of multiple forms of capitalism, the consequence is that different forms of entrepreneurship can exist. We formulate it here as a hypothesis: European entrepreneurship is more likely to be entrepreneurship by an entrepreneurial group and American entrepreneurship is more likely to be entrepreneurship by an entrepreneurial individual. In the next section, we will see this assertion is in line with a cultural perspective on innovation roles.

Ethnic Culture and Corporate Culture: Exploring the Concept of Fit within Innovation Processes

Corporate culture can be viewed as a behavioral control mechanism used to direct and coordinate organizational behavior (Schneider, 1988). Recently, some efforts have been made to examine the influence of corporate culture on innovative efforts within a company (Feldman, 1988). In order to understand the extent to which aspects of corporate culture affect innovative behavior, this research question must be embedded in a larger sociocultural environment (Feldman, 1988; Adler and Jelinek, 1986). At this point, the existing literature shows a gap. For instance, O'Reilly (1989) tries to answer the question of how culture helps or hinders the process of successfully developing new products or new processes by empirically examining the elements of the appropriate culture that promote creativity and implementation. However, while elements such as openness, autonomy, and risk taking may provide excellent inducements within American companies, this may very well not be applicable in other cultural environments. For instance, they do not explain the context in which successful innovation occurs within Japanese companies (Nonaka and Kenney, 1991). So far, no theoretical framework has been developed that explains the cross-cultural applicability of such concepts.

The following question can then be formulated: At a theoretical level, how are the elements of corporate culture and ethnic culture related to each other, and what are the effects on innovative behavior? Because of a lack of existing theoretical models that incorporate the impact of cultural dimensions on innovation processes, an inductive approach seems to be appropriate. We will investigate the very diverse perspectives that have been adopted to analyze the construct of culture. Furthermore, an analysis will be made of innovative behavior during innovation processes. These discussions lay the foundation for a more profound understanding of the effects of ethnic culture on innovative behavior.

The meaning of culture

The way the construct of culture has been defined depends largely on the intended use of the concept. From a functionalist viewpoint, the concept of culture can be seen as something an organization "has." From an anthropological viewpoint, culture can be defined as something an organization "is." The construct culture is then used as a general paradigm to describe organizational functioning (Smircich, 1983). However, researchers have tended to use these concepts divergently in a number of ways. Instead of increasing the conceptual confusion by using anthropological definitions to define ethnic differences and a functional approach to explain corporate culture, we will view both from a cognitive anthropological angle. In the tradition of this perspective, we will interpret the cognitive map that individuals use to structure and make sense of their actions. Ethnic and corporate culture explain the meaning of differences between those maps.

Ethnic culture has been defined as a "shared way of being, evaluating and doing that is passed from one generation to another" (Ronen and Shenkar, 1985, p. 17). Thus, culture implies a set of shared assumptions that distinguish the behavior of one group of people from another. At an individual level, ethnic culture gives a meaning to the underlying assumptions directing beliefs, values, norms, and concrete behavior. Based on the definitions offered by Schein (1985, p. 3) and Smircich (1983, p. 342), corporate culture can then be defined as the unique system of beliefs, values, and norms that members of an organization share (Gibson, Ivancevich, and Donnelly, 1988, p. 45).

Generally, authors state that beliefs are constructs about the desirable, whereas values are perceived statements of the truth (Wagner and Moch, 1986). Norms differ from values and beliefs because they express a concrete, specific set of standards defining socially appropriate or inappropriate behavior in certain situations (O'Reilly, 1989; Wagner and Moch, 1986). As norms are in essence a product of the prevailing beliefs and values in the organization (Smircich, 1983), they can be viewed as socially created standards of expected behavior. What differentiates beliefs and values from underlying assumptions is that the latter are situated on a less conscious and a less visible level, and they are inherently shared by a larger group of people (Schein, 1985). Thus, beliefs, values, and norms have a much more direct bearing on our daily behaviors (see Fig. 9.2.).

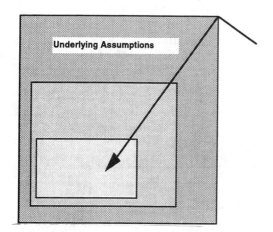

Figure 9.2 Levels of cultural influence at the individual level.

Dimensions of ethnic culture

Theoretical treatises and empirical cross-cultural inquiries have generally focused on four variables that characterize an ethnic culture, that is, the level of individualism (Wagner and Moch, 1986; Triandis, 1983; Hofstede, 1984; Schwartz, 1990), the degree of determinism (Kluckhohn and Strodtbeck, 1961; Maruyama, 1984; Adler and Jelinek, 1986), the level of hierarchical distance (Hofstede, 1984), and the degree of complexity (Shaw, 1990; Gamble and Ginsberg, 1981). Those variables of cultural heritage are supposed to determine the underlying assumptions shared by ethnic groups.

1. The *individualism of an ethnic culture* has been defined by Wagner and Moch (1986, p. 281) as "the condition in which cooperation is motivated by the contingent satisfaction of personal interest (whether this satisfaction stems from personal or group performance)." At the other end of the scale, it has been referred to as collectivism, that is, "the condition in which cooperation stems from the pursuit of interests shared among the members of a collectivity (even if this pursuit sometimes conflicts with members' immediate personal desires)."

2. The *determinism of an ethnic culture* refers to the perceived relationship between a people and its surrounding nature (Adler and Jelinek, 1986). Some people live in a very challenging sociocultural environment, others in a very peaceful environment. A high level of determinism means that people who belong to such a culture believe they are subjugated to nature because of its ambiguity (Maruyama, 1984). Cultures that are positioned at the low end of this dimension (i.e., the so-called free will cultures) believe they dominate the external environment. In addition, such cultures tend to view change as both feasible and good, and they hold a future orientation (Adler and Jelinek, 1986, p. 84).

3. The *distance perception of a culture* relates to the elements an ethnic culture adopts about the impact of status and power within that culture (Schneider, 1989). If the meaning of how people behave is highly dependent upon the status and power of the specific persons they are dealing with, they are considered to be high-distance cultures (Hall and Hall, 1988). A great emphasis on social class differences and status, a large power distance, and situationally related behavior patterns are characteristic features of such cultures (Triandis, 1983; Schneider, 1989).

4. Finally, the *complexity of an ethnic culture* refers to the number of formal roles one is allowed to perform and the analyzability of these roles (Gamble and Ginsberg, 1981). Some societies know

that certain activity patterns lead to a certain success. However, they do not know how these patterns can be explained. When many of the key activities cannot be explained, we call the culture relatively complex. Among the major indicators of this variable of cultural heritage are the level of understanding of how conduct leads to success in a particular domain and the importance of the activity to the group.

Thus, the construct of ethnic culture is deeply rooted in the assumptions members of one ethnic group hold about the relationship with other persons (individualism, hierarchical distance) and about the relationship with the environment (determinism, complexity). First, the individualism assumption influences the interrelationship between individual members of an organization (Schneider, 1989). Empirical research shows evidence that individualism stimulates competitive behavior within a team, whereas collectivism supports cooperative behavior as a group dynamic force (Cox, Lobel, and McLeod, 1991). Also, the hierarchical distance assumption within a culture may affect the shared perception of appropriate behavior in certain situations (Shaw, 1990). If behavior depends on the perceived status of the other, this may influence the internal integration within a group of people. Furthermore, the determinism dimension can be seen as influencing the external adaptation process. Indeed, empirical research provides evidence that people belonging to "mastery" cultures (i.e., cultures on the free will extreme) perceive uncertainty in a different way than people belonging to "subjugation" cultures (Adler and Jelinek, 1986). Furthermore, the type of complexity assumption that predominates within a given culture will affect relationships with other persons. Since complexity is related to role performance and activity analyzability, interrelational behavior may be more or less formalized according to these elements. Simple cultural environments allow more formalized respect for certain people performing a certain role.

Ethnos versus Organization: An Examination of Cultural Interactions

Before we examine how ethnic dimensions influence an individual's beliefs, values, and norms, let us first define the content of such a system (see Fig. 9.3). A first dimension, along which they can be classified, concerns *power orientation*. This dimension refers to the beliefs and values individuals hold about their ability to change things (Lee and Fox, 1993). Individual power-oriented people have confidence in their own ability to make and change things. Hence, they can cope

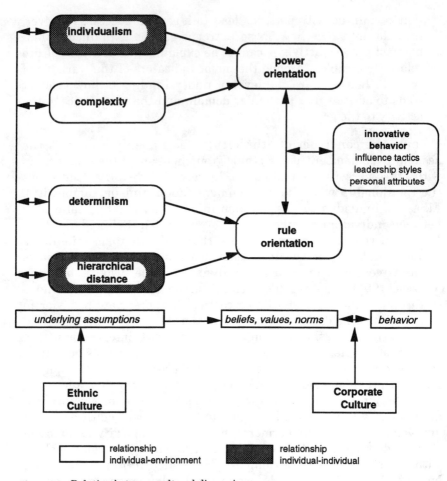

Figure 9.3   Relation between cultural dimensions.

with a significant amount of autonomy in their job performance. Authority and power-oriented people, at the other extreme of the dimension, are persuaded of the leader's ability to change things. They expect clear-cut guidelines by persons who are in the position and have the authority to formulate them. Without those guidelines, such individuals feel lost.

We expect a significant correlation between the level of individualism and the individual power orientation of a person. The pursuit of personal interests can be best understood within cultures that show a high level of individualism. Low individualistic cultures (i.e., collectivistic cultures), on the other hand, induce a group orientation on behalf of the members (Wagner and Moch, 1986). Therefore, they are expected to be more authority and power oriented. Another cultural assumption we expect to be related to an individual's power orientation concerns the complexity that prevails within an ethnic group. A culture that scores low on the complexity dimension can best be described as a "simple" culture. Simplicity then points to a small number of important roles for which there exists a quite thorough understanding of how the conduct of these roles will lead to success. A "complex" culture implies a large number of important roles. In addition, the link between conduct and performance is not well understood. Within an organizational context, the importance of a role is related to the perception of what type of behavior gets rewarded (Frohman, 1978). Since the number of roles that are performed in simple cultures is relatively small, the authority associated with those roles may be expected to be rather high (Shaw, 1990). However, persons who belong to a complex culture are expected to have more confidence in their individual power to succeed in a certain job or task. Whereas in the former culture, generalization is supported, in the latter there seems to be more need for specialization.

A second dimension postulated to classify beliefs and values has been called *rule orientation*. This dimension refers to the belief one has in his or her capacity to react against existing roles and procedures. An external rule-oriented individual expects externally imposed rules and guidelines. Without such directives, only chaos would remain (Lee and Fox, 1993). Internal rule-oriented individuals on the other hand, use externally imposed roles merely as a framework to structure their proper behavior. They give their own meaning to these external guidelines. People belonging to cultures characterized as deterministic see themselves as being subjugated to the surrounding environment (Shaw, 1990). Such subjugation is expected to increase the need to be in constant harmony with that environment. As a consequence, uncertainties will be considered as natural to certain environments. Within deterministic cultures, we expect people to be more externally rule oriented in their beliefs and values. At the opposite end of the determinism spectrum, a free will culture will trigger an internal rule orientation. A free will culture is permissive toward people who think they can depart from and challenge the cultural environment.

Finally, we expect the cultural assumption about hierarchical distance to have an effect on the beliefs concerning the internal rule orientation of persons. The formalization of expected behaviors and the development of formal rules generates a hierarchical structure. Interpretative inquiries about the nature of such hierarchies suggest that ranking systems are kept alive by a respect for formally established rules and procedures. Traditions and roles facilitate the development of a "wise-men" and status culture. According to Kagono et al. (1985), a high hierarchical distance explains who will be the preferred personal source of information in a Japanese company.

The relation between corporate culture and innovation effectiveness is expected to be dependent on the ethnic cultural environment. Indeed, some ethnic cultures may fit certain cultural profiles of an organization. For instance, the American individualistic free will culture is probably quite supportive toward the entrepreneurial style of many high-tech computer companies in the United States. By contrast, internal freedom may not be compatible with the cultural heritage of other cultures, for example, the French (Adler and Jelinek, 1986). Defining as a rigorous management tool which norms have to be developed to promote creativity and innovation implementation may be helpful in particular ethnic cultures. But we doubt its appropriateness for other ethnic cultural environments. According to O'Reilly (1989), norms like common goals, autonomy, and belief in action facilitate the process of innovation implementation. However, they require certain orientations in the belief and value systems of people. Thus, difficulties may arise when such norms are deployed in a context where shared underlying assumptions influence beliefs and values in an inappropriate way. A corporate culture, probably seen as "innovation stimulating," will most likely be very dysfunctional in those ethnic environments.

In addition, some ethnic cultural environments, such as Anglo, promote the development of group or corporate cultures whereas other ethnic contexts do not emphasize this, such as Latin (Adler and Jelinek, 1986). Consequently, underlying assumptions may moderate the effects of corporate culture on individual beliefs, values, and ultimately behavior. For this reason, the process of enculturation will be more successful in certain cultural environments than in others. *Enculturation* means the adjustment of one's belief and value system to the central norms of the present corporate culture. The degree of freedom one has in order to adjust depends on the ethnic underlying assumptions. Recent empirical research seems to further underscore this statement (Roger, 1990).

Assumptions experienced in common by the members of one ethnic group have been historically developed. Therefore, they can be expected to be inherently static in the short term. Beliefs, values, and norms maintained by members of one organization have emerged within a shorter term. Hence, they can be assumed to be inherently more dynamic. Thus, in the short term, the effects of corporate beliefs, values, and norms on corporate behaviors are codetermined by the ethnic culture that prevails in the organization. In the long term, organizational culture itself is influenced by the ethnic culture that prevails in the organization.

Innovative behavior during an innovation process

The definition of innovative behavior has spawned much confusion in the existing literature. For certain, there have been quite a few attempts to define the concept. However, these endeavors have shown a tendency to result in rather vague concepts (Howell and Higgins, 1990). Added to this confusion is the ambiguous specification and labeling of the diverse key roles played by individuals during innovation processes. Innovation research is indeed not devoid of name dropping. For instance, "product champions" (Howell and Higgins, 1990), "project champions" (Rothwell et al., 1974), "change advocates" (Ginsberg and Abrahamson, 1991), "change evangelists" (Beatty and Gordon, 1991), and "technological champions" (Maidique, 1980) all refer more or less to the same role in the innovation process.

In this chapter, we define the *innovative behavior* of a member of an organization as the actions and efforts he or she performs to initiate, motivate, and translate new ideas into action (Bass, 1985; Conger and Kanungo, 1987; Maidique, 1980; Van de Ven, 1986). Innovations require a network of individuals performing entrepreneurial, technological, and management roles. However, the characteristics of that network may be related to a company's size, structure, and culture (Maidique, 1980). The SAPPHO study has been one of the very first studies to specify a number of different roles during an innovation process (Achilladelis et al., 1991; Rothwell et al., 1974). Technical innovators work on the design and development process of an innovation. Business innovators assume the role of resource allocators for certain projects. Product champions transform creative ideas into activities. Finally, the chief executive provides the business context of the organization. The SAPPHO data provide evidence that product champions and business innovators are pivotal to the commercial success of innovation processes. Given the evidence on integrated project management (e.g., Moenaert, Deschoolmeester, De Meyer, and Souder, 1992), we extend the content of the technical innovator role to include marketing and production functions as well. Therefore we will alter the name "technical innovator" to "project innovator." The latter is expected to diminish potential confusion about its role content. Hence, whether project personnel belong to marketing, engineering, or production, they have to cooperate with the other team members in order to specify and develop the product. As has been suggested by

Maidique (1980), business innovators, as well as project innovators and product champions, perform *championing activities*. In fact, these three roles are a more specific elaboration of the champion concept generated by Schon (1963).

On the basis of the above typology, we can outline the innovation process as follows. Following the generation of the initial idea, the innovation process departs from putting emphasis on problem solving. At this stage in the process, the implementation of new products or processes starts. Therefore, product champions are expected to sell the idea throughout the organization (Quinn, 1978). In doing so, they seem to use transformational leadership styles to gain commitment from team members to support the innovative action (Howell and Higgins, 1990; Beatty and Gordon, 1991). Experience, personality, and organizational position enable them to develop transformational skills. This implies an ability to transform organizational goals, which are inherently a reflection of individual interests, toward the interests of other persons (Conger and Kanungo, 1987; Bass, 1991). At the same time, the product champion persuades the business innovator to invest firm resources in order to establish the further development of the idea. As the implementation stage proceeds, the role of the business innovator gains more and more importance. Formal project leaders need to be assigned to continue the project. In the meantime, project innovators work on the operational level of the project. Market orientation versus technical orientation of the top management team explains differences in the concrete composition of the group during the process (Frohman, 1978). The number of engineering, marketing, and production people working on the project differs as does their individual power to change things. The business innovator functions as a catalyst. He or she needs the experience, personality, and authority to support the innovation process and to keep the inhibiting forces within the organization off the innovation team (Morgan and Ramirez, 1983). These three innovation roles all include activities affecting the successful implementation of innovations. However, according to the SAPPHO data, the role of the business innovator seems to be the most critical one.

Howell and Higgins (1990) argue that the variables affecting the emergence of champion activities within an organization are personal attributes, leadership style, and influence tactics (Howell and Higgins, 1990). In order to have a better understanding of the identity of the persons performing championing activities, a better insight into their individual characteristics is needed. These *personal attributes* refer to the characteristics that differentiate them from nonchampions. Second, in order to understand how they motivate and persuade other people to cooperate, we are interested in their *leader-*

ship style. Third, as leadership styles are related to the formal authority of the person, they do not cover all means and methods for motivating and persuading other team members. Therefore, we are interested in the *influence tactics* these people use. Do they differ in variety and content between champions and nonchampions and are they situation related? Corporate cultures identified as "innovative" may carry norms that stimulate these three variables in the right direction.

Cultural constraints on championing activities

Again, we argue that an insight into the way underlying assumptions influence the beliefs, values, and norms fosters a better stimulation of the appropriate behaviors (Feldman, 1988). Norms of corporate culture that are widely accepted as stimulants to innovative behavior in American companies may initiate counterproductive results in other ethnic contexts. In order to promote innovative behavior throughout the organization, a supportive organizational culture is needed. However, since organizations are open systems, management should not develop that culture without taking the environment into account. If the local culture is rather collectivistic (e.g., Japan), which makes decision making quite consensus based, risk-taking efforts of one individual champion will be rather useless (Kagono et al., 1985). It may rather be better to promote facilitative group membership to foster change. The norms of corporate culture that are widely accepted to stimulate innovative behavior in American companies may not work very well within other ethnic contexts. Leadership styles, personal attributes, and influence tactics used in championing activities may differ between different ethnic cultural environments. Therefore, a blind reliance on American-based guidelines for innovation management in Europe may be as foolish as a blind implementation by American companies of Japanese management techniques.

Leadership style. The innovation literature seems to provide systematic evidence that people performing championing activities incarnate transformational *leadership qualities* (Howell and Higgins, 1990; Conger and Kanungo, 1987; Bass, 1991). However, transformational leadership is made possible when the belief and value system of the leaders is adopted by the followers (Kuhnert and Lewis, 1987). Transformational leaders raise the goals of their followers to a common goal, thereby promoting a change in their followers' beliefs, values, and norms. A new corporate culture is hence established that can either support or depart from the existing culture within the organization. Transformational leaders themselves seem to be characterized by individualized consideration, charisma, and resistance (Kuhnert and Lewis, 1987).

Is transformational leadership for innovation projects possible and desirable in every ethnic environment? The vision that people performing championing activities are in essence transformational leaders is the dominant one in the American innovation literature. Recently, other studies have emerged paying more attention to cross-cultural differences (Jaeger, 1986). Indeed, a change in corporate culture implies that individuals are expected to challenge underlying assumptions. As has been suggested above, the stimulation of innovative behavior through leadership is only possible if there is a fit between the culture of the corporation and the culture of the ethnic environment. Since transformational leadership implies individualized consideration, such leaders seem to be internal rule oriented. This orientation is expected to be more present in certain ethnic cultures (e.g., individualistic, complex) than in other cultures (e.g., collectivistic, simple). In collectivistic, simple environments other leadership styles may have more success in the implementation of innovations (e.g., transactional leadership styles). Individuals who are internal rule oriented prefer to make their own job rules. External rule-oriented people, on the other hand, need procedural control in order to perform well (Lee and Fox, 1993).

Also, a transformational leadership style implies a belief in the power of people to change beliefs, values, and goals. Indeed, an individual power orientation is necessary to make such a process feasible (e.g., free will, low hierarchical distance). If there is a prevalent belief about one's proper capabilities, an individualized consideration can persuade other team members to deviate from standard operational procedures and follow another person's values and beliefs. Since this process will be informal, an internal rule orientation has to empower it. Again, the development of central norms may be dependent upon the larger ethnic cultural environment. For example, there is empirical evidence that the transactional leadership style is more pervasive within Japan than the transformational leadership (Kagono et al., 1985; Nonaka and Johansson, 1985). Sharing of information and decision making by consensus affect the information processing capability of an organization. Within such a system, simple transactions between a superior and a subordinate are very applicable and quite useful. Authority power orientation establishes a team-oriented climate that enables innovation implementation. The development of norms that promote transformational leadership styles may not only be misunderstood but may even be impossible in the Japanese context.

Personal attributes. The American literature on the *personal characteristics* of people performing championing activities indicates that they are risk takers (Bass, 1991; Schon, 1963). This risk-taking propensity of innovation champions seems to be the only characteristic that has been systematically explored in the championing literature. Other research has also shown them to be persistent in the face of resistance (Quinn, 1978; Sinetar, 1985) and to be politically astute (Howell and Higgins, 1990). Frohman (1978) has pointed out that individual differences predispose individuals in such a way that they tend to perform only a limited number of innovation roles very well. Concerning the three innovation roles belonging to the broader category of championing activities, this means that personal attributes will differ between business innovators, project innovators, and product champions. The only personal characteristic they seem to have in common is their risk-taking behavior (Cox et al.,1991).

This suggests that in order to develop or reorient the corporate culture, the stimulation of risk-taking people will improve the organization's innovativeness. However, empirical cross-cultural research has postulated that individual risk-taking propensity is positively related to the underlying assumptions of individualism and hierarchical distance (Cox et al., 1991). Thus, efforts to found and promote a corporate culture that stimulates the tendency of the organizational members to take risks may be quite dysfunctional in ethnic cultures that are not supportive of individual risk taking (Feldman, 1988). In addition, the development of norms stimulating risk may be unnecessary in certain environments (e.g., subjugation cultures like the Japanese). Risk taking with the aim of reducing uncertainty is not perceived in the same way as it is in Western cultural environments (e.g., free will cultures such as the American). In those cultures, one performs innovative behavior by matching that uncertainty instead of facing it and reducing it (Schneider, 1988). When those cultures are also collectivistic (e.g., the Chinese), the risk-taking behavior is embedded in the group dynamic process, while "face saving" hinders the individual development of taking risks.

Influence tactics. A third variable characteristic of championing concerns the use of *influence tactics*. Champions are thought to make frequent use of a variety of influence tactics. Those influence tactics vary from shared communication and enthusiasm to rational reasoning (Howell and Higgins, 1990). Influence tactics can be described as techniques that are informally used to gain commitment and compliance from other organizational members at the same or a higher hierarchical level. Their informal character makes the effectiveness of this process quite susceptible to cultural influences. External rule-oriented people may react in a different way on the same persuasive elements than internal rule-oriented people. Frequency and variety in the use of influence tactics is supposed to differ according to the beliefs and values of people. For instance, forming norms within a corporate culture that stimulates assertiveness (frequent use of "hard" influence tactics) is expected to have a positive impact on individual power-oriented individuals but may well have a negative effect on authority power-oriented individuals (Sullivan and Taylor, 1991). The motivation and education of a polite friendliness (relatively infrequent use of hard influence tactics) may be a more appropriate strategy in the latter environment.

An integrative framework

As a result of the above theoretical analysis, we propose an integrated framework which links the different championing characteristics (i.e., leadership style, personal attributes, influence tactics) to variations in corporate culture. In Table 9.1, we have classified corporate cultures on the basis of the two dimensions we have extensively discussed above (i.e., rule orientation and power orientation). Four archetypes can be singled out. Since the espoused leadership styles and influence tactics can be seen as an extension of the prevailing culture, the imposition of certain styles or tactics by managers with a different cultural background (either corporate or ethnic) may result in very poor outcomes. Furthermore, persons successful in one culture may be quite unsuccessful in other ethnic cultures, given the differences in terms of personal attributes.

However straightforward Table 9.1 may seem, one should be very wary of stereotyping world regions. Again, the *diversity of Europe* leads us to believe that simple comparisons between the United States and Europe are not possible. Indeed, important variations are present within the European continent. A recent study by Schneider and De Meyer (1991) underscores this assertion. A cross-national study at a major European business school clearly indicates multiple cultures within Europe exist. In addition, the findings of that study demonstrate that interpretation of and responses to strategic issues vary significantly among these European subcultures.

Conclusion

This chapter began with an outline of the framework we have used to develop a U.S.-European analysis in the field of innovation manage-

TABLE 9.1 Corporate Culture, Championing Activities, and Fit with Ethnic Culture

Orientation of corporate culture	Internal rule individual power	Internal rule authority power	External rule individual power	External rule authority power
Leadership style	Transformational	Transactional	Charismatic	Low-level transactional
Personal attributes	Entrepreneurial, individualized consideration, risk takers, proactive behavior	Facilitative group members, consensus oriented, hierarchical	Chaotic persons, need for rules, individualized consideration, risk averse, reactive behavior	Bureaucratic personalities, highly resistant to change, punctual
Influence tactics	Task-related influence tactics, based on individual power progress, career development, and job rewards	Social-related influence tactics based on confidence, enthusiasm, and shared communications	Task-related influence tactics, administrative expertise	Social and authority related influence tactics
Fit with ethnic culture	Ethnic cultures characterized by individualism, complexity, low hierarchical distance, and free will	Ethnic cultures characterized by collectivism, simplicity, low hierarchical distance, and free will	Ethnic cultures characterized by individualism, complexity, high hierarchical distance, and determinism	Ethnic cultures characterized by collectivism, simplicity, determinism, and high hierarchical distance
Examples of nations	United States, Australia	Japan, Indonesia	France, Belgium	Poland, Czechoslovakia

ment practices. On the socioeconomic front, our analysis suggests that technological heritage, market heterogeneity, administrative heritage, and regional entrepreneurship may explain the different approaches used by firms in Europe and the United States. The cultural analysis leads us to believe that ethnic cultural dimensions such as individualism, determinism, distance perception, and complexity provide a second explanation of why product development projects are managed differently across the Atlantic.

Globalization of world competition compels organizations to acquire an understanding of differences between nations and cultures. In the field of technological innovation, research is needed to explain the influence of the socioeconomic and the cultural context on international differences in terms of radicalness of innovation, product and process renewal, formalization of innovation processes, and centralization of R&D operations. We believe the present analysis has enabled us to make a first step in that direction.

Many questions remain. As argued before, the present analysis is best seen as a nonexhaustive essay on potential relations between the socioeconomic and cultural context and the management and outcome of innovation projects. In relation to the success of innovation projects, the present scientific state of the art allows us to merely scratch the surface. For instance, we may wonder whether some of the ethnic cultural dimensions are associated with product development performance. In a recent study, Hofstede's framework (1984) has been used to analyze R&D productivity in 302 industrial and 506 academic R&D units from Austria, Belgium, Finland, and Sweden (Kedia, Keller, and Julian, 1992). Low-power distance and high masculinity were associated with higher R&D productivity. Uncertainty avoidance was not significantly related to R&D productivity. This study suggests that innovation performance does not solely depend on finding a fit between ethnic culture and corporate culture, but that some cultures are inherently better suited to deal with the challenging task of developing technological product innovations.

Such studies require replications across countries and across measures. Continuation of such empirical inquiries may also explain why the United States seems to lose its edge in the product development arena. Even in traditional strongholds like the information technology industry, American companies' competitiveness seems to decline (e.g., the case of Wang computers). Both the socioeconomic and the cultural context suggest that the entrepreneurial spirit is a common denominator in many American corporations. On this basis, we would expect the United States to be a major contributor to the development of new markets and products. On the other hand, however, the homogeneity of its domestic market, the strong position of corporate headquarters within most U.S. firms, and the individualism that prevails within these organizations may explain why many American companies have declined. For instance, many Belgian companies export more than 80 percent of their output, a statistic rarely achieved by American companies. Cut-throat competition on the European market and vigorous rivalry on the foreign markets are very strong motivators for the efficient management of new product development activities. European companies are in need of design platforms that help them to achieve scope and scale in their innovation efforts simultaneously. A useful avenue for understanding the international differences with regard to product development success concerns a cross-cultural validation of Van de Ven's framework (1986). Some cultures may indeed be better positioned to spot new ideas, to translate ideas into action, to achieve interfunctional integration, or to create organization structures that stimulate innovation.

The scientific and business literature on the management of technological innovation has traditionally ignored the concept of ethnic culture. However, there seems to be a growing concern for the powerful effects corporate culture has on organizational innovativeness and individual innovative behavior (Feldman, 1988; Peters and Waterman, 1982; Golden, 1992; Dake, 1991). Yet, no consensus exists regarding whether that culture must represent homogeneity (Tichy, 1982), heterogeneity (Rose, 1988), or even ambiguity (Meyerson and Martin, 1987) to support innovative behavior. Must the central norms characterizing corporate cultures be widely shared? Can there be conflict and space for multiple sets of norms? Must the central norms be specified, or does ambiguity foster a better motivational context? The attentive reader will have observed that we have not exhausted all ethnic configurations in Table 9.1. What about other configurations? Clearly then, we have only started to understand international variety in the product development arena.

References

Achilladelis, B. P., Jervis, P., and Robertson, A. *A Study of Success and Failure in Industrial Innovation,* Sussex England: University of Sussex Press, 1991.
Adler, N. J., and Jelinek, M. "Is Organization Culture Bound?" *Human Resource Management,* 25: 1 (Spring 1986), 73–90.
Albert M. *Capitalisme Contre Capitalisme.* Editions du Seuil, 1991.
Atkinson, A., and Stiglitz, J. "A New View of Technological Change," *Economic Journal,* 78, 1969, 573–578.
Bartlett, C., and Ghoshal, S. *Managing the Transnational Corporation,* Boston: Harvard Business School Press, 1989.

Bass, B. M. *Leadership and Performance beyond Expectations,* New York: Free Press, 1985.

———. "From Transformational to Transactional Leadership: Learning to Share the Vision," *Organizational Dynamics,* Winter 1991, 19–31.

Beatty, C. A., and Gordon, J. R. M. "Preaching the Gospel: The Evangelists of New Technology," *California Management Review,* Spring 1991, 73–94.

Clark, K. B., and Fujimoto, T. *Product Development Performance,* Boston: Harvard Business School Press, 1991.

Conger, J. A., and Kanungo, R. N. "Toward a Behavioral Theory of Charismatic Leadership in Organizational Settings," *Academy of Management Review,* 12:4, 1987, 637–647.

Cool, K. *Analysis of Patenting in the World's Pharmaceutical Industry (1945–1987),* Mimeo, Insead Presentation, 1992.

Cox, T. H., Lobel, S. A., and McLeod, P. L. "Effects of Ethnic Group Cultural Differences on Cooperative and Competitive Behavior on a Group Task," *Academy of Management Journal,* 34:4, 1991, 827–847.

Dake, K. "Orienting Dispositions in the Perception of Risk: An Analysis of Worldviews and Cultural Biases," *Journal of Cross-Cultural Psychology,* 22:1, 1991, 61–82.

De Meyer, A. "New Manufacturing Strategies: Taking Advantage of Uniform Standards and Distinctive Technologies." In S. G. Makridakis (ed.), *Single Market Europe,* Oxford: Jossey, Bass, 1991, 119–140.

Dvir, D., and Shenhar, A. "Success Factors of High-Tech SBUs: Toward a Conceptual Model Based on the Israeli Electronics and Computers Industry," *Journal of Product Innovation Management,* 7:4, December 1990, 288–296.

Feldman, S. P. "How Organizational Culture Can Affect Innovation," *Organizational Dynamics,* 17, 1988, 57–68.

Florida, R., and Kenney, M. *The Breakthrough Illusion,* New York: Basic Books, 1990.

Frohman, A. L. "The Performance of Innovation: Managerial Roles," *California Management Review,* 20:3, 1978, 5–12.

Gamble, T. J., and Ginsberg, P. E. "Differentiation, Cognition and Social Evolution," *Journal of Cross-Cultural Psychology,* 12:4, 1981, 445–459.

Gibson, J. L., Ivancevich, J. M., and Donnelly, J. H. *Organizations,* 6th ed., Plano, TX: Business Publications, 1988.

Ginsberg, A., and Abrahamson, E. "Champions of Change and Strategic Shifts: The Role of Internal & External Change Advocates," *Journal of Management Studies,* 28:2, 1991, 173–189.

Golden, K. A. "The Individual and Organizational Culture: Strategies for Action in Highly-Ordered Contexts," *Journal of Management Studies,* 29:1, 1992, 1–22.

Hall, E. T., and Hall, S. T. *Hidden Differences: Doing Business with the Japanese,* New York: Prentice-Hall, 1988.

Hofstede, G. *Culture's Consequences: International Difference in Work-Related Values* (abridged ed.), Beverly Hills, Calif.: Sage Publications, 1984.

Howell, J. M., and Higgins, C. A. "Champions of Technological Innovation," *Administrative Science Quarterly,* 35, 1990, 317–341.

Jaeger, A. M. "Organization Development and National Culture: Where's the Fit?", *Academy of Management Review,* 11:1, 1986, 178–190.

Kagono, T., Noaka, I., Sakaibara, A., and Okumura, O. *Strategic versus Evolutionary Management: A U.S.-Japan Comparison of Strategy and Organization,* Amsterdam: North Holland, 1985.

Kedia, B. L., Keller, R. T., and Julian, S. D. "Dimensions of National Culture and the Productivity of R&D Units," *High Technology Management Research,* 3:1, 1992, 1–18.

Kluckhohn, F., and Strodtbeck, F. L. *Variations in Value Orientations,* New York; Harper & Row, 1961.

Kuhnert, K. W., and Lewis, P. "Transactional and Transformational Leadership: A Constructive/Developmental Analysis," *Academy of Management Review,* 12 :4, 1987, 648–657.

Lee, M., and Fox, S. *Organizational Archetypes and Culture,* Lancaster University, 1993.

Maidique, M. "Entrepreneurs, Champions and Technological Innovation," *Sloan Management Review,* 21:2 (Winter 1980), 59–76.

Maruyama, M. "Alternative Concepts of Management: Insights from Asia & Africa," *Asia Pacific Journal of Management,* January 1984, 100–111.

Meyerson, D., and Martin, J. "Cultural Change: An Integration of Three Different Reviews," *Journal of Management Studies,* 24:6 1987, 623–647.

Moenaert, R. K., Deschoolmeester, D., De Meyer, A., and Souder, W. E. "Information Styles of Marketing and R&D Personnel during Technological Innovation Projects," *R&D Management,* 22, January 1992, 21–39.

Mokyr, J. *The Lever of Riches,* Oxford: Oxford University Press, 1990.

Morgan, G., and Ramirez, R. "Action Learning: A Holographic Metaphor for Guiding Social Change," *Human Relations,* 37:1, 1983, 1–26.

Nonaka, I., and Johansson, J. K. "Japanese Management: What About the 'Hard' Skills?" *Academy of Management Review,* 10:2, 1985, 181–191.

———and Kenney, M. "Towards a New Theory of Innovation Management: A Case Study Comparing Canon, Inc. and Apple Computer, Inc.," *Journal of Engineering & Technology Management,* 8, 1991, 67–83.

O'Reilly, C. "Corporations, Culture and Commitment: Motivation and Social Control in Organizations," *California Management Review,* Summer 1989, 9–25.

Patel, P., and Pavitt, K. "Is Western Europe Losing the Technological Race? " *Research Policy,* 15:2–4, 1986, 59–86.

Peters, T. E., and Waterman, R. H. *In Search of Excellence,* New York: Harper & Row, 1982.

Prahalad, C. K., and Hamel, G. "The Core Competence of the Corporation," *Harvard Business Review,* 68:3, 1990, 79–91.

Quinn, J. B. "Technological Innovation, Entrepreneurship and Strategy," *Sloan Management Review,* Spring 1978, 19–30.

Roger, A. "Culture Nationale ou Culture d'Entreprise? Une Comparaison Internationale sur les Perceptions et les Attentes de Chercheurs Industriels," *IAE,* September 1990, 1–14.

Ronen, S., and Shenkar, O. "Clustering Countries on Attitudinal Dimensions: A Review and Synthesis," *Academy of Management Journal,* 10:3, 1985, 435–454.

Rose, R. A. "Organizations as Multiple Cultures: A Rules Theory Analysis," *Human Relations,* 21:2, 1988, 139–170.

Rothwell, R., Freeman, C., Horlsey, V., Jervis, A. , Robertson, B., and Townsend, J. "Sappho Updated—Project SAPPHO Phase II," *Research Policy,* 3, 1974, 258–291.

Schein, E. H. *Organizational Culture and Leadership,* San Francisco: Jossey-Bass, 1985.

Schneider, S. C. "National versus Corporate Culture: Implications for Human Resource Management," *Human Resource Management,* 27:2, (Summer 1988), 231–246.

———. "Strategy Formulation: The Impact of National Culture," *Organization Studies,* 10:2, 1989, 149–168.

——— and De Meyer, A. "Interpreting and Responding to Strategic Issues: The Impact of National Culture," *Strategic Management Journal,* 12, 1991, 307–320.

Schon, D. A. "Champions for Radical New Innovations," *Harvard Business Review,* 41, March–April 1963, 77–86.

Schrader, S. "Informal Technology Transfer between Firms: Cooperation through Information Trading," *Research Policy,* 20:2, 1991, 145–152.

Schwartz, S. H. "Individualism—Collectivism: Critique and Proposed Refinements," *Journal of Cross-Cultural Psychology,* 21:2, June 1990, 139–157.

Shaw, J. B. "A Cognitive Categorization Model for the Study of Intercultural Management," *Academy of Management Review,* 15:4, 1990, 626–645.

Sinetar, M. "Entrepreneurs, Chaos and Creativity—Can Creative People Really Survive Large Company Structure?" *Sloan Management Review,* Winter 1985, 57–62.

Smircich, L. "Concepts of Culture and Organizational Analysis," *Administrative Science Quarterly,* 28, 1983, 339–358.

Sullivan, J., and Taylor, S. "A Cross-Cultural Test of Compliance Gaining Theory," *Management Communication Quarterly,* 5:2, November 1991, 220–239.

Tichy, N. M. "Managing Change Strategically: The Technical, Political, and Cultural Keys," *Organizational Dynamics,* Autumn 1982, 59–80.

Triandis, H. C. "Dimensions of Cultural Variation as Parameters of Organizational Theories," *International Studies of Management and Organization,* 12:4, 1983, 139–169.

Van de Ven, A. H. "Central Problems in the Management of Innovations," *Management Science,* 32:5, May 1986, 590–607.

Wagner, J. A., and Moch, M. K. "Individualism-Collectivism: Concept and Measure," *Group & Organization Studies,* 11:3, 1986, 280–304.

10

Strategic Management of Technology: An Emerging Field

Michael W. Lawless
University of Colorado at Boulder

The strategic management of technology is a major issue for the management of technology (MOT) movement. For example, in order for it to be fully successful, the new technology development process must be an integral part of the overall strategic mission of the organization. New product and service developments that are not congruent with the strategic missions of the developing organization will not have the required resource and personal support to succeed. Beyond these notions, the principles of strategic technology management have yet to be defined. How is the strategic management of technology (SMOT) different from strategic business management? What technology management problems should be considered strategic issues? How can a new-product-developing organization use strategic technology management concepts to successfully position itself against global competition? How can SMOT be applied to keep an organization technically competent? Are there natural tensions between the strategic management of new technology development and the strategic management of the business that managers must be alert to and aggressively work to alleviate? These are some of the questions posed in this chapter. Because SMOT is an emerging paradigm, the answers are more tentative than conclusive. But all new product and all new business managers should be aware of the issues and the nature of emerging thoughts about SMOT.

WM. E. SOUDER AND J. DANIEL SHERMAN

Introduction

What makes the study of SMOT different from the rest of strategic management research? It may only be the guiding belief among some scholars that technology plays a significant role in the conduct and success of an enterprise. Those who recognize technology's recurrent role in disrupting, reshaping, or reinforcing social systems may be more inclined than others to investigate its mutual effects on strategy and organizations. Their disposition to model technology as potentially important to explain conduct and performance variance has created a specialization differentiated by content from strategic management research as a whole. Many questions in the strategy mainstream have not had high priority in SMOT, and key SMOT issues are on the periphery of research of the strategic management agenda. For instance, correspondence between firms' technology and strategy types has been central to the SMOT community in recent years. But it is only one issue among many in general strategy formulation.

Since SMOT is differentiated principally on issue content, perhaps the best way to describe its current domain is through conditions facing general managers in technology-intense industries. The list is sure to include uncertainty in market outcomes, the potential for rapid, competence-destroying change, the pressure for fast decision making under uncertainty, short product and technology life cycles, value added from innovation, strategic value of knowledge, nontraditional value-chain arrangements, competencies distributed among different firms, and performance effects of information. These issues constitute an array of phenomena, managerial problems, concepts, and heuristics that appear for the moment to have their focus on technology in common, even if other connections are tenuous. They attract a diverse collection of scholars—economists, psychologists, and sociologists, among others. Lawless (1992) argued that the literature in MOT is eclectic in part because the content is attractive to individuals from many disciplines. Distinct world views among scholars add to the strains away from a coherent overall approach, so SMOT has a broad domain but no paradigm.

Future developments of SMOT research might be anticipated in two different patterns. Some work will be performed through an interdisciplinary approach. Issues may be decomposed according to their fit with the models of various disciplines, and the insights from each synthesized into a perspective that is more comprehensive for the diversity. Since this type of work would be organized around issues like those listed above, the results would be issue oriented as well. Individual questions could be resolved appropriately—especially for managers—if less focused from a scientific viewpoint. Progress may not necessarily be retarded by this mode, but the record for cumulative development of research in SMOT based on issue studies is mixed at best.

The second approach could be anticipated where important issues are embedded in the prevailing views of strategic management research in general and the scholarship that underlies it. The unique character of some SMOT issues notwithstanding, conceptualization and useful prescriptions have much in common with the rest of strategic management research and with its root disciplines of economics and sociology, among others. The cost is in trading off rigorous, systematic analysis by established methods for partial, rather than comprehensive, perspectives on SMOT problems.

An efficient structure within which to review a complex, diverse field is through its important issues and the status of research associated with them. This chapter is organized as a survey of firms' market relations as they are affected by technology. It is divided into three sections: technological change and market change, competitive positioning, and cooperative arrangements. The chapter's scope is a recognition that competitive strategy—from firms' adaptations to the constraints of rapid change to realignment of competitors and partners alike—is transformed as technology evolves.

Scope of SMOT

The SMOT subject matter might attract reader interest on several possible grounds. First are questions that sustain continuing attention, although they have been around for some time. The role of R&D management in the overall company strategy is an example. Second are recent developments that might challenge familiar assumptions. In strategy-technology integration, economic concepts like differentiation and sustainable advantage might advance our understanding more than the static analyses of fit that are familiar to students of the field. Third are emerging issues that could precede future contributions. Technology standardization is a strong candidate for further study for its influence on market conditions and on competitive and cooperative conduct.

This chapter focuses on the following SMOT issues. First is the matter of *technological change, markets,* and *management*. Some models of change patterns are summarized, prior to analyzing their effects on the conduct of firms. Technology standardization is a life-cycle phase with profound effects on the competitive conditions in a market, for example. The dominant wisdom since Abernathy and Utterback (1978) has been that competition becomes an efficiency footrace when a standard emerges with price as the main differentiating factor among rivals. The principal question for technology strategists is how to outpace others in process innovation. This perspective may be due for a reexamination. In particular, industries where technology is standardized may not have a deterministic bent toward cost and price competition. Incumbents may instead differentiate on other dimensions (packaging, distribution, service, etc.) that are uniquely supported by the firm's capabilities. The resulting picture allows for much more diversity and consistency with economics-based thought on differentiation, immutability, and the economic concept of rents.

Second is the issue of *integrating competitive strategy and technology*. Much of the literature relies on a mechanistic view of competitive strategies and their relationship to technology. Superficially, it seems consistent with Porter's conclusion that "technologies that should be developed are those that most contribute to a firm's generic strategy" (1985). In a typical formulation, integration is conceived as the intersection of a strategy type and a technology type. For example, there is poor fit between a cost-leader strategy and aggressive new product development, but investment in production efficiency is consistent with cost leadership. The process of strategy-technology integration is a direct extension. Recipes for integrating strategy and technology resemble a SWOT analysis, assessing the role of technology in the overall strategy as one might do to another resource or capability. This chapter proposes an alternative to this mechanistic view and many of its implications. The bedrock idea is that technology matters in strategy formulation to the extent that it allows firms to take competitive positions that lead to extraordinary profits. It fits within a dynamic model of competitive positioning as a potential source of differentiation and value. This approach systematizes technology's role in strategy and incorporates industry and technology change. It also introduces issues that have not been prominent in strategy-technology integration work despite their potential importance. Durability of competitive advantage from technology is an example.

Third are matters of *cooperation, complementary capabilities,* and *claimants.* Idiosyncratic knowledge, combined with scale effects, contributes to a propensity toward alliances in technology development and marketing. Incentives, costs, and evaluations of participating in alliances can be modeled with the logic of differentiation and value and with bargaining in the value chain. Participants in alliances are subject to losing rents to alliance partners and other claimants on the returns from alliances, but they gain advantage from specializing in their strengths and in reducing administrative burden relative to integration.

Punctuated equilibrium: Technological change, markets, and management

The interaction among markets, firms, and technology is a complex socioeconomic and technical process. The patterns by which firms interact with each other and with other environmental forces often change along with the introduction of new core technologies, new products, or changes in product-technology performance. On the other hand, technology cannot be regarded strictly as an exogenous force affecting market conditions. Firms' investments in R&D, formation of alliances for developing new technology, environmental scanning, and other activities can also affect the state of and changes in the technology.

Early change typologies were principally based on time lines, and their categories for ordering information were stages in various kinds of cycles. Sahal (1981) and Dosi (1982) offer their own insights within this type of framework. Bright's model (1969), for example, has eight stages, starting with discovery and ending with widespread adoption. These were extremely useful for ordering a process that appeared complex and stochastic but as a group tended to be static, perhaps a bit too orderly, and seldom (mostly by design) included any implicit strategy prescriptions. Their successors are contingency approaches where strategy and organization structure are associated with some prescribed time periods, for example, stages in a technology life cycle. Such an analysis might lead to a set of propositions regarding a fit between strategy and technology in each time period.

Cycles between phases of evolutionary and revolutionary growth are at the core of another generation of models. Change comes either as evolutionary (which mainly preserves the market's status quo) or revolutionary (which disrupts market parameters and changes the competitive rules of the game). Schumpeter's distinction (1950) between incremental and discontinuous change based on technology—"creative destruction"—is at the heart of these models. They take as their theme that the competitive rules of the game can actually be reinforced by relatively minor technology changes that occur within the context of the existing product. By keeping pace with technological developments, incumbents can retain or improve their advantage. On the other hand, the rules can be wholly disrupted by change. The effects of such discontinuous change could obsolete an existing product, change the value of complementary goods, deplete investments in manufacturing and other capabilities, eliminate entry barriers, or even create a whole new market to replace an existing one. The tension for managers is between profiting from technological impacts on the market and becoming displaced by them.

Some models involving the concept of punctuated equilibrium also draw on Kuhn's description (1970) of the evolution of scientific knowledge. His famous description of preparadigmatic and paradigmatic science parallels the experience of technological change. Teece (1988) uses it to describe technological change and the emergence of a standard by analogy to the development of scientific knowledge. The early trajectory of a technology, in both conceptual and empirical accounts, is from many competing alternatives to a few or even to a single standard. The preparadigmatic phase occurs early on, while there are many different versions of a new technology. Schumpeter's conceptualization is intuitively consistent. His model of economic change is also a platform for contemporary thinking on technology as well. There has traditionally been broad acknowledgment within the

TABLE 10.1   Punctuated Equilibrium: Technological Change Events and Periods

Event	Technological period
Emergence of a dominant design	Paradigmatic regular change
Discontinuous change	Preparadigmatic ferment

SMOT community of the role that technology plays in the changing firms' environments. Table 10.1 is a simple representation of the periods that follow two kinds of technology change events.

Past these simple premises, understanding of the relation between technological change and strategy formulation becomes more complicated. Other evolutionary models note the potential for continuous change in the form of "dematurity" (Abernathy and Clark, 1985) or "stepwise growth" (Sahal, 1981), leading to a relative advantage for some firms. Anderson and Tushman (1991) describe technological change as competence enhancing, which reinforces the capabilities of incumbents, or competence destroying, in which the capabilities of incumbents are devalued. Henderson and Clark (1990) make a recent addition to this discussion by distinguishing several types of technological change according to their scope: architectural, incremental, modular, and radical. Lawless and Anderson (1991) add to this a perspective based on the importance of a firms' differentiation to their rent-earning potential. They link technology change events to overall similarity in conduct among competitors in a market. They are interested in the degree to which technology similarity, which defines the period under a dominant technology, is associated with market similarity—or like conduct among firms. In brief, companies that are undifferentiated in their markets are more likely to face intense competition. The emphasis on competitive positioning draws on the research on strategy and market structure. Economic implications for performance become clearer. Conversely, the more alike different firms' competitive positions are, the less variance will be expected in their returns. In this approach, the essence of technology's relation to strategy is its influence on similarity or diversity in competitive conduct. Emergence of a dominant design in a technology is expected to reduce variety in competitive conduct in a market, *ceteris paribus*. Incumbents' potential to differentiate on technology-driven product characteristics diminishes if it does not completely disappear. Lawless and Anderson (1991) take as their baseline the premise that market-response periods of similarity and dissimilarity among incumbents are associated with periods of regular change or ferment. Table 10.2 summarizes these relationships among events, technological periods, and response periods.

TABLE 10.2   **Technological Periods and Market Response Periods**

Event	Technological period	Market response period
Emergence of a dominant technology	Paradigmatic regular change	Similarity
Discontinuous change	Preparadigmatic ferment	Dissimilarity

Firms tending toward like conduct in a market characterize a period of technological stability in part because their means to differentiate on products are reduced. Market dissimilarity is a period of technological heterogeneity and uncertainty (following a revolutionary change) during which firms are expected to be relatively differentiated along important strategic dimensions. The models of types of technological change and respective impacts on organizations and strategies are conceptually rigorous but still have some implications for managers in technology-intense industries. Anderson and Tushman's observations (1991) are a good example. They recommend that managers:

1. *Expect discontinuities.* Develop competencies that survive technological revolution because, for instance, they are not specific to a particular product. One could see a kinship between Anderson and Tushman's idea (1990) and Prahalad and Hamel's "core technology" (1990) in the general strategic management literature.

2. *Expect dominance.* Expect a *follower* design from incumbents to dominate other designs. The best technology from a technical perspective is by no means certain to become the new standard.

3. *Expect obsolescence.* Outside firms may initiate new product technologies that can outmode existing know-how. In their scanning and threat analysis, incumbents should be aware of new entrants whose main connection to their market is technological.

4. *Expect technology-driven shakeouts.* Industry shakeouts can come from technology change rather than exclusively from economic downturns.

Integrating competitive strategy and technology: Lessons for managers

Questions regarding the development of the field that interests scholars in SMOT—the emergence of a paradigm and the likelihood for rigorous cumulative research—are unresolved for now. At the same time, however, others continue to ask for "real-world" relevance: what they can take from the research on SMOT that is useful in the pre-

TABLE 10.3   Customary Management Priorities in Technological Periods

Event	Technological period	Market response period	Customary management priority
Emergence of dominant technology	Paradigmatic regular change	Similarity	Efficiency
Discontinuous change	Preparadigmatic ferment	Dissimilarity	Innovativeness

sent. From models based on technology cycles, we might deduce the main concerns for management according to the period of technological change, evolutionary or revolutionary. The themes, efficiency for periods of incremental change and innovativeness for periods of discontinuous change, are consistent with Utterback and Abernathy (1975), among other precedents. Table 10.3 shows the expected priorities for management in relation to technological change periods.

Evolutionary models of technological change lead in a direct way to heuristics for managing technology at the firm level. Three approaches to strategy development in individual firms explicitly attempt to accommodate some effects of technology described in the preceding section. They also serve to represent the current state of theory development in SMOT.

Foster (1986) proposed *S curves* as a means to generically describe the relationship between time and performance for individual technologies and to capture the timing of "revolutionary" change. Each technology is described with a curve that includes start-up, growth, and maturity phases. The last is paralleled by a new, displacing technology for some period. During that period, the initial lower performance of the displacing technology can be deceptive—particularly to those vested in the maturing technology. The old may look far superior in terms of current performance. If S curves really do convey the course of a technology in time, they may be used to anticipate revolutionary change and to prepare for it. Depending on the flexibility of their asset base and the extent of their commitment, incumbents might attempt to restructure or exit if they could anticipate the inflection point in performance and the onset of the new technology. This is, of course, a significant constraint on the utility of S curves. Problems persist both in anticipating the nature of a new technology (e.g., what the dominant design will be) and in forecasting changes in the performance of old and new technologies alike.

A second approach that may serve as a point of comparison are Clark's *success factors* (1989). This is a recipe for strategic management in competition under technological change. Based on observations of companies that have been successful, Clark concludes that the following measures are helpful in identifying the firm's *technological core*. First, understand the source of the firm's value-added advantage and cultivate it. Second, since concentrating on the core implies some specialization, develop competence at participating in alliances. Joint development, manufacturing, or distribution agreements, for example, require management skill to lead to relative net benefit. Third, emphasize manufacturable designs. Clark recognizes the importance of design, development, and production engineering to speed and quality manufacturing later in the life cycle. This leads to his fourth point: There is a premium to be earned by moving fast, especially in technology-intense industries.

Porter's generic strategies (1985) are a basic part of the strategy-technology formulation literature. They are the platform for a number of models where strategy types are overlaid with technology dimensions. Matrices result in which the cells are strategy-technology combinations with varying degrees of fit. Table 10.4 presents an example of such a matrix. It proposes that technology leadership is not compatible with a cost-leader generic strategy. But a follower approach to acquiring technology allows a cost leader to exploit the learning of technology first movers. A firm with a differentiation competitive strategy can strengthen its position by moving early on technological developments but finds no advantage in being a technology follower. In this construction, there is a symmetry between rows and columns that conveys simple strategy-technology choices.

TABLE 10.4   A Newtonian Model of Strategy-Technology Integration: Differentiation and Cost Leader Strategies' Fit with Technology Strategies

Generic strategy	Technology leader	Technology follower
Cost leadership	Not compatible	Acquire technology cheaply
	Process innovation	Exploit others' learning
Differentiation	Technology-based competitive-monopoly position	Weak position
	Product differentiation based on technology	Differentiate in value other than on technology; innovate outside the technological core

Porter (1985) presented some rules of thumb that presage this conceptual development. He described conditions that relate technological change to increased competitive advantage. First, advantage is gained if—consistent with Lawless and Anderson (1991)—the change differentiates, is valued by customers, and creates a sustainable lead. Second, first mover advantages matter. Being first leads to advantages other than from the technology itself (e.g., from an innovative reputation). Third, he says that some technology changes "lift all boats." They may improve overall industry conditions and give an advantage to all incumbents, particularly in relation to new entrants. These ideas are based on a combination of industrial organization economics and observation. They allude to the importance of a differentiated market position to competitive performance in general and make a start on a way of thinking that is new to the technology-strategy literature.

Such analysis could be said to have made a contribution in the exploration of strategy-technology interaction. It certainly gives an elegant structure to a complex set of interactions. However, Lawless and Anderson (1991) recognize that technology and strategy are linked at a more profound level, where subtlety leads to advantage. When, for all its special characteristics, technology is incorporated into the foundation strategic management issue of gaining competitive advantage, strategy-technology interaction takes on the aura of the "invisible hand." Simple contingency models of their fit appear newtonian and unnecessarily artificial. Lawless and Anderson's model of similarity and dissimilarity (1991) proposes that the emergence of a standard increases pressure toward similarity in the relative positions of incumbents. They also argue that firms resist becoming more similar since their market comes to resemble perfect competition, and their profits can theoretically be reduced to zero. Efforts toward developing unique capabilities are a rational expectation. Their propositions are consistent with those of Rumelt (1987) among others, where the differentiated position of a firm leads to rents, even in a competitive market. The durability of those returns depends on a firm's abilities to deter imitation. Technology affects competition and advantage as part of the value chain, integrated with complementary assets and know-how. Technology is strategically important since it contributes to a differentiated, valued market position that leads to rents. It can be evaluated as a potential source of *sustained* competitive advantage according to the same general analysis that is applied to other assets and know-how. Barney (1989), for instance, argues that resources are more likely a source of sustained advantage to a firm if they are rare, valuable, hard to substitute, and hard to imitate. Technology adds to duration of advantage if it deters rivals from approaching the market position through acquisition, imitation, substitution, or improvement. Similar concepts, having gained acceptance bordering on voguishness in the strategic management community, have not yet claimed a place in the discussion of strategic implications of technology cycles.

This thinking might be distilled in a new way to evaluate the appropriateness of management priorities, given technological change periods. Competition after the emergence of a dominant design might amount to a footrace based on efficiency and speed as in the prevailing wisdom. To recognize that firms earn rents by differentiating in a way valued by customers, however, is to anticipate that they would invest in distribution, marketing, service, or other strategic dimensions. Incumbents assume diverse market positions according to their capabilities and the durability of returns based on their investments. Thus, the period of regular change expected after the emergence of a dominant design could be one of differentiation on many dimensions rather than solely on efficiency. These notions are summarized in Table 10.5.

Cooperation, complementary capabilities, and claimants: Suggestions for managers

In some industrial settings, evaluating the state of the art in competitive strategy would be sufficient to describe strategic management as a whole. There simply is little going on otherwise. In the strategic management of technology, however, cooperation among various firms in a "competitive" market is routine yet important conduct. It warrants specific attention to modeling costs and benefits because the impacts are too important for management practice to be spontaneous and unrehearsed.

If cooperative strategies do not typically have the same priority as competitive strategies among technology companies, they are prominent in both frequency and impact. Being good at one or a few activities in the value chain may be all that many firms choose or can man-

TABLE 10.5 Differentiation Management Priorities in Technological Periods

Event	Technological period	Market response period	Differentiation management priority
Emergence of dominant technology	Paradigmatic regular change	Similarity	Differentiation on dimensions other than technology
Discontinuous change	Preparadigmatic ferment	Dissimilarity	Innovativeness

age. Managers turn to cooperation to gain access to assets, know-how, and other capabilities that complement and lever their own. The complexity and scope of technology development appear to drive tendencies to cooperate. It may also be all that is needed to be satisfactorily profitable. Cooperation in the supply of products with significant value added from technology could be explained as firms' concentrating on core competencies and accessing other resources and activities in markets.

Technology attracts returns by creating value: reducing costs, enhancing performance, or increasing returns of customers in subsidiary markets. In concept, a "bucket of profits" is created by all of the member companies. Providers of goods and services create more value together, and therefore a bigger bucket, than they could individually. Their relations with suppliers, customers, and rivals are then a matter of their contributions to the total profit created and the efforts of each company to keep as much total profit as they can.

These claims by members of alliances, joint ventures, and other cooperative forms might be clarified by invoking the framework of property rights. The bucket of profits is finite, and claims can be mutually exclusive, even though they are related to strategies and organization structures that are nominally cooperative. Competitive conduct may persist in cooperative arrangements—and may in fact be a reasonable anticipation, other things equal—because of scarcity in benefits and conflicting expectations among participants. This approach does not necessarily exclude, and may indeed complement, interpreting cooperation as a cultural or emotional good. A rational expectation is that, because cooperation is so often observed, firms benefit from it, at least under some conditions. Additionally, intangibles like goodwill and trust play a role.

By assessing value added in their negotiations to develop a cooperative arrangement, and in their operations while the project is ongoing, participants can advance their claims on returns. They can also increase the chances that they obtain a price for their involvement that reflects its true value to other members.

Hamel, Doz, and Prahalad (1989) actually describe cooperation with other companies, market rivals or not, as a continuation of competition under somewhat different circumstances. Profits from the sales of a product, the value of partners' contributions, and the costs and benefits of participation that first lead to cooperation can change quickly in technology-intense environments. Partners on one generation of a technology can later become rivals. Others may operate in other activities within the value chain and be jeopardized by forward or backward integration in derived demand relationships. Other changes in the scope of a company's activities—for example, market diversification or line extensions—could lead to a disadvantage to alliance partners. Finally, some cooperation occurs with rivals, and these may have more obvious risks from changing market conditions.

If property rights are the conceptual foundation of a cooperative relationship, the value of exchange takes on several different dimensions from conceptual and managerial viewpoints. First is a partner's immediate benefit from cooperation. It could include access to proprietary information, distribution channels, or manufacturing capacity. These are tangibles and may be the bottom line for some firms simply because they can fill an immediate need and are easy to measure. Still, the future value of technology, particularly in periods before it has passed market or even technical tests, can be difficult to incorporate into a decision before the fact. Second, cooperative arrangements are defined in part by the perspectives that parties bring to the initial negotiations and to their actions during implementation.

Hamel, Doz, and Prahalad (1989) point to differences between terms of agreements and the scope of exchange that actually occurs. Their article interprets joint ventures between U.S. and Japanese firms, for example, as access to tangible capabilities with immediate, but limited, benefit to the Americans. Access to skills, know-how, and information is integrated into operations for the long-term benefit of the Japanese partner. All partners are subject to the terms of such arrangements as set out in contracts. However, one is equipped to extract more value under provisions that are equally binding. Third, a cooperative arrangement is characterized by the value of the exchange to the overall position of each company, both as a member of the alliance and in its other activities. Consider that information revealed in an alliance may be more valuable to the receiving firm than it is to the sender because of differences in their capabilities. It seems a reasonable empirical question to ask whether such value is accounted for in the price and incentive structure of the arrangement. And an appropriate conceptual question concerns whether such value should be included.

In managerial terms, we expect that members of an alliance or joint venture can keep the largest possible portion of profits by managing cooperative relationships effectively. Measures toward such an objective fall directly into the agency model. The agency framework has been applied to a wide range of organizational issues, including governance. It raises a host of issues that otherwise would not systematically find a place in discussions of alliances: membership requirements and fees, preference revelation and alignment, monitoring, incentives and penalties, and more. Other approaches have considerable potential as well. First, capabilities and the conduct of firms in a market can be divided into *maintenance* dimensions (areas where all firms must have strengths in order to participate in an industry but which do not lead to differentiation in proportion to investment) and *leverage* dimensions (on which companies are able to differentiate). Cooperating on maintenance dimensions is not particularly threatening. Managers should be especially diligent, however, regarding cooperation where their leverage capabilities are at stake. Second, the discussion of cooperation can be framed as a matter of sourcing. If companies did not specialize and sourced all of their needs internally, cooperation would be an empty question. But we have noted that scope, complexity, and the need to be on the cutting edge of several different technologies simultaneously are intense pressures. Managers can consider the value chain that links them with factor sources and with their markets. It might lead to insights regarding the value that they can realize from alternative forms and the value they create for their other partners by their involvement in alliances. Third, the term "cooperative advantage" might help to convey that parties should keep the largest possible portion of profits for themselves by managing cooperative relationships effectively. They start with an appropriate vision, for example, distinguishing between learning and more immediate objectives like access to assets. In choosing partners, they may also consider value *recovery* or the ability of members, based on capabilities and other unique characteristics, to capture benefits from cooperation that are nominally available to all. Without discrimination in the distribution of some shared good (like a patent or formula), greater value may still be realized by some members relative to others if they are uniquely equipped to exploit it. There are two remedies where such differences are a barrier to cooperating. First, develop the firm's own capabilities to decrease this advantage. Second, set acceptable initial terms based on the best available information. Entry "fees," for instance, could be set in relation to the expected benefit of the partners.

Research on cooperative conduct and technology is in a state similar to the strategic management of technology in competition. Common forms have been described and evaluated with varying degrees of rigor—consortia, joint ventures, and licensing, for example. Some experience-based heuristics have appeared, including valuable insights like the way that differential benefits in alliances are partially based on the partners' relative perceptions of the arrangements (Hamel, Doz, and Prahalad, 1989). But a large portion of the body of work so far consists of classifications and anecdotes. Cooperative strategies that are especially prominent in technology-intense industries—for example, adherence to standards for product components and know-how trading—are in the early stages of critical study. Cooperation in general, and familiar cooperative forms in particular, can be approached more systematically and rigorously than the record to date shows.

Summary

In this chapter, several interesting issues in SMOT were not treated because of the focus here on describing a few important questions that lend themselves to theory-based inquiry in a straightforward way. These issues include first mover disadvantages and advantages specifically where products have high value added from technology; diffusion of new technology, with attention to effects of installed base; switching costs; and demonstrated value added, gaining acceptance of new product technologies in the market and inside organizations and coping with costs of new technology that do not fall appropriately on users and other beneficiaries. These are important issues to many, and economics-based approaches to SMOT can add considerably to their study.

While the literature on SMOT has grown rapidly, signs of systematic advances have been relatively few. SMOT scholars as a whole have not attended as well as we might to examining, challenging, and building on precedent. It may indeed be early even to outline a paradigm for this field. Still, core ideas that relate technology to market forces (e.g., differentiation and rents, durability of advantage based on the difficulty of imitation) put some important issues in a context that could advance innovative, cumulative research on SMOT.

Diversity among SMOT scholars complicates prospects for the future. Rather than a coherent field, a community of loosely coupled disciplines may be a better model of research in MOT today. A market for ideas may also describe the current stage of theory development more accurately than an emerging paradigm. For the present, midrange theory—rigorous scholarship that is still concrete enough to assist managers—seems both a worthy objective and a reasonable expectation.

Some implications, lessons, and rules for managers in SMOT have been suggested (e.g., Tables 10.3 and 10.4). However, the field has not yet advanced to the point where many profound, detailed operating rules can be formulated. It seems clear that SMOT must be the underpinning of effective new technology development processes. However, the new technology development process has developed out of necessity, ahead of its theory base. As the theory of SMOT develops, it can serve as both a check on the viability of the existing new technology development process and a means of rationalizing and fortifying existing processes.

References

Abernathy, W., and Clark, K. "Innovation: Mapping the Winds of Creative Destruction," *Research Policy*, 14, 1985, 3–22.

Abernathy, W., and Utterback, J. "Patterns of Industrial Innovation," *Technology Review,* June–July, 1978, 40–47.

Anderson, P., and Tushman, M. "Managing through Cycles of Technological Change," *Research and Technology Management,* 1991, 26–31.

Barney, J. "Asset Stocks and Sustained Competitive Advantage: A Comment," *Management Science,* 35, 1989, 1511–1513.

Bright, J. "Some Management Lessons from Technological Innovation Research," *Long Range Planning,* 2: 1, September 1969, 36–41.

Clark, K. "What Strategy Can Do for Technology," *Harvard Business Review,* November–December 1989, 94–98.

Dosi, G. "Technological Paradigms and Technological Trajectories," *Research Policy,* 11: 3, 1982, 147–162.

Foster, S., *Innovation: The Attackers Advantage,* New York: Summit Books, 1986.

Hamel, G., Doz, Y., and Prahalad, C. "Collaborate with Your Competitors—and Win," *Harvard Business Review,* January–February 1989, 13–139.

Henderson, R., and Clark, K. "Architectural Innovation: The Reconfiguration of Existing Product Technologies and the Failure of Established Firms," *Administrative Science Quarterly,* 35, 1990, 9–30.

Kuhn, T. *The Structure of Scientific Revolutions,* Chicago: University of Chicago Press, 1970.

Lawless, M. "A Market for Ideas: Diversity and Progress in the Management of Technology," *Organization Science,* 3: 3, 1992, 301 and 442.

——— and Anderson, P. "Competitive Conformity Under Regimes of Standardization," Academy of Management National Meeting, Miami, 1991.

Porter, M. *Competitive Advantage,* New York: Free Press, 1985.

Prahalad, C., and Hamel, G. "The Core Competence of the Corporation," *Harvard Business Review,* May–June 1990, 79–91.

Rumelt, R. "Theory, Strategy, and Entrepreneurship." In D. Teece (ed.), *The Competitive Challenge,* Cambridge: Ballinger, 1987, 137–158.

Sahal, D. *Patterns of Technological Innovation,* Reading, MA: Addison-Wesley, 1981.

Schumpeter, J., *Capitalism, Socialism, and Democracy,* 3d ed., New York: Harper and Row, 1950.

Teece, D., "Capturing Value from Technological Innovation: Integration, Strategic Partnering, and Licensing Decisions," *Interfaces,* 18: 3, 1988, 46–61.

Utterback, J., and Abernathy, W. "A Dynamic Model of Product and Process Innovation," *Omega,* 3: 6, 1975, 639–656.

11

Directions for Future Research

J. Daniel Sherman
Wm. E. Souder
University of Alabama in Huntsville

The Domain of Research in the Management of Technology

An implicit assumption in the preceding chapters has been the notion that the management of technology (MOT) is not a field in itself but rather a broad domain of research which crosses multiple disciplines. The primary issues in this domain are based on the types of problems and conditions faced by managers in technology-intense industries and particularly by those who are either directly or indirectly involved with the management of new technology development. Thus, research on MOT crosses such diverse disciplines as marketing research, new product development, strategic management, organizational behavior, organizational theory, entrepreneurship, organizational sociology, industrial engineering, and management science. In this chapter we seek to explore directions for future research in the management of new technology development.

Idea Generation, Concept Development, and Market Definition

Managing idea generation

As noted by Professor Rubenstein in Chap. 2, the initial stage of idea generation is critical to the whole new product innovation process.

This is the crucial starting point for technical activities or projects which are intended to provide new or improved products or processes. As suggested by Professor Rubenstein, if poor ideas form the basis for the firm's project portfolio, the results are not likely to be successful in technical or commercial terms. Furthermore, good technical work on poor commercial ideas has resulted in wasted resources, missed opportunities, and declining performance in many industrial R&D labs.

One area in which future research may provide useful findings for technical managers and strategic planners is that of decisions regarding the outsourcing of new product concepts. There has been a trend in recent years among many firms to look outside for new technical ideas and product concepts as opposed to encouraging and fully exploiting internally generated ideas. Questions emerge as to when and to what degree such external reliance is advisable. Also, issues like the types of contingency factors which operate as moderating factors in such decisions require the attention of researchers.

A second issue raised by Professor Rubenstein centered around the increasing trends toward a short-term focus in U.S. research funding and efforts. The problem is compounded by the fact that generally larger portions of budgets in corporate research labs are provided by funding from operating units. Rubenstein argues that the reduction in long-term research emphasis on major new innovations may have long-term negative ramifications. He suggests that incentive systems must be created which provide incentives for divisional and other operating unit managers to set aside and protect a minimum level of resources (personnel and funding) for long-term projects. This should tend to ensure the organization's longer-term survival and prosperity. The research question emerges as to how such incentive systems should operate. For example, could a percentage of each operating unit manager's personal bonus be set aside and paid out on a contingency basis as a bonus for longer-term results (i.e., longer than the current fiscal year)? Issues such as this pose interesting research questions for researchers in organizational behavior and human resource management.

Another research question which emerges from the study of reward systems in R&D is the potential inhibiting effect of certain types of incentive systems on the acceptance and development of new technical ideas. As noted by Rubenstein in Chap. 2, reward systems can sometimes generate barriers to acceptance. Perceptions of a zero-sum nature of rewards (in the form of bonuses, promotions, or other financial incentives) for successful innovations may lead some individuals or groups in the organization to ignore or even oppose new ideas which will provide rewards to others. This problem may even extend beyond incentive systems. If the allocation of support in terms of personnel or funding is at issue, some units may be less than cooperative with the unit in which the innovative concept originated in order to protect scarce resources. These types of problems reinforce the need for MOT research that crosses the fields of organizational behavior and human resource management on incentive systems.

As Rubenstein suggests in Chap. 2, an important potential research area relevant to developing new product concepts is the area which has been labeled intrapreneurship. While much has been written in the past 10 years on this topic, the literature is primarily case oriented and based on anecdotal information. Very little systematic hypothesis testing (empirical research) has been published. Yet informal evidence suggests that the role of internal entrepreneurs, or intrapreneurs, and product champions is critical to the survival and development of new product concepts. As noted by Professors Leonard-Barton and Smith in Chap. 6, specific research questions might focus on what factors, special arrangements, or management practices are necessary to facilitate intrapreneurship and the role of product champions within large firms. Equally important is the need for research on new venture teams and the factors which influence or inhibit their success. Finally, as suggested by Professors Moenaert, DeMeyer, and Clarysse in Chap. 9, the need exists for studies on intrapreneurship which cross cultures.

As Professor Rubenstein intimated in Chap. 2, the generation of new product ideas or incremental improvements in existing ideas is a fertile area for research. A relatively new methodology is emerging which has utility in this area of research. This is the methodological technique known as network analysis. Originally developed in sociology, this method is finding increased utility in the field of organizational theory. This method is very well suited to studying the network patterns of information flow between engineers in the R&D lab, marketing people, manufacturing people, suppliers, customers, and competitors. Such investigations may facilitate understanding of the patterns which characterize successful versus failed new product ideas.

Concept and market development

As Gupta notes in Chap. 3, expenditures by U.S. firms have generally increased since the 1960s. However, formal market research studies are conducted for only a small fraction of new products (Mahajan and

Wind, 1992). In contrast, the Japanese invest more resources in these early stages of concept generation, screening, and market research. Gupta suggests that this may be one reason why the Japanese respond more effectively to customer needs and waste less time on postlaunch debugging. This observation suggests another potential avenue for future research in the cross-cultural area. Investigations on Japanese management practices have proliferated in the past 15 years. However, studies of Japanese practices in the areas of concept generation, screening, and market research have not emerged significantly in the literature.

One concept which ostensibly has not been utilized to its full potential in guiding technology development, forecasting, and market opportunity identification is the concept of the product life cycle. In Chap. 4, Professor di Benedetto observed that American firms have too often managed their products as if they were in the maturity or decline stages of the product life cycle rather than revitalizing them through incremental improvements for renewed growth. He noted that in many industries with slowing growth rates (i.e., mature markets), some of the common traits were declining prices, marketing mix strategies centered around advertising differentiation, lower investment in product and process research, and low emphasis on market share growth. He suggests that these traits characterize firms which are managing mature products for the short term rather than attempting to revitalize them for the long term. Thus, di Benedetto argues that greater research on the role of the product life cycle can benefit managers through the development of analytic methods. Multiple product life cycles can be used to identify strategic windows of opportunity for gaining competitive advantage. Such information would aid managers in decisions to invest in R&D or, in the event that the projected window of opportunity will have passed, to invest in joint ventures or technology licensing.

Based on the practical benefit of improved knowledge and analytic methods derived from the product life-cycle concept, in Chap. 4 di Benedetto identified several needs for future research. They include research which identifies factors which cause differential patterns in product life cycles. Because most previous research on product life cycles has focused on consumer markets, research on industrial markets is needed. Previous research has also focused on minor product changes with insufficient attention to technology changes. Finally, di Benedetto suggests that future research should explore the effects of characteristics of firms and management practices on the length and pattern of product life-cycle stages.

Critical front-end decisions and trade-offs

As a potential answer to the problem of reducing cycle time, rapid prototyping has attracted considerable attention. The notion is appealing: Reducing prototyping time directly reduces cycle time. A day saved in prototyping translates into at least a day saved in the entire new technology development cycle time. But how can rapid prototyping be achieved? Won't it necessarily require the assumption of greater risk? Some answers to these questions are already at hand. Prototyping time can be reduced without increasing the risk to the developer by the use of three-dimensional modeling, applications of computer-assisted design, and stereolithography. This latter technique is a computer-assisted design methodology in which plastic models are created directly from the inputs of a computer keyboard operator. The operator can see the prototype as it is being created in real time, make changes on the spot, and experiment with shape and configuration changes in real time. One can easily see how having a user present to make inputs at this design session could improve the design process and actually reduce the developer's risk of creating an unwanted product. But these are rapid design applications. How can the user actually experience the product in its intended application? How can a prototype be placed in the user's application more quickly, and how can user feedbacks be obtained more rapidly? These are major questions for further research. Though many aspects of rapid prototyping are being considered, the field is completely open to new approaches.

There is a complex combinatorial issue of risk-design-cost-performance trade-offs that needs considerable study. For example, high-performance products are often more costly than their lower-performance cousins. These designs may or may not be more risky (e.g., the higher-performing/higher-cost product may have greater reliability and therefore be more or less prone to failure). Alternatively, the higher-performing product may be more prone to misuse and failure due to its complexity. All product developers want to minimize their costs and exposure to the risks of product failure and user injury. But they also want to design and deliver the highest performance and user satisfaction possible. Clearly, these are conflicting goals. Although it may not be possible to satisfy all of these goals simultaneously, it seems clear that some of them can be satisfied while others are traded off against them. There is a need for models that can assist the developer in making these types of trade-offs. Although some progress has been made in developing and using such models (e.g., see Souder and Bethay, 1993), much more study and research is needed on these aspects.

Managing Development, Manufacturing, and Launch

Integration between R&D and marketing

An issue which is critical to the effectiveness of the early stages of idea generation and market definition is the general problem of integration between R&D and marketing. This issue was addressed in Chaps. 2, 3, and 4 by Professors Rubenstein, Gupta, and di Benedetto, respectively. It is an issue of longstanding concern (Gupta, Raj, and Wilemon, 1985; Souder, 1987). The separation which exists between R&D and marketing is based on functional specialization. But this specialization tends to lead to functional loyalties and organizational cultural differences which inhibit the cross-functional collaboration so critical to the early stages of new product innovation. Furthermore, achieving R&D and marketing integration will tend to become even more difficult as technologies and markets become more complex in the future. Thus, this problem has important practical implications and the nature of this problem raises a number of potential research questions which cross the disciplines of organizational theory, organizational behavior, strategic management, and organizational development (Souder, 1988).

One direction for future research which was noted by Professor Rubenstein in Chap. 2 centered on reward systems which facilitate rather than inhibit R&D and marketing collaborative efforts. Such incentive systems should not only include material rewards for successful cooperation but also censure for noncooperation. As noted by Rubenstein, marketing personnel are typically less likely to receive formal recognition or material rewards for ideas leading to new or significantly improved products. The design and testing of such incentive systems is a subject which requires future research. Such research would not only encompass the compensation paradigms in areas such as bonus plans but also the paradigm of performance evaluation systems.

One of the major problems in achieving adequate integration between R&D and marketing is based on the perceptual differences that exist between these two areas. Ironically, most R&D and marketing managers agree on the importance of developing high levels of integration. However, perceptual differences tend to detract from effective coordination. This problem was shown clearly in a study by Gupta, Raj, and Wilemon (1985). In this investigation, differences in perceptions were found in areas like marketing's involvement with R&D in setting new product goals and priorities, integration in establishing product development schedules, the role of marketing in generating new product ideas, marketing's effectiveness in providing feedback to R&D from customers regarding product performance, the role of marketing in providing information to R&D on competitors, and the role of R&D in analyzing customer needs. These findings and others by Souder (1977, 1987, 1988) indicate the importance of perceptual differences stemming from different frames of reference as a contributor to the problem of effective cross-functional integration.

The nature of this problem suggests a number of potential directions for future research that crosses the areas of organizational development, conflict resolution, and organizational design. In the area of organizational development, experimentation with different methods of improving R&D and marketing cooperation and joint understanding of customer needs could provide useful approaches for practicing managers. An example of this would be a study of the longitudinal effects of involving individuals in the channels of distribution in new product development decisions as suggested by Professors Calantone and Montoya in Chap. 7. Other examples include the effects of sending both R&D and marketing personnel to visit customers, bringing customer groups to the R&D facility to discuss customer needs, recruiting new technical personnel who were previously employed by the customer, and training for R&D personnel utilizing case histories of products which involved collaborative efforts with marketing. These are all potential methods which can be subjected to empirical testing.

Whenever there are high levels of interdependence, competition over limited resources, or the possibility for incongruence of goals, the potential for conflict exists. In the marketing and R&D interface these preconditions often exist. Failure to effectively manage the inevitable interdepartmental conflicts will often result in breakdowns in understanding and communication, increased tensions between marketing and R&D, and thus, failed coordination between the two areas. This will adversely affect the new product development process. Research on a variety of conflict resolution methods and organizational development interventions germane to conflict at the R&D and marketing interface may provide future useful guidance to managers involved in new product development. Such research may focus on when and how to employ intergroup confrontation meetings, process consultation, survey feedback (to diagnose sources or causes), team building interventions, and the creation of superordinate goals. In addition to research identifying which method should be used in varying situations (i.e., identification of contingency factors), research should also focus on the development of new types of interventions or resolution methods particularly applicable to interdepartmental conflict.

A final area in which future research may provide helpful guidance in determining how to create optimal integration between marketing and R&D is in the area of organizational design. A variety of forms of integration commonly exist. These include decentralized informal direct contact (with formalized approval), the creation of liaison positions to facilitate cross-functional communication or coordination, and the creation of temporary or permanent cross-functional task forces or teams. Some options include the creation of new product review boards (with cross-functional representation) or even a new products department which is separate from R&D and marketing but is responsible for integrating their efforts. Final forms include the creation of a product form of departmentation (each with a product or project manager) or in some cases a matrix structure which pulls personnel from the functional areas of marketing and R&D to form cross-functional new product development teams. The questions which are important to future research efforts pertinent to the problem of R&D and marketing integration focus on determining which form or forms of integration are appropriate under varying conditions. These may differ with the type of new product being developed, the stage in the development life cycle, and possibly the size of the effort (in terms of the number of personnel involved). Future research should focus on the development of contingency models which will provide the kind of guidance managers in new product or new technology development need in order to make the organizational design decisions that will provide the optimal levels of integration across R&D and marketing.

Organizational design and new product innovation

The issue of research on organizational design extends beyond the marketing and R&D interface. In Chap. 5, Professors Jelinek and Litterer discuss the structural inadequacies associated with so-called traditional scale-based organizations. They argue that increased global competition, customer demands for high levels of quality and variety, the development of sophisticated information technology, the increasing level of environmental turbulence, and the accelerating rate of technological innovation have all contributed to the need for organizations to become increasingly responsive and adaptive. Jelinek and Litterer suggest that many traditional assumptions about organizational design are becoming obsolete. While many of their recommendations are based on well-established research in organizational design, one important need which they identify focuses on an area which is less well developed. This again centers on the need to create optimal cross-functional integration. Jelinek and Litterer present the need to reduce cycle times, speed up design, and involve both manufacturing and marketing in early design decisions in the interest of market responsiveness, improved efficiency, and better quality. Like Wilemon and Millson (see Chap. 8), they point to concurrent or simultaneous engineering with the use of cross-functional teams to create a fluid iterative mode of organizing which contrasts with the traditional sequential approaches.

Such methods have allowed the Japanese to excel in manufacturing quality based on the fact that they commonly do not have completely separate design and manufacturing functions. By considering manufacturing and assembly in the initial design process, both the part and the manufacturing process can be engineered simultaneously. Hence, the goal is to engineer quality into the manufacturing process rather than relying on assembly line production inspections to eliminate defects. In the broadest applications, R&D, manufacturing, marketing, and procurement functions can coordinate concurrently rather than sequentially. Such concurrent engineering applications can reduce production costs, improve product quality, and reduce cycle time. Concurrent engineering is an emerging paradigm, and the challenge for future research is to develop knowledge about concurrent engineering methods and the types of organizational structural modifications that will lead to more effective new product design and manufacturing. For example, design for manufacturability has attracted great attention among U.S. managers during the past decade. The notion here is that manufacturing obstacles should be overcome at the design stage. Thus, the product comes to market much faster, at lower cost, with higher quality and greater value. This notion is appealing, especially since many manufacturing issues originate at the design stage. Concurrent engineering and interfunctional team management approaches have been applied at many U.S. firms in attempts to achieve design for manufacturability goals. The obstacles are the fundamental ones that have always plagued organizations: achieving cross-organizational team commitment to speed of delivery, low cost, and quality outputs. More research is needed on these aspects of the new product development and delivery process.

A second structural issue which emerges in Jelinek and Litterer's discussion (Chap. 5) of "real-time organizations" is the use of autonomous work teams. The concept of autonomous work teams has existed for approximately 20 years. In recent years, industrial experiments with self-managing teams have proliferated. Unfortunately, empirical research has lagged behind, with most published articles documenting the use of autonomous work teams appearing in professional or trade journals presenting anecdotal information. Unlike the empirical research literature on job design at the individual level (which is very well developed in the academic research journals), the literature on autonomous work teams is not well developed. This represents another challenge for future research. However, here the opportunity is less in the area of organizational design and more in the areas of group dynamics, incentive systems (motivation), performance evaluation, training, etc.

A third issue which relates to organizational design was raised by Wilemon and Millson in Chap. 8. They observed that more firms are moving toward flatter organizations with fewer hierarchical levels and larger spans of control. While it is true that historically many firms have operated with too many hierarchical levels and spans of control which often were too narrow, the reaction to this problem may also produce suboptimal structures. In some cases the pendulum may be swinging too far (i.e., organizations may develop suboptimally flat structures with spans of control which are too wide). This suggests the need for renewed research investigations in the area of hierarchical levels and spans of control. While these variables have been studied for many years in organizational theory, the research in this area remains inadequately developed to provide practicing managers the level of guidance which is needed in important organizational design decisions.

A fourth area which relates to organizational design is the use of matrix organizations in new product and new technology development. In recent years, the use of matrix structures has been widely criticized. Popular writers like Peters and Waterman (1982) have argued that matrix organizations should be abandoned altogether. There have been numerous cases where matrix structures have failed. Many of these cases may be attributable to poor management implementation or the creation of a matrix structure in an environment where a simpler product or functional form may have been more appropriate. In some cases, matrix structures may have resulted in failure simply because the size of the organization or the number of project assignments (on average) were too large. Most of the published literature presenting critical reviews utilizes anecdotal information without empirical data. This is of some concern because there may be many cases where the integration needs and the optimal use of engineering talent would suggest the need for some type of matrix system.

The challenge for future research is to determine with greater precision the conditions which require the use of matrix structures. Furthermore, questions must be answered regarding when a balanced matrix, a functional matrix, or a project matrix system should be utilized. This issue was noted by Professors Leonard-Barton and Smith in Chap. 6. Finally, research which provides more specific guidance to managers on how to overcome the common implementation problems associated with matrix structures is needed.

Organizational culture and technological innovation

Organizational culture is an area that has received increasing attention in recent years. Unfortunately, most of the empirical research on organizational culture has been general in nature. Relatively little research has focused exclusively on R&D or on the effects of different characteristics of organizational culture on innovation and effectiveness at various stages in the process of new product development.

In Chap. 5, Professors Jelinek and Litterer discuss a range of organizational cultural characteristics which should encourage high performance. These include factors like autonomy, responsibility, initiative, a shared strategic vision, responsiveness to the customer, highly decentralized authority based on expertise more than position, leadership by example, clearly articulated and pervasively held core values, and an emphasis on continuous learning. Similarly, Wilemon and Millson in Chap. 8 emphasize a culture of continuous improvement to foster innovation.

In addition to factors such as these, there are other organizational cultural characteristics which may influence responsiveness to the customer and innovation. For example, tolerance for risk should be a desirable characteristic for the early stages of innovation. This does not necessarily suggest that successful innovators gamble more than less successful innovators. Rather, innovative firms are more willing to look at high-risk opportunities and to include them in a balanced portfolio of high- and low-risk proposals (Souder, 1983).

Openness of communication is another factor which can influence innovation (Gupta, Raj, and Wilemon, 1987; Kozlowski and Hults, 1987). This not only includes internal organizational communication (Shrivastava and Souder, 1985) but also communication external to the organization. Contacts with scientists, customers, and others outside the organization can be critical sources of information and expertise in the new product development process.

Support from senior management is another characteristic which can positively influence innovation (Abbey and Dickson, 1983; Kozlowski and Hults, 1987; Shrivastava and Souder, 1985). Without executive sponsorship, excellent technical concepts can fail due to insufficient resources. A final organizational culture factor which can influence performance in new product innovation is the system of rewards. The issue extends beyond the existence of formal rewards for exceptional performance and incorporates a climate of recognition or informal reward for excellent performance.

These factors have received varying levels of support in the empirical research literature. What is less clear are the combined effects of such factors on outcomes like innovation. This is partially due to inadequate measurement development. In the next decade one of the major challenges to researchers in the area of organizational culture is to develop reliable measures with high construct, convergent, and discriminant validity. In addition, the need exists for research analyzing the combined effects of such organizational culture factors using multivariate methods like LISREL.

Another important direction for research is to examine the potentially differing effects of organization cultural characteristics at different stages in the product or project life cycle. Of the studies cited in this section on organizational culture, all but one did not consider stages in the life cycle and thus inherently assumed that the effect of organizational culture on new product development would be homogeneous across these stages. To illustrate why this may be an important research question with practical implications for R&D managers, note that an organizational cultural factor like tolerance for risk may be a strong determinant of performance in the early stages of new product development. But it may be less important in later stages. Similar patterns may exist for factors such as warmth in the work group environment or support from senior management. Thus, various configurations of organizational cultural characteristics may have differential influences on R&D performance at different stages of development. Identification of the dimensions which have the greatest influence at each stage in the innovation process would enable managers to develop strategies to facilitate R&D performance in new product innovation.

Emerging Cross-Cultural Issues and Strategic Management Issues

Cross-cultural issues

Another issue which relates to organizational culture is based on the observations of Moenaert, De Meyer, and Clarysse in Chap. 9. Here the macroeffects of national or societal cultures are shown to influence the organization cultures of firms within these societies. Hence, based on cultural dimensions which differentiate Japanese, American, and different European societies, we find characteristics which may positively (or negatively) influence new product or new technology innovation in one culture but may not affect performance in another culture. The observations presented by Moenaert, De Meyer, and Clarysse reinforce the need for more cross-cultural investigations to improve our understanding of the role of cultural differences on new product innovation.

Strategic management of technology

An area in which current research is underdeveloped is the important area of strategic alliances for new technology development. As discussed in Chaps. 6, 8, and 10, strategic alliances for purposes of new technology development have proliferated during the 1980s. The types of alliances range from research grants and R&D contracts to licensing to joint ventures to mergers or acquisitions. One direction for future investigations would be to determine the kinds of conditions or strategic objectives which would identify the optimal type of strategic alliance. Such research could lead to the development of contingency models which could potentially be useful in guiding managerial decision making. As Professors Leonard-Barton and Smith point out, many strategic alliances for new technology development result in failure. This raises important research questions regarding the causes of failure and the determination of the kinds of managerial actions which can reduce the probability of failure.

Conclusion

It is apparent that there are many promising directions for future research in MOT. These new directions for research cross numerous paradigms in several fields. In particular, in the area of developing new product concepts, research is needed on outsourcing, incentive systems, intrapreneurship, screening, and the role of the product life cycle. In the area of R&D and marketing integration, future research is needed which focuses on conflict resolution, reward systems, organizational development applications, and structural modes of achieving optimal integration. In the area of organizational design, future work is needed which focuses on concurrent engineering, autonomous work teams, decentralization, and the optimal utilization of matrix structures. In the areas of organizational culture and human resource management, needs ostensibly exist for investigations which identify the types of characteristics which facilitate new product innovation. Studies are needed on the differential role of such characteristics, depending on stages in the product life cycle or on the wider cultural context. Finally, in the area of strategic management, future research is needed which will provide guidance to managers on the utility of varying types of strategic alliances and the integration and application of the general strategic management literature to the management of new technology development.

References

Abbey, A., and Dickson, J. "R&D Work Climate and Innovation in Semiconductors," *Academy of Management Journal,* 26, 1983, 362–368.

Gupta, A., Raj, S., and Wilemon, D. "The R&D Marketing Interface in High Technology Firms," *Journal of Product Innovation Management,* 2: 1, 1985, 12–24.

——, ——, and ——. "Managing the R&D Marketing Interface," *Research Management,* 30: 2, 1987, 38–43.

Kozlowski, S., and Hults, B. "An Exploration of Climates for Technical Updating and Performance," *Personnel Psychology,* 40, 1987, 539–563.

Mahajan, V., and Wind, Y. "New Product Models: Practice, Shortcomings and Desired Improvements," *Journal of Product Innovation Management,* 9: 2, 1992, 128–139.

Peters, T., and Waterman, R. *In Search of Excellence: Lessons from America's Best Run Companies,* New York: Harper & Row, 1982.

Shrivastava, P., and Souder, W. "Phase Transfer Models for Technological Innovation," *Advances in Strategic Management,* 3, 1985, 135–147.

Souder, Wm. E. "An Exploratory Study of the Coordinating Mechanisms between R&D and Marketing as an Influence on the Innovation Process, Final Report to the National Science Foundation," University of Pittsburgh, 1977.

——. "Effectiveness of Nominal and Interacting Group Decision Processes for Integrating R&D and Marketing," *Management Science,* 23: 6, February 1977, 595–605.

——. "Organizing for Modern Technology and Innovation: A Review and Synthesis," *Technovation,* 2, 1983, 27–44.

——. *Managing New Product Innovations,* Lexington, MA: Lexington Books, 1987.

——. "Managing Relations Between R&D and Marketing in New Product Development Projects," *Journal of Product Innovation Management,* 5: 1, 1988, 6–19.

—— and Bethay, D. "The Risk Pyramid for New Product Development: An Application to Complex Aerospace Hardware," *Journal of Product Innovation Management,* 10: 3, 1993, 3–16.

Index